高等学校计算机教材

Android 实用教程

（第2版）（含视频分析）

郑阿奇　主编

电子工业出版社

Publishing House of Electronics Industry

北京·BEIJING

内 容 简 介

本书以 Android Studio 3.x 作为平台，系统介绍 Android 平台 App 开发。本书秉承前一版以应用开发为主线，在应用中理解 Android 开发基本内容及相互关系的优点，贴近 5G 智能手机，又大幅增加最新的应用体系，还增加系统介绍移动 App 高级界面开发、数据库和网络编程、多媒体和图形图像编程以及各种常用第三方开发与设备操作等内容。全书尽量采用（或模拟）当前移动互联网实际应用的真实场景，学以致用。配套的实例尽量做到独立完整，方便读者试做；同时又增强了启发性，提示读者可以根据需要进行更改和完善。

本书配套的习题和实验均进行了扩充，更加系统化，同时紧密贴近当前 5G 智能手机上真实的 App 案例。扫描书中二维码可在线观看微视频，另外还提供配套的教学课件和全部应用实例代码（工程文件）。

本书可作为大学本科和高职高专院校有关专业的教材和教学参考书，也可作为 Android 自学用书和开发参考书。

未经许可，不得以任何方式复制或抄袭本书之部分或全部内容。
版权所有，侵权必究。

图书在版编目（CIP）数据

Android 实用教程：含视频分析 / 郑阿奇主编. —2 版. —北京：电子工业出版社，2020.7
ISBN 978-7-121-39021-0

Ⅰ. ①A… Ⅱ. ①郑… Ⅲ. ①移动终端－应用程序－程序设计－高等学校－教材 Ⅳ. ①TN929.53

中国版本图书馆 CIP 数据核字（2020）第 082022 号

责任编辑：程超群　　文字编辑：赵云峰
印　　刷：三河市良远印务有限公司
装　　订：三河市良远印务有限公司
出版发行：电子工业出版社
　　　　　北京市海淀区万寿路 173 信箱　邮编　100036
开　　本：787×1 092　1/16　印张：23.25　字数：671 千字
版　　次：2017 年 6 月第 1 版
　　　　　2020 年 7 月第 2 版
印　　次：2020 年 7 月第 1 次印刷
定　　价：69.00 元

凡所购买电子工业出版社图书有缺损问题，请向购买书店调换。若书店售缺，请与本社发行部联系，联系及邮购电话：（010）88254888，88258888。

质量投诉请发邮件至 zlts@phei.com.cn，盗版侵权举报请发邮件至 dbqq@phei.com.cn。
本书咨询联系方式：（010）88254577，91034@qq.com。

前　言

　　Android 是目前流行的智能手机操作系统之一，其中文名称为"安卓系统"。随着 5G 时代的来临，基于 Android 的应用开发将出现更多新的应用场景。

　　Android 官方推荐的开发环境——Android Studio，基于 IntelliJ IDEA，集成了 Android 平台的诸多组件，并提供完善的 Android 开发和调试工具，无论安装还是使用都十分简便，成为当下用于 App 开发的主流利器。2017 年 10 月之后发布的 Android Studio 3.x 版本，谷歌（Google）官方从中剥离了 Android SDK，改由用户通过环境提供的 SDK Manager 或 AVD 映像工具下载，再根据实际开发需求定制安装，这么做极大地精简了原 Android Studio 安装包的体积，降低了对用户计算机系统配置的要求。新版 Android Studio 增加了默认的约束布局（ConstraintLayout），为用户设计布局优雅的 App 界面提供了便捷之处。

　　本书以 Android Studio 3.x 作为平台，系统介绍 Android 平台 App 开发，删减了过时和简单常识性的内容，贴近当前 5G 智能手机，又大幅增加新的应用体系和读者重点关注的内容。全书内容包括 Android 开发入门、Android 用户界面、界面布局与活动页、移动 App 高级界面开发技术、Android 服务与广播程序设计、Android 数据存储与共享、Android 数据库和网络编程、Android 多媒体和图形图像编程，以及 Android 第三方开发与设备操作。本书秉承前一版以应用开发为主线，在应用中理解 Android 开发基本内容及相互关系的特点，并尽量采用（或模拟）当前移动互联网实际应用的真实场景，学以致用。本书配套的实例尽量做到独立完整，方便读者单独试做；同时又增强了启发性，读者可以根据自己的需要进行更改和完善。

　　本书的习题和实验均在上一版的基础上进行了扩充，更加系统化的同时更贴近当前 5G 智能手机上真实的 App 案例。习题用于消化知识，实验则用于完成应用性实例。

　　扫描书中二维码，可在线观看相应的微视频（建议在 WiFi 环境下操作），这些视频将主要内容联系起来讲解，分析文件关系和代码之间的相互联系，介绍解决问题的过程和要点，回答读者关心的问题。

　　本书提供配套的教学课件和全部应用实例代码（工程文件），需要者可从华信教育资源网（www.hxedu.com.cn）免费下载。

　　本书由郑阿奇（南京师范大学）主编，参加本书编写工作的还有刘美芳、周何骏、孙德荣等。

　　由于编者水平有限，错误在所难免，敬请广大师生、读者批评指正。

　　意见建议邮箱：easybooks@163.com

编　者

本书视频目录

（建议在 WiFi 环境下扫码观看）

序　号	视频文件名
1	1.3.1 创建 Android Studio 工程
2	1.3.7 事件处理的 4 种编程范式
3	1.4 Android SDK 的安装与管理
4	2.2.1 例 2.1 TextViewDemo（文本视图：TextView）
5	2.2.3 例 2.3 ImageViewDemo（图像视图：ImageView）
6	2.2.7 例 2.7 SpinnerDemo（下拉框：Spinner）
7	3.1.1 约束布局：ConstraintLayout
8	3.1.3 例 3.1 LoginPage（界面布局的应用）
9	4.2.1 例 4.1 MyTabHost（标签栏）
10	4.2.2 例 4.2 MyBanner（轮播条）
11	4.2.3 例 4.3 MyRecyclerGrid（频道栏）
12	4.3.1 例 4.4 MyListView（列表视图）
13	4.3.2 例 4.5 MyGridView（网格视图）
14	4.4 EasyFruit（整合为完整 App）
15	5.1.2 例 5.1 CircleStartService（启动方式使用 Service）
16	5.1.3 例 5.2 CircleBindService（绑定方式使用 Service）
17	5.1.4 例 5.3 PowerMonitorNormal（多 Service 交互及生命周期）
18	5.2.2 例 5.4 PowerMonitorNormalBroadcast（普通广播举例）
19	5.2.3 例 5.5 SmsFilterOrderedBroadcast（有序广播举例）
20	6.1.2 例 6.1 SharedReg-SharedLog（SharedPreferences 举例）
21	6.2.2 例 6.2 InterReg（文件存储举例）
22	6.3.2 例 6.3 SqliteReg（SQLite 应用举例）
23	6.4.3 例 6.4 SqliteReg-SqliteLog（ContentProvider 应用举例）
24	7.2.2 例 7.1 MySqlTest（Android JDBC 直连 MySQL）
25	7.3.3 例 7.3 MyServlet（Web 应用开发和部署）
26	7.4.3 例 7.5 MyWebService（开发 WebService）
27	7.5.2 例 7.7 MyServlet（Web 端开发）
28	8.1.1 例 8.1 MyVideoView（视频播放）
29	8.1.2 例 8.4 MyMediaPlayer（音频播放）
30	8.2.1 例 8.6 MyScale（图像倾斜缩放）
31	8.3 例 8.8 MyGallery（手机相册功能）
32	9.1.4 例 9.1 MyQRCodeDemo（二维码生成）
33	9.2.3 例 9.3 MyEasyPayDemo（支付功能实现）
34	9.3.1 例 9.4 MyGd（使用高德地图）

代码段二维码说明：

　　为了节省本书篇幅和保持应用实例的完整性，扫描部分应用实例附近的二维码，可查看省略的部分代码段内容。本书提供的配套教学资料包中包括了应用实例程序的完整工程文件，可从华信教育资源网（www.hxedu.com.cn）免费下载。

目 录

第1章 Android 开发入门 1
1.1 Android 平台简介 1
1.2 Android Studio 3.x 安装 2
1.2.1 安装前的准备 2
1.2.2 安装 Android Studio 3
1.3 第一个 Android 应用程序 5
1.3.1 创建 Android Studio 工程 6
1.3.2 设计应用程序界面 8
1.3.3 添加程序代码 9
1.3.4 Android 应用程序运行 12
1.3.5 Android Studio 工程结构 17
1.3.6 应用程序代码解析 19
1.3.7 事件处理的 4 种编程范式 25
1.4 Android SDK 的安装与管理 28
1.4.1 通过下载 AVD 映像安装 SDK 29
1.4.2 通过 SDK Manager 安装 SDK 29
1.4.3 两种安装方式的区别与联系 30

第2章 Android 用户界面 31
2.1 用户界面基础 31
2.1.1 用户界面基本要求 31
2.1.2 控件概述 34
2.2 基本的界面控件 39
2.2.1 文本视图：TextView 39
2.2.2 按钮和图像按钮：Button/ImageButton 42
2.2.3 图像视图：ImageView 45
2.2.4 复选框：CheckBox 48
2.2.5 单选按钮及其容器：RadioButton 和 RadioGroup 50
2.2.6 文本编辑框：EditText 53
2.2.7 下拉框：Spinner 58
2.2.8 自动完成文本视图：AutoCompleteTextView 61
2.2.9 日期时间选择器：DatePicker/TimePicker 63
2.3 界面事件 67
2.3.1 按键事件 67
2.3.2 触摸事件 69

第3章 界面布局与活动页 73
3.1 界面布局 73
3.1.1 约束布局：ConstraintLayout 73

	3.1.2 自定义布局	76
	3.1.3 界面布局的应用	82
3.2	Activity 活动页	85
	3.2.1 Activity 概述	86
	3.2.2 页面间的数据交互	86
	3.2.3 页面生命周期	92

第4章 移动 App 高级界面开发技术 98

4.1	App 通用界面元素	98
4.2	界面元素开发	100
	4.2.1 标签栏	100
	4.2.2 轮播条	106
	4.2.3 频道栏	111
4.3	内容的呈现	116
	4.3.1 列表视图	116
	4.3.2 网格视图	122
	4.3.3 类别标签列表	128
4.4	整合为完整 App	134
	4.4.1 界面元素集成	135
	4.4.2 通知消息计数	138

第5章 Android 服务与广播程序设计 143

5.1	Service（服务）程序设计	143
	5.1.1 Service 概述	143
	5.1.2 启动方式使用 Service	145
	5.1.3 绑定方式使用 Service	150
	5.1.4 多 Service 交互及生命周期	155
5.2	广播（BroadcastReceiver）	162
	5.2.1 BroadcastReceiver 概述	162
	5.2.2 普通广播举例	165
	5.2.3 有序广播举例	166

第6章 Android 数据存储与共享 171

6.1	SharedPreferences（共享优先）存储	171
	6.1.1 SharedPreferences 概述	171
	6.1.2 SharedPreferences 举例	172
6.2	内部文件存储	183
	6.2.1 Android 系统文件访问	183
	6.2.2 文件存储举例	184
6.3	SQLite 数据库存储与共享	187
	6.3.1 SQLite 概述	187
	6.3.2 SQLite 应用举例	188
6.4	ContentProvider 数据共享组件	191
	6.4.1 ContentProvider 组件	191

	6.4.2 ContentProvider 创建	192
	6.4.3 ContentProvider 应用举例	195

第 7 章 Android 数据库和网络编程 · 199

- 7.1 数据库准备 · 199
- 7.2 Android JDBC 编程 · 200
  - 7.2.1 基本原理 · 200
  - 7.2.2 Android JDBC 直连 MySQL · 202
  - 7.2.3 Android JDBC 直连 SQL Server · 205
  - 7.2.4 Android JDBC 直连 Oracle · 206
  - 7.2.5 以表格形式显示数据库表数据 · 206
- 7.3 Android HTTP 编程 · 213
  - 7.3.1 基本原理 · 213
  - 7.3.2 环境安装 · 215
  - 7.3.3 Web 应用开发和部署 · 217
  - 7.3.4 移动端 Android 程序开发 · 222
- 7.4 Android 与 WebService 交互 · 226
  - 7.4.1 基本原理 · 226
  - 7.4.2 配置 IIS 服务器 · 227
  - 7.4.3 开发 WebService · 229
  - 7.4.4 发布 WebService · 233
  - 7.4.5 移动端 Android 程序开发 · 234
- 7.5 网上商城 JSON 数据操作 · 237
  - 7.5.1 基本原理 · 237
  - 7.5.2 Web 端开发 · 238
  - 7.5.3 移动端开发 · 240

第 8 章 Android 多媒体和图形图像编程 · 246

- 8.1 媒体播放器的开发 · 246
  - 8.1.1 视频播放 · 246
  - 8.1.2 音频播放 · 256
  - 8.1.3 录像功能 · 258
- 8.2 图形图像处理 · 263
  - 8.2.1 图像倾斜缩放 · 263
  - 8.2.2 图像扭曲 · 267
- 8.3 手机相册功能 · 270
- 8.4 OpenGL 图形库 · 276
  - 8.4.1 OpenGL 简介 · 276
  - 8.4.2 构建 OpenGL 环境 · 276
  - 8.4.3 定义和绘制图形 · 279

第 9 章 Android 第三方开发与设备操作 · 282

- 9.1 生成和扫描二维码（ZXing 库） · 282
  - 9.1.1 ZXing 概述 · 282

    9.1.2 整合 ZXing 框架 ·············· 282
    9.1.3 界面设计 ·············· 286
    9.1.4 二维码生成 ·············· 286
    9.1.5 二维码扫描 ·············· 289
  9.2 接入支付宝（alipaySdk 库） ·············· 291
    9.2.1 支付接口 ·············· 291
    9.2.2 集成支付功能 ·············· 294
    9.2.3 支付功能实现 ·············· 295
  9.3 地图应用开发（高德地图开放平台） ·············· 300
    9.3.1 配置地图环境 ·············· 300
    9.3.2 地图基本检索应用 ·············· 306
    9.3.3 GPS 定位和周边搜索 ·············· 314
    9.3.4 驾驶路径规划 ·············· 320
    9.3.5 百度地图应用开发 ·············· 327
  9.4 Android 设备操作 ·············· 328
    9.4.1 计步器 ·············· 328
    9.4.2 摇一摇 ·············· 330
    9.4.3 蓝牙设备发现 ·············· 332
    9.4.4 手电筒 ·············· 335

习题和实验 ·············· 338
  第 1 章　Android 开发入门 ·············· 338
  第 2 章　Android 用户界面 ·············· 340
  第 3 章　界面布局与活动页 ·············· 343
  第 4 章　移动 App 高级界面开发技术 ·············· 346
  第 5 章　Android 服务与广播程序设计 ·············· 348
  第 6 章　Android 数据存储与共享 ·············· 349
  第 7 章　Android 数据库和网络编程 ·············· 350
  第 8 章　Android 多媒体和图形图像编程 ·············· 352
  第 9 章　Android 第三方开发与设备操作 ·············· 354

习题参考答案 ·············· 356
  第 1 章　Android 开发入门 ·············· 356
  第 2 章　Android 用户界面 ·············· 357
  第 3 章　界面布局与活动页 ·············· 358
  第 4 章　移动 App 高级界面开发技术 ·············· 359
  第 5 章　Android 服务与广播程序设计 ·············· 360
  第 6 章　Android 数据存储与共享 ·············· 361
  第 7 章　Android 数据库和网络编程 ·············· 361
  第 8 章　Android 多媒体和图形图像编程 ·············· 362
  第 9 章　Android 第三方开发与设备操作 ·············· 363

第 1 章 Android 开发入门

1.1 Android 平台简介

Android 是目前主流的两大智能手机操作系统之一。Android 一词的本义是指"机器人"，最初是由 Andy Rubin（安迪·鲁宾）开发的一款专用于移动终端设备的操作系统软件。2003 年 10 月，Andy Rubin 等人创建 Android 公司。2005 年 8 月 17 日，Google（谷歌）收购了成立仅 22 个月的 Android 公司，Andy Rubin 成为 Google 工程部副总裁，继续负责对 Android 系统进行开发运营，此后 Android 版本不断升级，功能不断增强，使 Android 得到了迅速发展。2019 年 9 月的预测报告显示，Android 在手机操作系统市场的份额将从 2018 年的 85.1%上涨到 87%，在中国更是家喻户晓，其中文名称为"安卓系统"，最新版本为 Android 10.0。

Android 系统采用了分层的架构，如图 1.1 所示，从高到低分为应用程序层（Applications）、应用框架层（Application Framework）、系统运行时库（Libraries/Android Runtime）和 Linux 内核（Linux Kernel）四个层次。

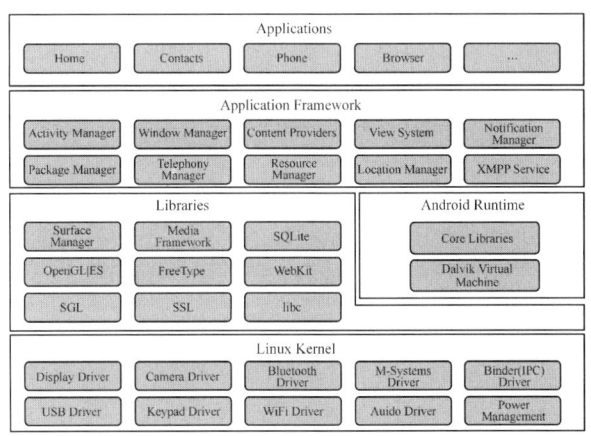

图 1.1 Android 系统的分层架构

- **应用程序层**（Applications）：包括随 Android 系统一起发布的核心应用程序（如 SMS 短消息、日历、计算器、地图、浏览器、联系人管理簿等）以及手机开发商开发的应用程序。所有应用程序又称为 App。
- **应用框架层**（Application Framework）：它为 App 提供系统服务，主要有活动管理器（Activity Manager）、内容提供器（Content Providers）、视图系统（View System）、通知管理器（Notification Manager）以及资源管理器（Resource Manager）等，App 开发人员可以完全访问和使用它们。
- **系统运行时库**（Libraries/Android Runtime）：包含一些 C/C++库，这些库能被 Android 系统中不同的组件使用。
- **Linux 内核**（Linux Kernel）：Android 是基于 Linux 的操作系统，故其架构的最底层是 Linux 内核，但 Google 为达商业目的，移除了内核中被 GNU/GPL 授权证所约束的部分（例如驱动程序）。

Android 平台具有开源开放、多硬件兼容、支持丰富的第三方 SDK 二次开发等诸多优点,深受广大程序员的青睐,已然成为当今互联网应用开发的最主要平台,其官方推荐的集成开发环境(IDE)为 Android Studio,编写本书时的最新版本是 Android Studio 3.x。

1.2　Android Studio 3.x 安装

1.2.1　安装前的准备

1. 硬件要求

因为 Android 各个版本的 SDK 包普遍较大,且运行 Android 模拟器也要耗费大量内存,对开发 Android 应用程序的计算机系统的配置要求较高,强烈建议 CPU 采用酷睿 i7(或相当档次的)及以上;内存至少有 8GB 及以上;硬盘(尤其是 C 盘)最好是大容量固态硬盘;具备宽带高速上网条件。

2. 安装 JDK

Android Studio 是使用 Java 工具构建的软件,因此它只能在有 JDK 的计算机上安装和运行。如果读者的计算机上没有 JDK,需要先安装 JDK。可以到 Oracle 官网(https://www.oracle.com/technetwork/java/javase/downloads/index.html)下载最新版本的 JDK,注意选择适合自己操作系统的版本。笔者使用的是 JDK12,安装包文件名为"jdk-12.0.2_windows-x64_bin.exe"。

双击安装包文件,启动安装向导,按照向导的指引进行操作,过程很简单,这里不展开介绍。

安装完 JDK 后,还要配置环境变量,以使 Android Studio 能够找到这个 JDK。以 Windows 10 为例,在桌面上右击"此电脑"图标,从弹出的快捷菜单中选择"属性",打开"系统"窗口,单击"高级系统设置"选项,出现"系统属性"对话框,单击"环境变量"按钮,显示当前系统中环境变量的情况,如图 1.2 所示。

图 1.2　当前系统中的环境变量

在图 1.2 底部列出的"系统变量"列表中,如果 JAVA_HOME 项不存在,则单击"新建"按钮创建它。单击"系统变量"列表下方的"新建"按钮后,系统显示"新建系统变量"对话框,在"变量名"文本框中输入 JAVA_HOME,在"变量值"文本框中输入我们安装 JDK 的位置(笔者将 JDK 安装在"C:\Program Files\Java\jdk-12.0.2"目录中),单击"确定"按钮,如图 1.3 所示。

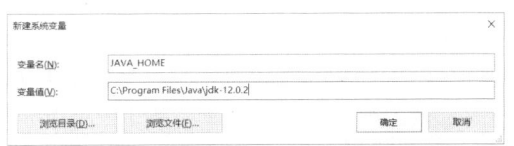

图 1.3　新建 JAVA_HOME 系统变量

接下来添加系统 Path 变量，在"系统变量"列表中选中"Path"，单击列表下方的"编辑"按钮，出现"编辑环境变量"对话框，如图 1.4 所示。

图 1.4　添加系统 Path 变量

图 1.4 列出了系统中已有的 Path 变量，在行的末尾添加以下行：
%JAVA_HOME%\bin

连续 3 次单击"确定"按钮，Windows 接受这些修改并返回到最初的"系统"窗口。至此，系统就在原来的 Path 路径上增加了一个指向新安装 JDK 的查找路径。

3. 下载 Android Studio

访问 Android 官网（https://developer.android.google.cn/studio），下载 Android Studio 的安装包（编写本书时已经更新到 Android Studio 3.5），单击页面上的"DOWNLOAD ANDROID STUDIO"按钮并接受许可条款，开始下载。

Android Studio 自 3.0（2017 年 11 月 20 日发布）版本开始，其安装包中已不再包含 Android SDK，此举极大地减小了 Android Studio 安装包的体积，Android Studio 3.5 的安装包文件大小仅 710MB，文件名为"android-studio-ide-191.5791312-windows.exe"。Android SDK 可在用户安装完 Android Studio 后再根据需要选择相应的版本下载和集成。

因为在 Android Studio 的安装过程中需要时刻到网络上去获取所需的各种文件，为避免产生麻烦，建议读者先关闭 Windows 防火墙和杀毒软件。

1.2.2　安装 Android Studio

1. 执行安装向导

双击下载的 Android Studio 安装包文件，启动安装向导，如图 1.5 所示。

单击"Next"按钮向前推进，每一步都采用默认设置，直至安装完成到达"Completing Android Studio Setup"界面，如图 1.6 所示。其中，"Start Android Studio"复选框能够让 Android Studio 在单击"Finish"按钮之后启动。确保选中了该复选框，接着单击"Finish"按钮，Android Studio 将会启动。

图 1.5　启动 Android Studio 安装向导　　　　　　图 1.6　完成 Android Studio 的安装

2. 第一次启动

当 Android Studio 第一次启动时，它会检查用户的系统之前是否安装过早期版本，并询问用户是否要导入先前版本 Android Studio 的设置，如图 1.7 所示。一般初学者建议使用初始设置，选中下面一个单选按钮（Do not import settings）后，单击"OK"按钮。

图 1.7　Android Studio 的初始设置

接着出现如图 1.8 所示的启动画面，在弹出的"Data Sharing"对话框中单击"Don't send"按钮拒绝 Google 对个人隐私信息的采集，在接下来弹出的"Android Studio First Run"提示框中单击"Cancel"按钮忽略系统对 Android SDK 的检查。

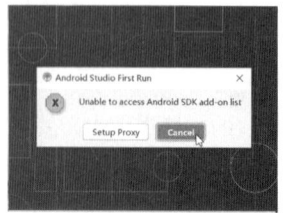

图 1.8　启动画面及对话操作

接着出现"Welcome"（欢迎）界面，如图 1.9 所示。单击"Next"按钮，在接下来的"Select UI Theme"（选择 UI 主题）界面选择淡蓝色较为清新的"Light"风格主题，如图 1.10 所示。单击"Next"按钮，

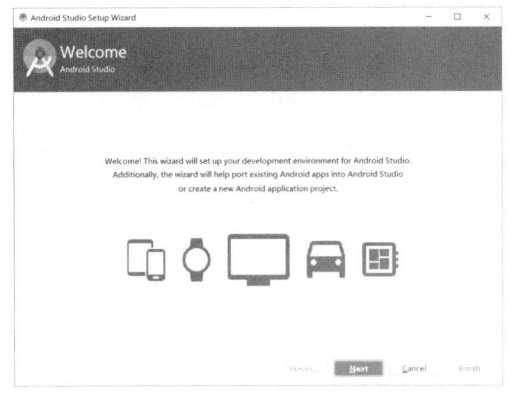 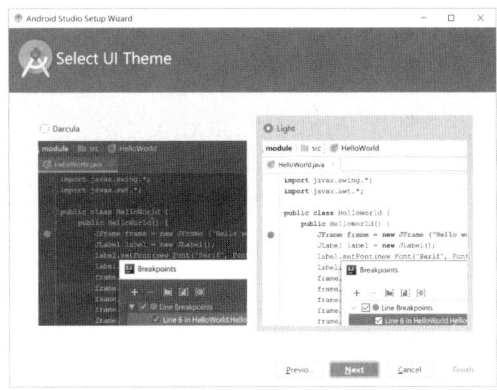

图 1.9 "Welcome"(欢迎)界面　　　　　　图 1.10 选择 UI 主题风格

在出现的 "Verify Settings"(核查设置)界面上,汇总显示了在 Android Studio 中开发应用所需要的全部 Android SDK 组件的信息,单击 "Finish" 按钮,向导就开始下载并安装这些组件,如图 1.11所示。稍等片刻,待完成后单击 "Finish" 按钮,结束安装向导。

图 1.11 向导下载安装所需 Android SDK 组件

至此,Android Studio 安装完成。

1.3 第一个 Android 应用程序

本节我们开发第一个 Android 应用程序(称为 App),界面如图 1.12 所示。

图 1.12 第一个 Android 应用程序界面

界面上有一张"安卓小机器人"的图片和一个"开始课程"按钮，点击"开始课程"按钮，将改变上方原来显示的"Hello World!"文字内容为"Hello World! 我爱 Android 编程！"。

1.3.1 创建 Android Studio 工程

要开发一个 Android 应用程序，必须先创建 Android Studio 工程。创建工程的步骤如下：

（1）启动 Android Studio 后出现如图 1.13 所示窗口，单击"Start a new Android Studio project"项来创建新的 Android Studio 工程。

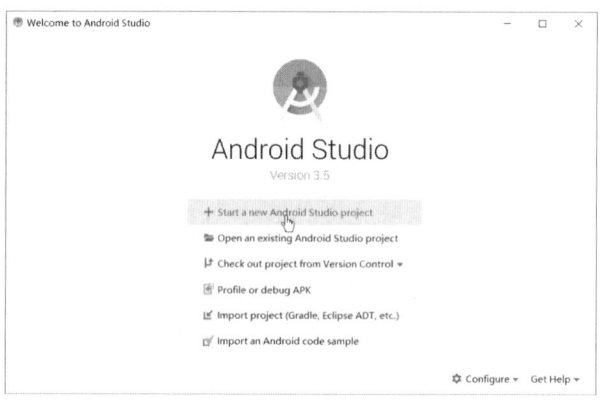

图 1.13　创建一个新的 Android Studio 工程

（2）在"Choose your project"页选择 Activity（活动页面）类型，如图 1.14 所示。这里选择"Basic Activity"类型，单击"Next"按钮进入下一步。

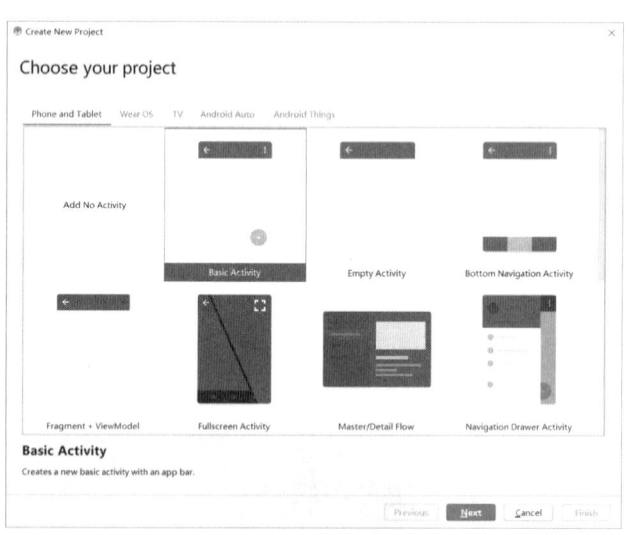

图 1.14　选择 Activity 类型

> 👀说明：
> 选择"Basic Activity"（基本活动页）类型，工程创建后系统会生成一个显示"HelloWorld!"的最简单的应用程序；如果选择"Empty Activity"（空活动页）类型，则仅仅创建应用程序框架，运行时并不显示任何内容。

（3）在"Configure your project"页填写应用程序相关的信息项，如图 1.15 所示。

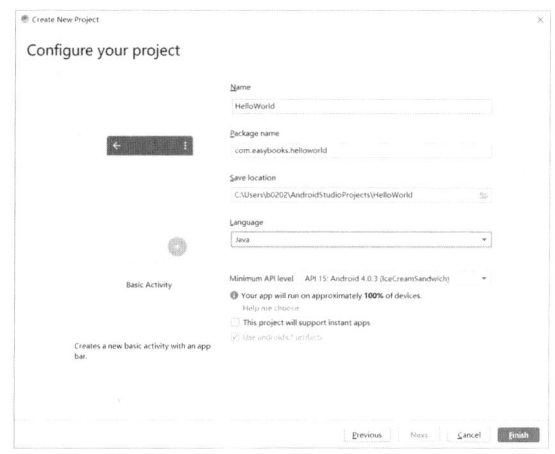

图 1.15　填写应用程序相关的信息项

这里需要填写的信息项如下：
- Name：应用程序工程的名称，默认为"My Application"，这里我们改为"HelloWorld"。
- Package name：工程中 Java 程序所在的包名，默认为"com.example.myapplication"，这里改为"com.easybooks.helloworld"。
- Save location：工程存放的路径，默认为"C:\Users\<用户名>\AndroidStudioProjects"（其中，<用户名>是当前 Windows 系统的登录用户名），可以修改，这里保持默认。
- Language：Android 开发所使用的编程语言，这里选"Java"。另外一个选项"Kotlin"是新版 Android Studio 支持的另一种新的编程语言，有兴趣的读者也可以尝试用一用。但建议初学者还是用我们最熟悉的 Java 作为开发语言。
- Minimum API level：设置为"API 15: Android 4.0.3 (IceCreamSandwich)"，表示要安装此应用程序的目标设备，其操作系统最低要求的版本为 Android 4.0.3。由于智能手机更新换代很快，现在市面上绝大部分手机的 Android 操作系统都已经是 Android 9 及以上，因为高版本的 Android 一般向下兼容，这样选择基本上能够兼容运行在所有的手机上。

填写完毕，单击"Finish"按钮，稍等片刻，系统显示如图 1.16 所示的工程开发界面。开发界面右上部是主开发区，显示要编辑的源程序文件内容，初始显示工程的 content_main.xml 文件源码。

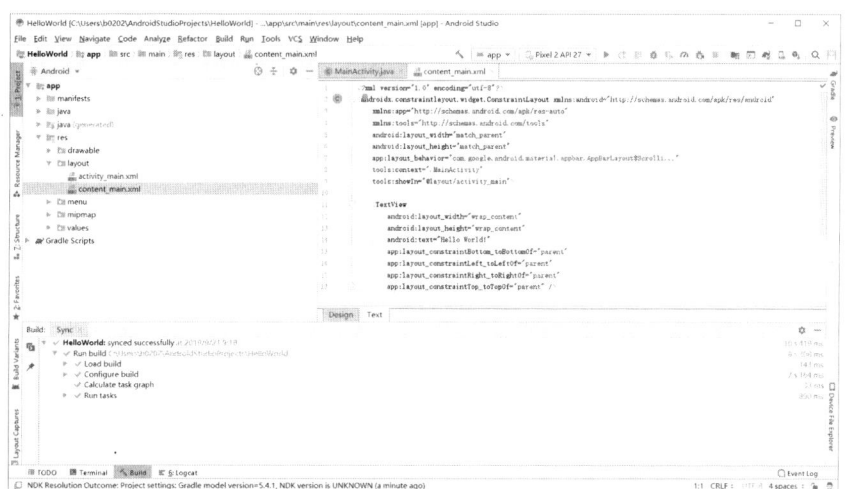

图 1.16　Android Studio 工程开发界面

1.3.2 设计应用程序界面

前面创建工程的名字是 HelloWorld，它也是应用程序的名称。如果运行，界面会显示"Hello World!"，可以将其修改、重新设计成自己想要的界面。下面我们来给界面上增加显示一张图片和一个按钮，设计步骤如下。

（1）进入可视化设计模式。单击 content_main.xml 文件 Editor 窗口底部左下角的"Design"选项卡，进入可视化设计模式，如图 1.17 所示。左侧是一个工具箱面板（Palette），上面有各种带图标的 UI 控件项；中央区域是智能手机屏幕（模拟）；右侧是设置控件属性的"Attributes"面板。

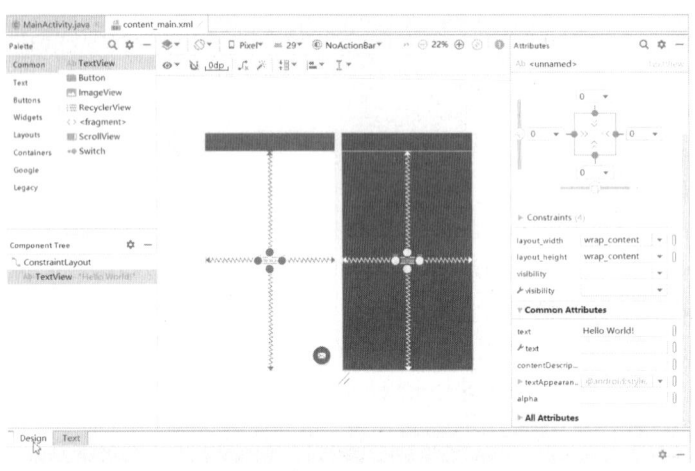

图 1.17　进入可视化设计模式

设计程序界面时用户只需用鼠标选中工具箱内的某个控件，将其拖曳到手机屏幕界面上进行布局即可。选中手机屏幕上的某个控件，可在"Attributes"面板上设置其属性。

（2）让"Hello World!"字号变大、字体变红。原工程的"Hello World!"文字太小，现通过修改 TextView 控件的属性将其放大变醒目。选中"Hello World!"（TextView 类型）控件，在"Attributes"面板中修改其"textSize"和"textColor"属性即可。将"Hello World!"文字颜色（"textColor"属性）设为醒目的酒红（#E91E63），字体（"textStyle"属性）设为加粗（勾选"bold"为"true"），如图 1.18 所示。

图 1.18　修改 TextView 控件的属性

（3）添加图片。要用图像视图 ImageView 控件显示图片，需要把图片资源加载到工程中。方法是：将需要显示的图片复制到当前工程的 drawable 目录下，笔者计算机上该目录的路径为"C:\Users\b0202\AndroidStudioProjects\HelloWorld\app\src\main\res\drawable"（读者可参考）。图片名"androidlover"为一个正在读书的可爱小机器人，读者也可使用自己喜欢的其他图片。

接下来设置 ImageView 的"srcCompat"属性，操作方式与前面类似，只不过这里是在属性设置的"Pick a Resource"（资源选择）窗口中选择由用户自己载入的图片资源"androidlover"，如图 1.19 所示，单击"OK"按钮。

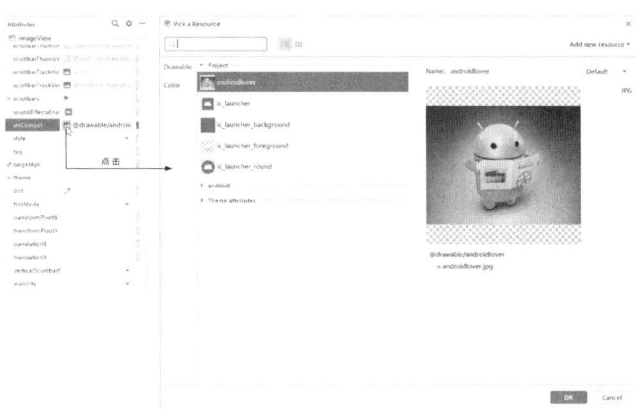

图 1.19　选择要显示的图片资源

（4）添加按钮。从工具箱面板拖曳一个 Button 控件至界面，考虑到与"Hello World!"字体大小相配且美观，这里将按钮的文字大小（"textSize"属性）设为 36dp，并调整界面布局。

最终设计后的程序界面如图 1.20 所示。

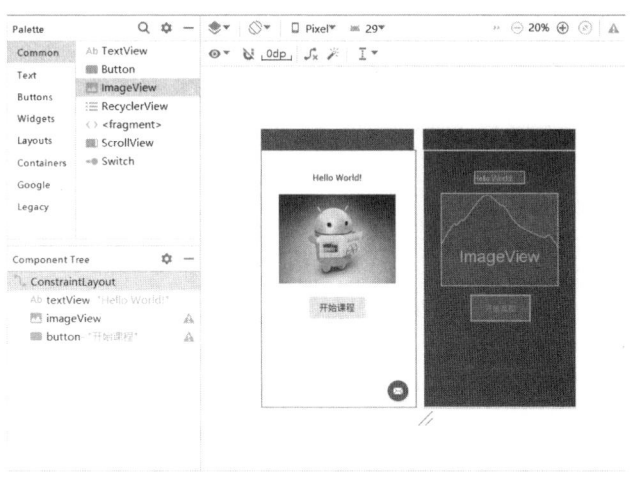

图 1.20　设计后的程序界面

1.3.3　添加程序代码

设计好应用程序界面后，接下来就要为 Android 程序编写代码以实现其应用功能。Android Studio 开发的工程，其代码编写遵循一定的模式，在既有的程序框架中的特定位置添加代码即可。此处，我们先给出需要添加的程序代码及其位置，至于这些代码的解析将在后面 1.3.6 小节再详细给出。

单击 Android Studio 工程开发界面右上方主开发区 "MainActivity.java" 切换至主程序的 Java 源文件编辑模式，如图 1.21 所示。

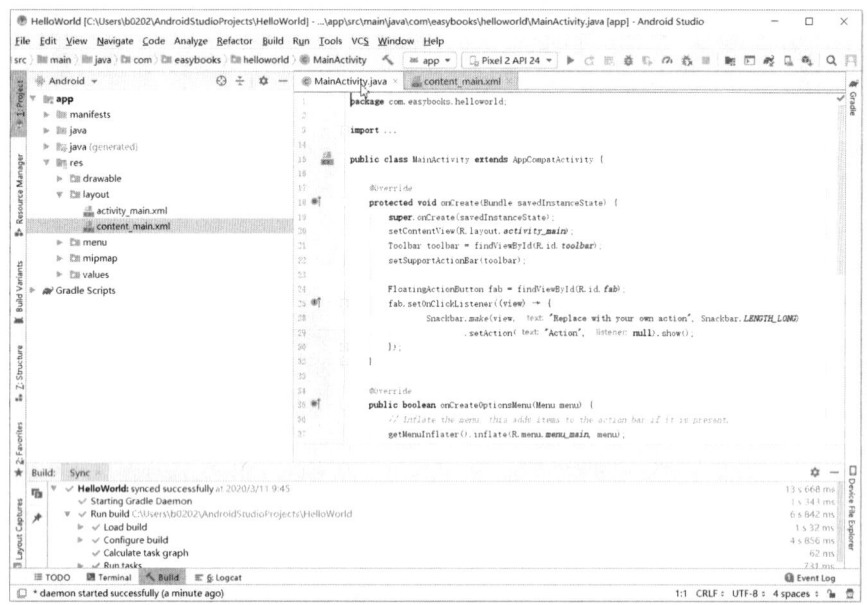

图 1.21 切换至主程序的 Java 源文件编辑模式

在 MainActivity.java 现成的程序框架基础上添加代码如下（加粗语句为添加的内容）：

```java
package com.easybooks.helloworld;

import android.os.Bundle;

import com.google.android.material.floatingactionbutton.FloatingActionButton;
import com.google.android.material.snackbar.Snackbar;

import androidx.appcompat.app.AppCompatActivity;
import androidx.appcompat.widget.Toolbar;

import android.view.View;
import android.view.Menu;
import android.view.MenuItem;

import android.widget.Button;
import android.widget.ImageView;
import android.widget.TextView;

public class MainActivity extends AppCompatActivity {
    private TextView textView;
    private ImageView imageView;
    private Button button;

    @Override
    protected void onCreate(Bundle savedInstanceState) {
        super.onCreate(savedInstanceState);
        setContentView(R.layout.activity_main);
```

```java
        findViews();

        Toolbar toolbar = findViewById(R.id.toolbar);
        setSupportActionBar(toolbar);

        FloatingActionButton fab = findViewById(R.id.fab);
        fab.setOnClickListener(new View.OnClickListener() {
            @Override
            public void onClick(View view) {
                Snackbar.make(view, "Replace with your own action", Snackbar.LENGTH_LONG)
                        .setAction("Action", null).show();
            }
        });
        button.setOnClickListener(new View.OnClickListener() {
            @Override
            public void onClick(View view) {
                String msg = textView.getText().toString().trim();
                msg += "    我爱Android编程！";
                textView.setText(msg);
            }
        });
    }

    private void findViews() {
        textView = (TextView) findViewById(R.id.textView);
        imageView = (ImageView) findViewById(R.id.imageView);
        button = (Button) findViewById(R.id.button);
    }

    @Override
    public boolean onCreateOptionsMenu(Menu menu) {
        // Inflate the menu; this adds items to the action bar if it is present.
        getMenuInflater().inflate(R.menu.menu_main, menu);
        return true;
    }

    @Override
    public boolean onOptionsItemSelected(MenuItem item) {
        // Handle action bar item clicks here. The action bar will
        // automatically handle clicks on the Home/Up button, so long
        // as you specify a parent activity in AndroidManifest.xml.
        int id = item.getItemId();

        //noinspection SimplifiableIfStatement
        if (id == R.id.action_settings) {
            return true;
        }

        return super.onOptionsItemSelected(item);
    }
}
```

1.3.4 Android 应用程序运行

Android 应用程序运行分为模拟运行和真机运行。

1. 模拟运行

Android Studio 自带仿真器（emulator，或称为 Android Virtual Device，简称 AVD）来模拟通常的智能手机屏，供开发人员随时运行和测试 App。用仿真器执行 Android 应用程序的方式称为**模拟运行**，以区别于在真实物理移动设备（如手机、平板电脑等）上的**真机运行**，所以在运行第一个 Android 应用程序前，先要创建一个 Android 仿真器。步骤如下：

（1）在 Android Studio 内，选择主菜单"Tools"→"AVD Manager"命令，或在工具栏上单击相应的图标按钮即可开启 AVD Manager（仿真器管理器），弹出"Your Virtual Devices"对话框，如图 1.22 所示。

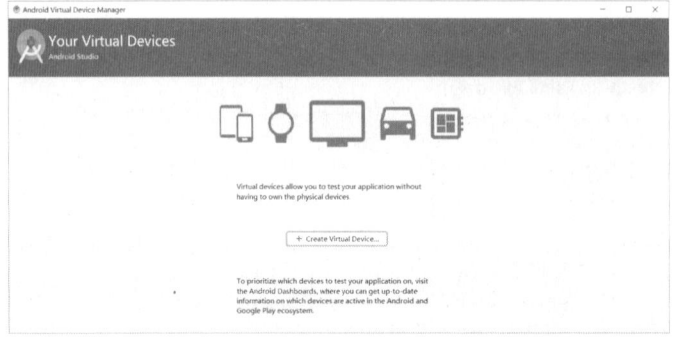

图 1.22 "Your Virtual Devices"对话框

（2）在对话框中单击"+ Create Virtual Device..."按钮，会显示可供选择的仿真器硬件类型，如图 1.23 所示，这里我们选一款名为"Pixel 2"的手机，单击"Next"按钮。

图 1.23 选择仿真器硬件类型

（3）选择仿真器要安装的操作系统版本。"Recommended"选项页下罗列出了所有建议选择的 ABI 字段值为 x86、Target 带有 Google Play 的系统映像，如图 1.24 所示。x86 代表模拟 Intel x86 Atom CPU，配合 Intel x86 仿真器加速器 1（Android Studio 默认会安装）可以让仿真器执行更顺畅；而带有 Google

Play 才能执行带有 Google Map 功能的应用程序。

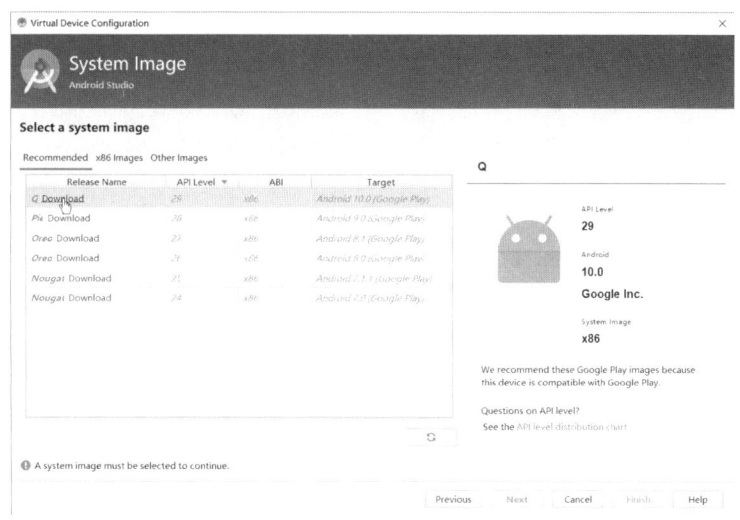

图 1.24　建议选择的系统映像

（4）安装所选映像对应版本的 SDK。由于 Android Studio 3.0 以上版本已不再默认集成 Android SDK，故要使用某个系统映像前必须先下载安装其对应版本的 SDK。这里我们想使用最新版 Android 10.0（Release Name 为 Q）操作系统，所以单击图 1.24 中该条目上的 "Download" 链接。在 "License Agreement" 页选中 "Accept" 单选钮接受许可条款，单击 "Next" 按钮随即进入下载页面，如图 1.25 所示。

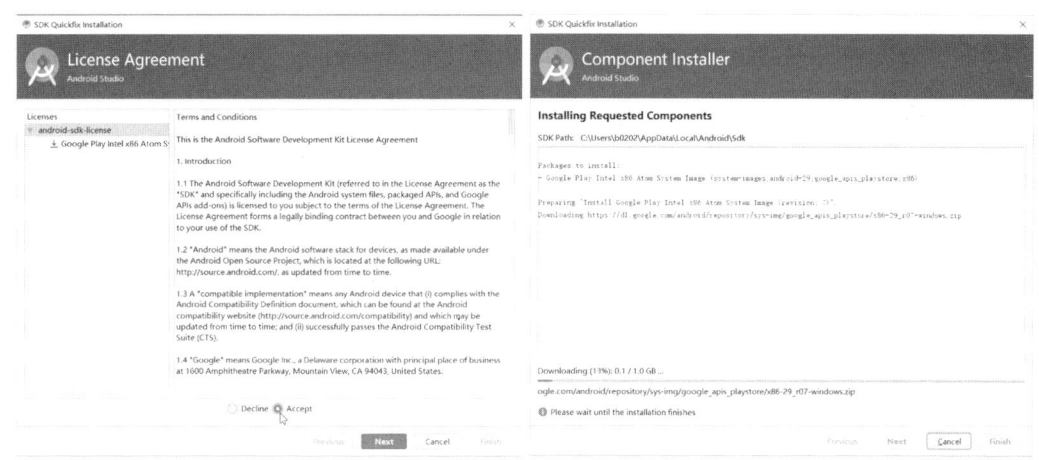

图 1.25　开始下载 SDK

完成后单击 "Finish" 按钮回到 "System Image" 页，可以看到 Q 版 Android 10.0 条目上的 "Download" 链接已经不见了，说明此 SDK 已经安装成功。这个时候就可以选中该条目，单击 "Next" 按钮进入下一步了。

> ◉◉注意：
> 某个条目映像的 SDK 只须安装一次，下次使用时就直接选中该条目，无须重复安装。读者也可将多个映像的 SDK 预先下载下来，留作以后测试 App 在不同版本 Android 上的兼容性之用。

（5）接下来确认仿真器的相关配置，如图 1.26 所示。用户还可以单击"Show Advanced Settings"按钮进行高级设置，设置完毕后单击"Finish"按钮。

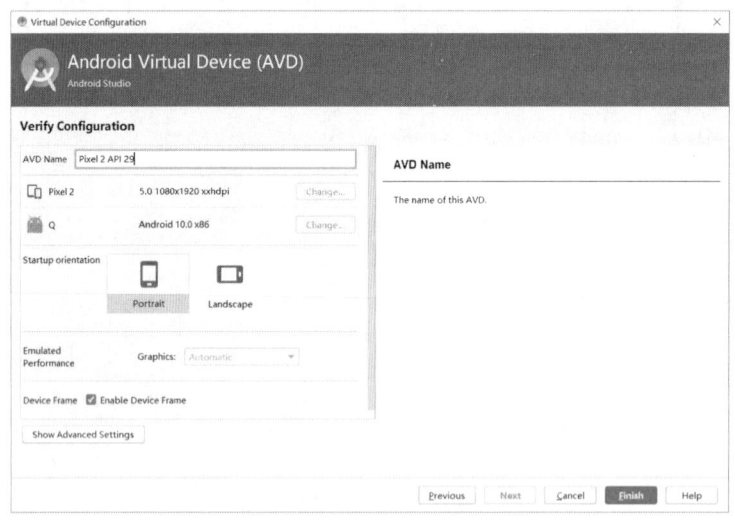

图 1.26　确认仿真器的相关配置

仿真器创建好之后，在"Your Virtual Devices"页可以看到刚刚创建的仿真器，如图 1.27 所示。用户可直接单击其后的"Actions"（▶）按钮来启动它，或者也可以在运行程序的时候由系统来自动启动。

图 1.27　创建完成的仿真器

仿真器创建后会自动生成 AVD 映像文件，在 Windows 10 系统中仿真器映像文件默认所在的路径为"C:\Users\<用户名>\.android\avd"，其中<用户名>为当前 Windows 系统的登录用户名。

创建好仿真器，就可以在其上运行 Android 程序了。选择 Android Studio 主菜单"Run"→"Run 'app'"命令（或单击工具栏上相应的图标按钮），等待片刻就会出现一个手机屏幕的界面，效果如图 1.28（a）所示。点击"开始课程"按钮，显示效果如图 1.28（b）所示。单击工具栏"Stop 'app'"（■）按钮可停止程序运行。

第 1 章 Android 开发入门

（a）

（b）

图 1.28 运行效果

> **注意：**
> 仿真器的启动要耗费大量系统资源，所以在开发程序时如无特殊情况不要关闭仿真器，直到工作完成后将要退出 Android Studio 前再关闭。仿真器开着的情况下，照样可以打开和关闭 Android Studio 工程，仿真器与工程是独立的，互不影响。

2. 真机运行

如果读者拥有一部智能手机，不妨尝试一下在真实的实体机上来运行 HelloWorld 程序，在此笔者以手机 vivo Z3i（型号为 V1813T/Android 9.0）为例，介绍在其上安装和运行 HelloWorld 程序的具体操作。步骤如下：

（1）将手机以 USB 数据线连接到开发环境所使用的计算机。

（2）下载安装 Google 驱动程序。选择 Android Studio 主菜单"File"→"Settings"命令，打开"Settings"窗口，如图 1.29 所示。切换至"SDK Tools"选项页，勾选列表中的"Google USB Driver"项，然后单击底部"Apply"按钮开始下载安装。

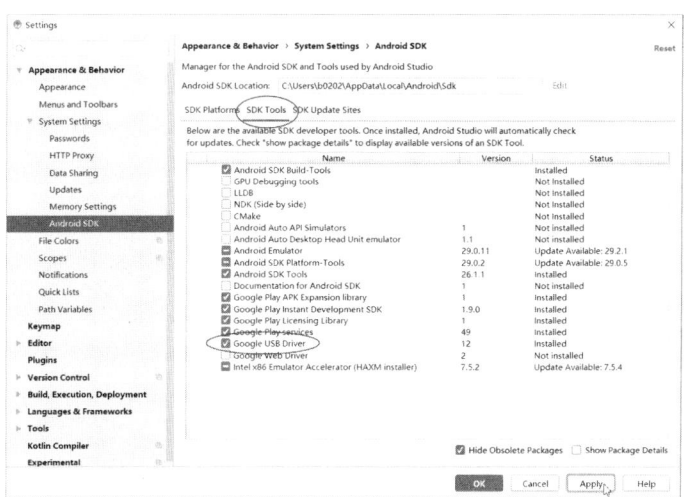

图 1.29 下载安装 Google 驱动程序

（3）更新手机设备驱动程序。打开 Windows 设备管理器，展开设备列表，找到手机对应的设备项后右击，在弹出的快捷菜单上单击"更新驱动程序软件"命令，在弹出对话框中单击"自动搜索更新的驱动程序软件"，如图 1.30 所示。稍等片刻，系统会自动找到刚刚下载安装的 Google 驱动程序并将它作为手机的驱动。

图 1.30　更新手机设备驱动程序

（4）打开手机开发者权限并允许 USB 调试。这一步在不同品牌和型号的手机中操作不尽相同，但大体上都是先进入手机设置界面，找到"开发者选项"打开，并开启"USB 调试"项即可。笔者的手机上开启权限的界面如图 1.31 所示，请读者参考着在自己的手机上进行操作。完成这一步后，在 Android Studio 工具栏选择 App 运行设备的下拉列表里就会多出一个对应该手机的设备选项，如图 1.32 所示，选中该项后单击旁边的"运行"(▶)按钮即可在真实手机上运行程序，运行效果如图 1.28 所示。

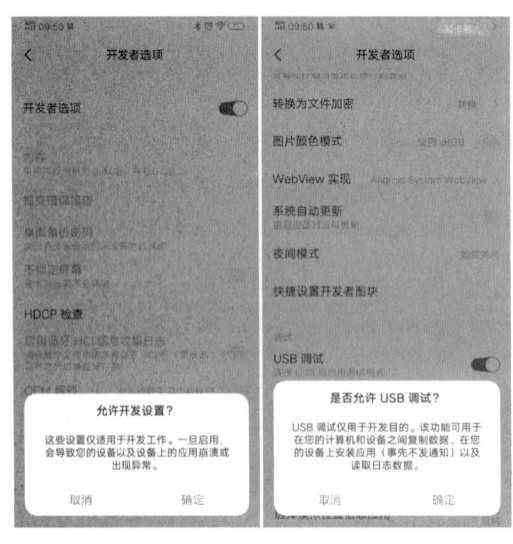

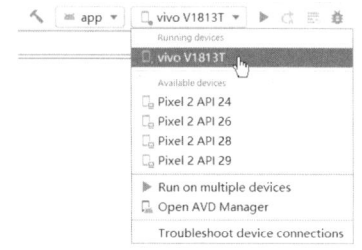

图 1.31　打开开发者权限并允许 USB 调试　　　图 1.32　对应真实手机的设备项

1.3.5 Android Studio 工程结构

接下来以前面这个 Android 应用程序为例来分析 Android Studio 工程的结构。

1. 工程结构概况

一个完整的 Android Studio 工程中包含的文件和资源很多，非常庞杂。为此，Android Studio 提供了 Project 工具窗口来辅助用户管理和查看工程的结构。Project 工具窗口位于开发界面的左侧，它有很多种显示模式（展开窗口左上方的下拉列表可以看到），如图 1.33 所示，其中"Android"和"Project"是最有用的两种模式。

默认情况下，Android Studio 会将模式设为"Android"，它为用户展示了工程中所有的包、目录和文件的概况。前面开发完成的 HelloWorld 工程在 Project 工具窗口中的 Android 模式视图如图 1.34 所示。

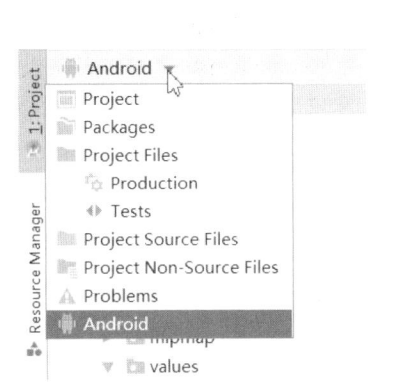

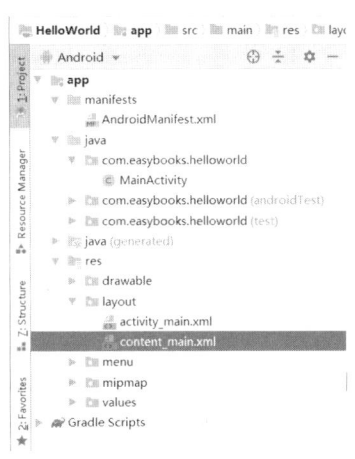

图 1.33　Project 工具窗口的显示模式　　　图 1.34　HelloWorld 工程结构

2. Android Studio 工程目录

由图 1.34 可见，一个典型的 Android Studio 工程的目录可分为 3 个部分：manifests 目录、java 目录与 res 目录。

（1）manifests 目录。每一个 Android 应用程序都对应一个 manifests 类型的文件，它存储着该应用程序的重要信息，系统生成的默认文件名为 AndroidManifest.xml，初学者一般不需要关注其中的内容。

（2）java 目录。java 目录内存放的是工程所有的 Java 源程序文件，Android Studio 工程默认创建的 Java 源文件为 MainActivity.java，它是 Android 开发者编写 Java 代码存放的主要文件。

（3）res 目录。工程所需的非程序资源大多放在 res 目录内，其中的文件名只能为小写字母、数字、_（下画线）、.（点）。res 目录的具体结构如图 1.35 所示。

res 目录中各个子目录的作用分别介绍如下：

① drawable 目录。drawable 目录存放工程需要的图片文件（.png、.jpg 等）资源。开发应用时，程序员需先将工程要用到的图片资源复制到该目录下，然后才能在代码中引用，设计界面上才能显示出想要呈现的图片。本应用程序界面上显示的"小机器人"图片正是放在该目录下面的。

图 1.35　res 目录的具体结构

② layout 目录。layout 目录专门存放用户界面的布局（layout）文件。在之前创建 Android Studio 工程时，第（2）步选择"Basic Activity"，系统会自动生成两个文件——activity_main.xml 与 content_main.xml，其中 activity_main.xml 是工程的主显示控制文件，而 content_main.xml 则是主设计文件。初学者一般不用过多关注 activity_main.xml 的内容，因为 content_main.xml 才是主设计界面，是用户编写 App 界面设计代码存放的主要文件。但若在创建工程的第（2）步选择了"Empty Activity"，则只生成一个 activity_main.xml，此时它就是主设计文件，用户直接在 activity_main.xml 中设计界面即可。

Android Studio 支持两种界面设计方式，除了之前介绍的可视化设计，用户也可以直接编写 XML 代码来描述和设计程序界面，单击 content_main.xml 文件 Editor 窗口底部的"Text"选项卡可切换到代码视图，如图 1.36 所示，在其中编辑 content_main.xml 文件的源码。

图 1.36　代码视图

从图 1.36 中可看到之前设计界面时拖曳放置到其上的各控件的标签，如<TextView ... />、<ImageView ... />等（图中圈出）。用户可通过直接修改这些标签的属性来改变对应控件的外观，或添加新的标签来实现自己需要显示的界面元素。

③ menu 目录。menu 目录只是用于菜单的界面设计，现在 App 开发并不流行菜单，这里就不具体介绍了。

④ mipmap 目录。与 drawable 目录的作用类似，也用于存放工程的图片资源，可以按照屏幕分辨率将图片分成 hdpi（高分辨率）、mdpi（中分辨率）、xhdpi（超高分辨率）等。实际开发中，可将相同图片制作成多份不同分辨率的版本分别存放在对应的目录内，以适应多种不同档次屏幕分辨率的手机。

⑤ values 目录。values 目录存放用户界面用到的颜色（colors.xml）、文字（strings.xml）和样式（styles.xml）等的定义文件，便于集中统一管理。以文字为例，对于要求同时支持多国语言的 App，将整个应用的界面文字全都集中在 strings.xml 内，翻译人员只要直接翻译 strings.xml 中的文字即可，而无须到工程 layout 目录下多个 XML 文件的源代码中去四处寻找要翻译的文字，便于应用程序本地化。不过，对于一般不需要同时支持多国语言的 App，笔者还是建议将界面文字直接写在 layout 目录下的设计文件内，这样可保持界面源码的可读性，使界面功能一目了然，便于调整和维护。

3. Android Studio 工程架构信息

单击打开 Project 工具窗口中的"Gradle Scripts"→"build.gradle (Module: app)"文件，会展示如图 1.37 所示的内容，这是本工程的架构信息，可用于查阅，初学者一般不要修改它。

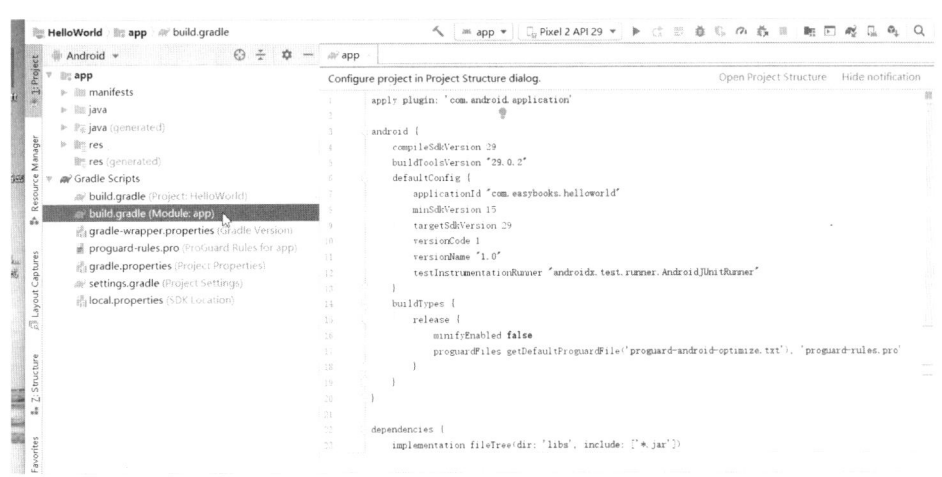

图 1.37　查看工程架构信息

1.3.6　应用程序代码解析

本小节就前面开发的 HelloWorld 工程的代码进行详细的解析说明。

1. 应用程序框架

应用程序框架是在创建工程时由 Android Studio 自动生成好了的，创建工程时所选择的 Activity（活动页面）类型不同，所生成的程序框架会有差异。本例程序的 Activity 类型为"Basic Activity"，生成的框架代码如下：

```
package com.easybooks.helloworld;                                    //（a）
//系统生成的导入组件代码
```

```java
import android.os.Bundle;                                                    // (b)

import com.google.android.material.floatingactionbutton.FloatingActionButton;
import com.google.android.material.snackbar.Snackbar;

import androidx.appcompat.app.AppCompatActivity;
import androidx.appcompat.widget.Toolbar;

import android.view.View;
import android.view.Menu;
import android.view.MenuItem;

public class MainActivity extends AppCompatActivity {                        // (c)
    @Override                                                                // (d)
    protected void onCreate(Bundle savedInstanceState) {                     // (d)
        super.onCreate(savedInstanceState);                                  // (d)
        setContentView(R.layout.activity_main);                              // (e)
        Toolbar toolbar = findViewById(R.id.toolbar);                        // (f)
        setSupportActionBar(toolbar);                                        // (g)

        FloatingActionButton fab = findViewById(R.id.fab);                   // (h)
        fab.setOnClickListener(new View.OnClickListener() {                  // (i)
            @Override
            public void onClick(View view) {
                Snackbar.make(view, "Replace with your own action", Snackbar.LENGTH_LONG)
                        .setAction("Action", null).show();     //页面底部提示信息条（用户可编程定制）
            }
        });
    }

    @Override
    public boolean onCreateOptionsMenu(Menu menu) {                          // (j)
        // Inflate the menu; this adds items to the action bar if it is present.
        getMenuInflater().inflate(R.menu.menu_main, menu);                   // (k)
        return true;
    }

    @Override
    public boolean onOptionsItemSelected(MenuItem item) {                    // (l)
        // Handle action bar item clicks here. The action bar will
        // automatically handle clicks on the Home/Up button, so long
        // as you specify a parent activity in AndroidManifest.xml.
        int id = item.getItemId();

        //noinspection SimplifiableIfStatement
        if (id == R.id.action_settings) {
            return true;
        }

        return super.onOptionsItemSelected(item);
    }
}
```

其中：

（a）**package com.easybooks.helloworld;**：此为 Android Studio 工程所在的包名，以"package"声明，后跟完整的包路径，这个包名就是之前创建工程时由用户指定的（见图 1.15）"Package name"信息项的内容。

（b）**import…**：系统会在程序框架开头以一系列的 import 语句自动导入一些程序运行必备的组件，这里主要包括：

- Bundle（android.os.Bundle）：用于 App 中页面间的数据传递。
- FloatingActionButton（com.google.android.material.floatingactionbutton.FloatingActionButton）：App 页面顶部导航栏的悬浮按钮。
- Snackbar（com.google.android.material.snackbar.Snackbar）：App 底部提示信息条。
- AppCompatActivity（androidx.appcompat.app.AppCompatActivity）：主程序类继承的 Activity 父类。
- Toolbar（androidx.appcompat.widget.Toolbar）：导航栏。
- View（android.view.View）：视图基类。
- Menu（android.view.Menu）和 MenuItem（android.view.MenuItem）：菜单和菜单项。

（c）**public class MainActivity extends AppCompatActivity**：MainActivity 是工程默认创建的主程序类，它代表了一个 App 的主页面。Android Studio 开发的应用程序，其 MainActivity 都继承自 AppCompatActivity；而过去用 Eclipse 开发的老 Android 程序的 MainActivity 继承的则是 Activity。比之传统的 Activity，AppCompatActivity 提供了更好的向下兼容性，确保各种基本控件在不同类型的设备上都有着统一的外观风格。

（d）**protected void onCreate(Bundle savedInstanceState)**：onCreate()方法是一切 Activity（AppCompatActivity）生命周期的第一个方法，所有 App 页面在启动时首先执行的就是这个方法。在 MainActivity 中对它进行了覆盖重写，故在方法前带"@Override"注解表示是重写的方法。一般开发中，都会通过重写这个方法来完成应用程序的初始化工作。该方法带一个 Bundle 类型的参数 savedInstanceState，它的作用是保存应用程序状态信息。当一个 App 退出时，系统会自动调用 onSaveInstanceState()方法来保存当前页面（Activity）的状态，状态数据以键值对的形式存入 savedInstanceState 中。App 下次启动时，执行 onCreate()方法一开始调用 super.onCreate(savedInstanceState) 就能从该参数中获取程序的状态数据。savedInstanceState 在实际 App 开发中非常有用，例如，很多手游 App 在启动时能够复原上次游戏的画面，让玩家可以接着继续玩下去；电子书 App 能够"记住"用户之前阅读到的章节等。

（e）**setContentView(R.layout.activity_main);**：通过 setContentView()方法加载主页面的内容视图，本例中就是程序主界面的视图，以资源 id（R.layout.*）的形式从工程中引用，activity_main 也就是设计界面的 UI 文件 activity_main.xml 的 id 标识，实际开发中还可以根据需要载入事先设计好的其他内容视图。

（f）**Toolbar toolbar = findViewById(R.id.toolbar);**：根据 id 关联应用程序导航栏控件对象，"Basic Activity"类型的工程默认都会包含导航栏，位于 App 页面顶端。导航栏控件对象定义在工程的 activity_main.xml 文件中，代码如下：

```
<com.google.android.material.appbar.AppBarLayout
    android:layout_width="match_parent"
    android:layout_height="wrap_content"
    android:theme="@style/AppTheme.AppBarOverlay">
```

```xml
<androidx.appcompat.widget.Toolbar
    android:id="@+id/toolbar"                              //就是通过这个id进行关联的
    android:layout_width="match_parent"
    android:layout_height="?attr/actionBarSize"
    android:background="?attr/colorPrimary"
    app:popupTheme="@style/AppTheme.PopupOverlay" />

</com.google.android.material.appbar.AppBarLayout>
```

（g）**setSupportActionBar(toolbar);**；将这个导航栏设为主页面的导航栏。

（h）**FloatingActionButton fab = findViewById(R.id.fab);**；根据 id 关联导航栏上的悬浮按钮，并为按钮绑定点击事件以响应用户操作。悬浮按钮控件对象也是定义在 activity_main.xml 文件中，代码如下：

```xml
<com.google.android.material.floatingactionbutton.FloatingActionButton
    android:id="@+id/fab"                                  //就是通过这个id进行关联的
    android:layout_width="wrap_content"
    android:layout_height="wrap_content"
    android:layout_gravity="bottom|end"
    android:layout_margin="@dimen/fab_margin"
    app:srcCompat="@android:drawable/ic_dialog_email" />
```

（i）**fab.setOnClickListener(new View.OnClickListener(){ ... });**；调用悬浮按钮上的 setOnClickListener()方法设置绑定其上的监听器对象，有关监听器绑定及事件处理的机制将在稍后展开介绍。

（j）**public boolean onCreateOptionsMenu(Menu menu)**：该方法用于初始化菜单，其中的参数 menu 就是即将要显示的 Menu 实例，方法返回 true 表示显示该 menu，返回 false 则不显示。

（k）**getMenuInflater().inflate(R.menu.menu_main, menu);**；调用 Activity 的 getMenuInflater()方法得到一个 MenuInflater，再通过 inflate()方法来把布局文件中定义的菜单加载给第 2 个参数所对应的 menu 对象。定义菜单的布局文件是位于工程 menu 目录下的 menu_main.xml，其中定义的菜单元素如下：

```xml
<menu xmlns:android="http://schemas.android.com/apk/res/android"
    xmlns:app="http://schemas.android.com/apk/res-auto"
    xmlns:tools="http://schemas.android.com/tools"
    tools:context="com.easybooks.helloworld.MainActivity">
    <item
        android:id="@+id/action_settings"
        android:orderInCategory="100"
        android:title="@string/action_settings"
        app:showAsAction="never" />
</menu>
```

用户可在其中设计菜单、子菜单及菜单项的具体内容。

（l）**public boolean onOptionsItemSelected(MenuItem item)**：该方法是当菜单项被点击时调用的，也就是菜单项的监听方法。对于一个 Activity 来说，同一时刻只能显示和监听一个 Menu 对象。

以上（j）、（k）、（l）皆是与系统菜单功能相关的方法，用户可通过修改或重写它们来对应用的菜单进行定制开发，但通常 App 开发中菜单用得并不多，故对此读者只须了解即可。

如果在前面创建工程的时候，用户选择的 Activity 类型是"Empty Activity"，则生成的代码框架中既不会包含与菜单相关的这几个方法，也不会有导航栏控件对象和悬浮按钮，此时的程序框架将变得极为简单（仅一个需要重写的 onCreate()方法）。这种情况下用户只要在"setContentView(R.layout.activity_main);"语句的后面紧接着往下编写页面的初始化代码段，并在 onCreate()方法外部编写实现程序各具体功能的方法代码即可，整个 App 程序的结构形如：

```
package 包名;
import ...;                                          //导入组件库
...
public class MainActivity extends AppCompatActivity {
    private ...;                                     //定义变量
    ......
    @Override
    protected void onCreate(Bundle savedInstanceState) {
        super.onCreate(savedInstanceState);
        setContentView(R.layout.activity_main);
        //添加初始化代码
           ......
    }
    ......
    //各功能方法的定义及实现
       ......
}
```

2. 添加代码详解

在以上既有程序框架中的特定位置添加代码，下面我们来对本例所添加的代码（加粗语句）进行详细分析：

```
package com.easybooks.helloworld;
……                                                  // （a）系统生成的导入控件代码
import android.widget.Button;                        // （a）导入命令按钮控件
import android.widget.ImageView;                     // （a）导入图片显示控件
import android.widget.TextView;                      // （a）导入文本显示控件

public class MainActivity extends AppCompatActivity {
    private TextView textView;                       // （b）定义显示"Hello World!"等文字的控件对象
    private ImageView imageView;                     // （b）定义显示"安卓小机器人"图片的控件对象
    private Button button;                           // （b）定义"开始课程"按钮控件对象

    @Override
    protected void onCreate(Bundle savedInstanceState) {
        super.onCreate(savedInstanceState);
        setContentView(R.layout.activity_main);
        private findViews();                         // （c）
        ……                                           //系统生成的其他代码
        button.setOnClickListener(new View.OnClickListener() {   // （d）
            @Override
            public void onClick(View view) {         // （e）
                String msg = textView.getText().toString().trim();   // （f）
                msg += "    我爱Android编程！";      // （g）
                textView.setText(msg);               // （h）
            }
        });
    }

    private void findViews() {                       // （c）
        textView = (TextView) findViewById(R.id.textView);
        imageView = (ImageView) findViewById(R.id.imageView);
        button = (Button) findViewById(R.id.button);
```

```
        }
        ……                                          //系统生成的其他代码
    }
```

其中：

（a）**import android.widget.Button;**、**import android.widget.ImageView;** 和 **import android.widget. TextView;**：一个 Android 应用程序的 UI 组件（Palette 面板上）可分成 widget 与 layout 两大类。widget 组件是 UI 的最基本单位，换句话说，不能在这类组件内再放入其他组件，如本例用到的 TextView、ImageView 和 Button 就都是 widget 组件。而 layout 组件是布局用组件，在它们的内部还可以放入其他组件。widget 组件相关的类都放在 android.widget 包内，当用户在编程过程中使用到这些组件时，系统自动生成"import android.widget.<组件类名>;"语句将相应的组件类引入程序，无须用户手工编写导入语句。

（b）**private TextView textView;**、**private ImageView imageView;** 和 **private Button button;**：Android 应用是基于 Java 语言的，已经设计好的界面控件如果要在代码中引用，需要首先定义成 Java 对象。通行的做法是声明成主程序类（MainActivity）的私有成员，并在初始化时将它们各自关联到界面上对应的控件。

（c）**private void findViews()**：findViews()是我们定义的一个方法，它的主要用途是将 UI 组件与 Java 对象关联起来，紧接在 setContentView()语句之后调用执行。如果界面上的控件不多，也可以直接将 findViews()中的语句写在 onCreate()方法中（也是位于 setContentView()语句之后）即可。在 findViews()方法内调用 findViewById()并指定 id 来获取界面源文件（content_main.xml）内相应的 UI 组件，然后赋值给相应的控件对象，将它们关联起来，之后这些控件对象就分别代表了界面上的 UI 组件，可以在程序中直接引用了。如果关联的 UI 组件类型很普通，并不是程序中的重要控件，在编程时也可省略 findViewById()返回的强制类型转换，而简写作：

```
textView = findViewById(R.id.textView);
imageView = findViewById(R.id.imageView);
button = findViewById(R.id.button);
```

这里关联 UI 组件对象时所指定的 id 也就是界面源文件（content_main.xml）中各 UI 元素的"android:id"属性值。

（d）**button.setOnClickListener(new View.OnClickListener(){ ... });**：为界面上的按钮添加点击事件处理功能。View.OnClickListener 代表 OnClickListener（即监听器），它是 View 的内部类（inner class）。View 可以调用 setOnClickListener()来向监听器注册点击事件，而 View 类又是所有 Android UI 组件的共同父类，故所有 UI 组件都可以这么做，因此，这里 button 才能顺利地调用 setOnClickListener()方法来注册事件。

（e）**public void onClick(View view)**：利用匿名内部类实现 OnClickListener 的 onClick()方法，当按钮被按下时，onClick()方法会自动被调用并执行。onClick()方法必须传入一个 View 类型的参数 view，它代表的是触发事件的 UI 组件，这里也就是被按下的 button 按钮。

> ● ● Android 事件处理：
> 在 Android 程序中，普遍采用注册事件监听器的方式来处理事件，它的原理是：将要求响应的事件预先注册到系统监听器（俗称"看门狗"）中，同时编程实现该事件对应的处理方法。当程序运行时，系统监听器时刻处于监听状态，一旦监听到该事件发生，随即执行其对应的方法代码。而对于未注册的事件，即便它发生了，监听器也会对其一概忽略，不做任何反应。以 HelloWorld 程序中的按钮 button 为例，其事件处理的工作流程如图 1.38 所示。

① button 向监听器注册事件：button 必须先调用 setOnClickListener()向 OnClickListener（监听器）注册点击事件，这样在程序运行时，OnClickListener 才会去监听 button 是否被单击。

② 实现点击事件的处理方法：事件处理方法用来实现用户单击 button 后希望程序执行的功能，处理方法的代码写在 OnClickListener 的 onClick()方法内。

③ 产生点击事件：当界面上的 button（"开始课程"按钮）被用户触击时，会产生一个点击事件，OnClickListener 监听器就会知道。

④ 执行处理：由于该点击事件事先已经向 OnClickListener 注册过了，故 OnClickListener 就会对它进行处理，于是自动调用用户编程实现好的 onClick()方法来响应用户触击按钮的操作。

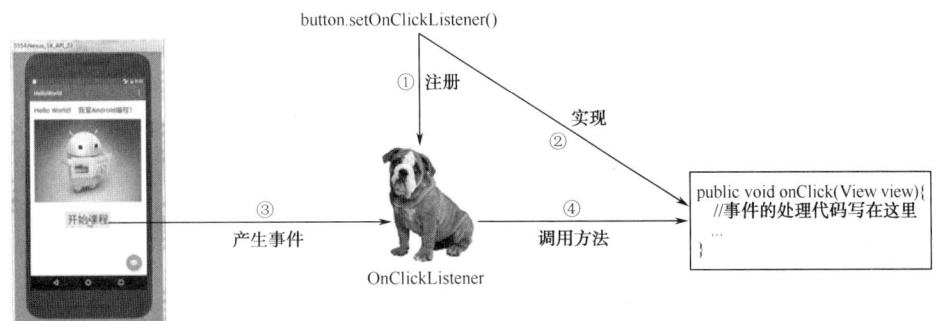

图 1.38 Android 事件处理的工作流程

（f）**String msg = textView.getText().toString().trim();**：通过 TextView 组件的 getText()方法获取其上的文字后，调用 trim()去除不必要的空格符号。

（g）**msg += "   我爱 Android 编程！";**：定义字符串类型的变量 msg，将"   我爱 Android 编程！"串接在原"Hello World!"之后。

（h）**textView.setText(msg);**：通过调用 TextView 组件的 setText()方法将新组合的文字显示在界面上。

1.3.7 事件处理的 4 种编程范式

Android 编程中的事件处理代码可以有 4 种不同的编写范式，为方便读者今后阅读和理解他人所写的程序代码，下面来系统地介绍这几种编程范式。

1. 范式一：UI 组件直接创建监听器

由 UI 组件对象在调用 setOnClickListener()方法时直接创建一个监听器对象，并在其中当场实现重写事件处理的代码，其程序结构形如：

```
UI 组件对象.setOnClickListener(new View.OnClickListener() {
    @Override
    public void onClick(View view) {
        ......                                          //事件处理代码
    }
});
```

前面 1.3.6 小节程序中的事件处理就是这样实现的，这里的"UI 组件对象"也就是界面上的"开始课程"按钮，对应于程序中的控件对象 button，于是就写成：

```
button.setOnClickListener(new View.OnClickListener() {
    @Override
```

```
        public void onClick(View view) {
            String msg = textView.getText().toString().trim();
            msg += "    我爱 Android 编程！";
            textView.setText(msg);
        }
    });
```

这种方式将事件处理代码内嵌于注册监听器的语句之中，当事件处理的代码量较大时会使程序的上下文结构不甚明晰，容易出错。改进的办法是将事件处理代码单独封装于一个类，于是得到下面的范式二。

2. 范式二：UI 组件实例化监听类

这种方式是在外部自定义一个监听类，它实现了 View.OnClickListener 接口，然后在需要监听和处理事件的 UI 组件上注册并实例化这个监听类，程序结构形如：

```
UI 组件对象.setOnClickListener(new 监听类() );
……
public class  监听类  implements View.OnClickListener {
    @Override
    public void onClick(View view) {
        ……                                              //事件处理代码
    }
}
```

其中，"监听类"是用户自定义的类，可写在程序中的任何地方，这样就将事件处理的代码与注册事件的语句分离开来，优化了程序结构。用这种方式改写后的程序代码如下：

```
……
@Override
protected void onCreate(Bundle savedInstanceState) {
    super.onCreate(savedInstanceState);
    ……
    button.setOnClickListener(new MyListener() );              //注册事件的语句
}
……                                                            //省略无关代码
public class MyListener implements View.OnClickListener {
    @Override
    public void onClick(View view) {
        String msg = textView.getText().toString().trim();
        msg += "    我爱 Android 编程！";
        textView.setText(msg);
    }
}
```

范式一和范式二编程所依赖的内部类（如这里的 OnClickListener）是内置于所有 UI 组件的共同父类 View 中的，除了 OnClick 点击事件，开发者同样可以灵活地运用这两种范式来实现对 Android 中各种 UI 组件任意类型事件的处理，差异仅在于注册监听器时用的 set 方法名，以及需要实现的监听器接口的名字不一样。例如，长按事件的注册方法为 setOnLongClickListener()，要实现的接口是 View.OnLongClickListener；CheckBox（复选框）控件选择状态改变事件的注册方法为 setOnCheckedChangeListener()，要实现的接口是 CompoundButton.OnCheckedChangeListener，等等。

3. 范式三：layout 界面设置事件方法

上述两种范式都要求用户对 Java 语言的事件模型、工作原理和机制有一个比较清楚的了解，编程存在一定难度。不过，Android Studio 还提供了另一种更简易的方式，就是开发者可以在 layout 文件内设置 Button 组件被单击时要调用的方法，该方法可以自定义，作用等同于 onClick()。如此一来，就省去了向监听器注册的步骤，简化了编程。具体操作如下：

（1）在主程序类中添加一个自定义的 onStartClick()方法（加粗处）如下：

```
public class MainActivity extends AppCompatActivity {
    ……
    @Override
    protected void onCreate(Bundle savedInstanceState) {
        super.onCreate(savedInstanceState);
        setContentView(R.layout.activity_main);
        ……                                    //系统生成的其他代码
        //注释掉原来的代码段
        /*
        button.setOnClickListener(new View.OnClickListener() {
            ……
        });
        */
    }
    private void findViews() { … }
    //添加的方法
    public void onStartClick(View view) {
        ……                                    //事件处理代码
    }
}
```

> **注意：**
> 用户自己添加定义的方法也必须传入一个 View 类型的参数，且该方法必须为"public void"类型。

（2）切换到设计视图，选中手机屏上的"开始课程"按钮，在右侧"Attributes"面板中找到"onClick"属性，可以看到其下拉列表选项中多了一个"onStartClick"项——正是上面程序中添加定义的方法，如图 1.39 所示。选中该项，将其设为按钮的"onClick"属性值。

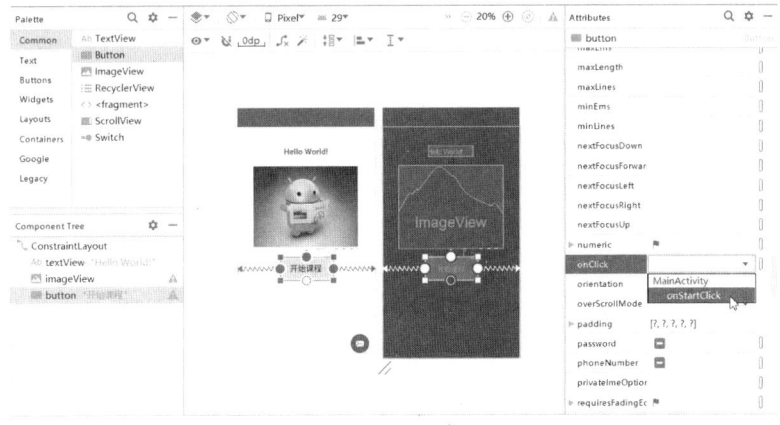

图 1.39 设置按钮的"onClick"属性

这样设置之后，打开设计界面的源文件 content_main.xml，可以看到按钮 button 的"android:onClick"属性已经与用户定义的事件处理方法 onStartClick 相关联了，如下（加粗语句）所示：

```xml
<?xml version="1.0" encoding="utf-8"?>
<androidx.constraintlayout.widget.ConstraintLayout xmlns:android="http://schemas.android.com/apk/res/android"
    …>
    <Button
        …
        android:onClick="onStartClick"
        … />
</androidx.constraintlayout.widget.ConstraintLayout>
```

运行程序，单击"开始课程"按钮就可以执行 onStartClick()方法实现功能。但这种方式有很大的局限性，它仅适用于单击类型的事件，而不像前两种编程范式那样可用于对任意类型 UI 事件的处理。

4. 范式四：主程序类实现监听接口

它的编程方式如下：

```java
public class MainActivity extends AppCompatActivity implements View.OnClickListener {
    ......
    @Override
    protected void onCreate(Bundle savedInstanceState) {
        super.onCreate(savedInstanceState);
        setContentView(R.layout.activity_main);
        ……
        button.setOnClickListener(this);                   //以 this 引用主程序本身
    }
    ……
    @Override
    public void onClick(View view) {                       //重写的事件处理方法
        String msg = textView.getText().toString().trim();
        msg += "    我爱 Android 编程！ ";
        textView.setText(msg);
    }
}
```

在这种范式中，是由主程序类直接实现监听器接口的，而重写的事件处理方法（要带@Override 注解）就作为主程序类中的一个成员方法，UI 组件对象在注册监听器的时候以 this 引用主程序本身即可，无须再另外定义和实例化监听类，使得程序代码逻辑清晰、简洁易读。

在 Android 程序规模比较大、涉及事件数目较多时，采用范式四编程利于对所有事件执行统一的注册和管理，便于程序维护，故在一般的 App 开发中用得最多的就是范式四（本书实例开发中也较多地使用了这种范式），它既简单，又能适应大多数不同类型的事件。但某些 UI 组件所特有的事件类型却并不在 MainActivity 主程序类能够实现的功能范围之内，故对这些特殊情况还是要由 UI 组件自己来注册和创建事件监听器。

1.4 Android SDK 的安装与管理

为使 Android 开发工具走向轻量化，Android Studio 自 3.0 以后不再与 SDK 集成发布，改由用户根据自己开发的需要下载和升级 SDK，Android Studio 仅提供管理下载的工具。因此，要用好 Android Studio，就必须对 Android SDK 的下载、安装、升级、管理和维护等相关的知识有个全面的了解。

Android Studio 提供了 SDK Manager（SDK 管理器）来下载和管理 SDK，选择主菜单"Tools"→"SDK Manager"命令，或单击工具栏上相应的图标按钮即可开启 SDK Manager。

1.4.1 通过下载 AVD 映像安装 SDK

通过下载 AVD 映像安装 SDK 是最简单便捷的方式。在之前运行第一个 Android 工程开启 AVD Manager（仿真器管理器）时，看到每一个条目前都有一个"Download"链接，单击后会下载安装对应版本 Android 系统的映像文件，在安装映像的同时，系统就会自动将该版本 Android 系统的 SDK 安装好，如图 1.40 所示，在 SDK Manager 的"Android SDK"→"SDK Platforms"选项页列表中带对钩的就是已经安装好的 SDK。与此同时，我们在"C:\Users\<用户名>\AppData\Local\Android\Sdk\platforms"目录下也能看到对应该版本 SDK 的目录。

图 1.40　安装映像的同时已经将 SDK 安装好

这时候就可以直接使用这个 SDK 及包含它的映像 AVD 来运行我们编写的 Android 程序。

1.4.2 通过 SDK Manager 安装 SDK

SDK Manager 的列表中显示了所有可用的 SDK，之前运行 HelloWorld 程序前已经安装了 Android 10.0，现在我们再选择安装一个 Android 8.1 版的 SDK，勾选对应条目，单击窗口底部的"Apply"按钮，如图 1.41 所示。

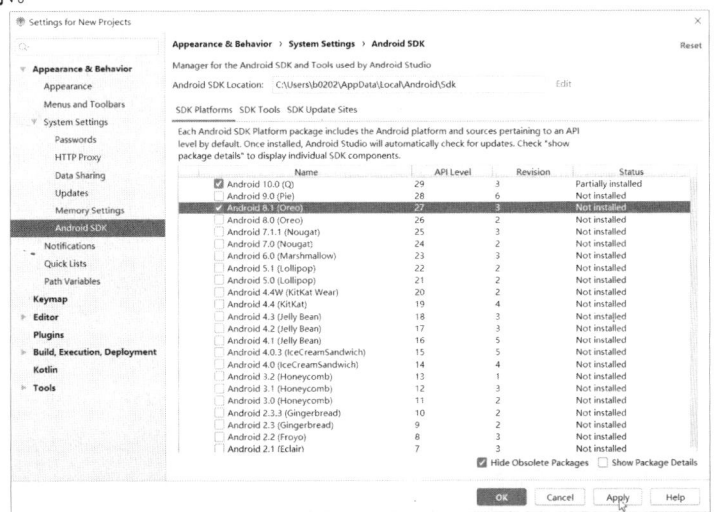

图 1.41　安装 Android 8.1 版的 SDK

弹出对话框，单击"OK"按钮，系统开始自动下载和安装 SDK，如图 1.42 所示。

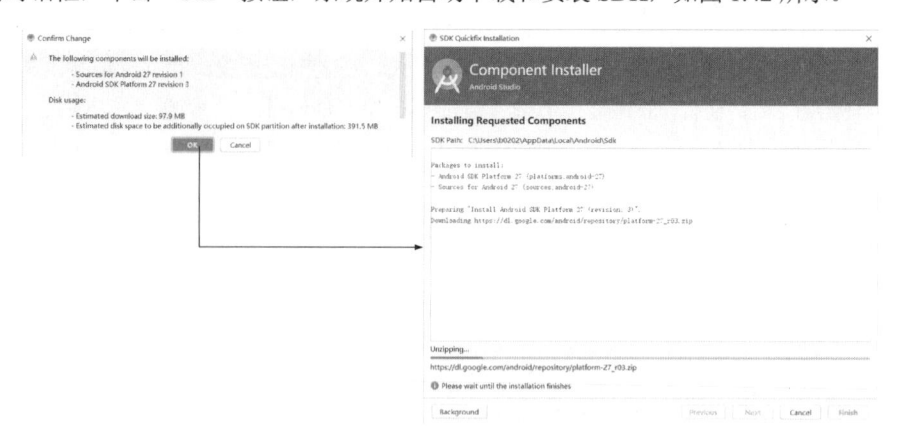

图 1.42　系统开始自动下载和安装 SDK

完成后，同样在"C:\Users\<用户名>\AppData\Local\Android\Sdk\platforms"目录下能看到此 SDK 对应的目录。

1.4.3　两种安装方式的区别与联系

前述两种方式虽然都能安装 SDK，但它们是有区别的。使用 AVD 映像安装的 SDK 直接就能用；而用 SDK Manager 下载安装的 SDK 并不能马上投入使用，因为还没有载入使用它的映像系统，在用 AVD Manager 创建仿真器时，其对应条目前依然有"Download"链接，只有在单击下载安装了映像系统后才能正常使用这个 SDK。

下载安装完映像之后，在"C:\Users\<用户名>\AppData\Local\Android\Sdk\system-images"目录下出现对应版本设备的 Image 映像目录，这个时候再用 AVD Manager 创建仿真器，可以看到对应 Android 8.1 版前面的"Download"链接已经不见了，如图 1.43 所示，此条目变为可选，才能用它来创建仿真器运行 Android 程序。

图 1.43　SDK 变为可用

综上所述：用 AVD Manager 下载安装映像的同时就会下载安装 SDK，可直接用来运行程序；而通过 SDK Manager 下载安装的只是单独一个 SDK（不含映像系统），若要在其上运行程序，还要再用 AVD Manager 安装映像。

Android Studio 中可同时安装多个版本的 SDK。实际开发中，常常安装多个映像（对应不同版本的 SDK）来测试同一个 App 在不同版本 Android 操作系统上的兼容性。

第 2 章 Android 用户界面

用户界面（UI）是系统和用户之间进行信息交换的媒介，实现信息的内部形式与人类可以接受的形式之间的转换。用户界面设计是应用程序开发的重要组成部分，决定了应用程序是否美观、易用。

本章介绍 Android 界面设计常用的基本控件及其事件的处理方法。

2.1 用户界面基础

2.1.1 用户界面基本要求

1. Android 应用界面设计

在手机上设计用户界面，与传统的桌面应用程序界面设计有所不同，它必须满足以下两点基本要求。

（1）界面与程序分离。界面设计与程序逻辑完全分离，这样修改界面时不需要改动程序功能实现的逻辑代码。Android 系统使用 XML 文件对用户界面进行描述，而实现程序逻辑为 Java 源文件，两者是完全分离的，各种资源文件分门别类地独立保存于各自专有的文件夹中。

（2）自适应手机屏幕。不同型号手机的屏幕可视参数（如解析度、尺寸和长宽比等）各不相同，Android 允许模糊定义界面控件的位置和尺寸，通过声明界面控件的相对位置和粗略尺寸，使界面控件能够根据屏幕尺寸和屏幕摆放方式动态地调整显示方式。

2. Android 框架模式：MVC

MVC 全名是 Model View Controller，是模型（Model）—视图（View）—控制器（Controller）的缩写，是一种框架模式。它是将业务逻辑、数据、界面分离的一种代码组织方式，修改界面时无须修改业务逻辑。

- 视图：接受用户的请求，然后将请求传递给 Controller。
- 控制器：进行业务逻辑处理后，通知 Model 去更新。
- 模型：数据更新后，通知 View 去更新界面显示。

控制器、视图和模型的关系如图 2.1 所示。

Android 中一般布局的 XML 文件就是 View 层，Activity 则充当了 Controller 的角色，数据模型 Model 通常是由开发者自定义类来实现的。

下面通过一个简单的例子来进一步说明三者之间的联系，这个例子的功能是在第 1 章 HelloWorld 程序基础上稍加改造而成的，实现每点击一次"开始课程"按钮就对课程数加 1 然后显示出来，效果如图 2.2 所示，这是用户点击按钮 3 次后的显示结果。

（1）Model 层。

在工程中新建 Java 类 CourseModel 作为数据模型，记录并保存开课数，其中有一个更新的方法，数据更新完之后会通知 UI 去刷新显示的内容。

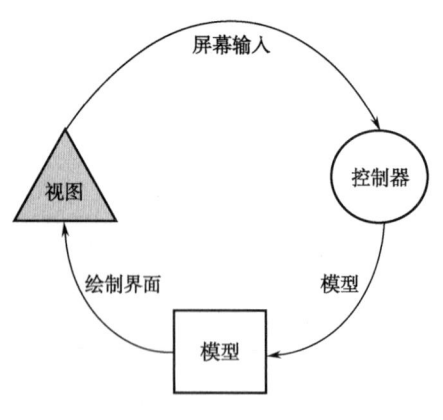

图 2.1　MVC 模型　　　　　　　图 2.2　MVC 实现对开课数计数显示

CourseModel.java 代码如下：

```
package com.easybooks.helloworld;
public class CourseModel {
    private int cnum = 0;                              //记录课程数

    public void addCourse(MainActivity activity) {     //更新的方法
        cnum++;                                        //增加开课数
        activity.updateUI(cnum);                       //刷新 UI
    }
}
```

（2）View 层。

View 层在 Android 中对应的就是布局的 XML 文件，在本工程中为 content_main.xml。
content_main.xml 代码如下：

```
<?xml version="1.0" encoding="utf-8"?>
<androidx.constraintlayout.widget.ConstraintLayout xmlns:android="http://schemas.android.com/apk/res/android"
    xmlns:app="http://schemas.android.com/apk/res-auto"
    xmlns:tools="http://schemas.android.com/tools"
    android:layout_width="match_parent"
    android:layout_height="match_parent"
    app:layout_behavior="@string/appbar_scrolling_view_behavior"
    tools:context=".MainActivity"
    tools:showIn="@layout/activity_main">

    <TextView                                                    //刷新显示开课数的文本视图
        android:id="@+id/textView"
        android:layout_width="wrap_content"
        android:layout_height="wrap_content"
        android:text="Hello World!"
        android:textColor="#E91E63"
        android:textSize="24sp"
        android:textStyle="bold"
        app:layout_constraintBottom_toBottomOf="parent"
        app:layout_constraintLeft_toLeftOf="parent"
```

```xml
        app:layout_constraintRight_toRightOf="parent"
        app:layout_constraintTop_toTopOf="parent"
        app:layout_constraintVertical_bias="0.085" />

    <ImageView
        android:id="@+id/imageView"
        android:layout_width="314dp"
        android:layout_height="243dp"
        android:layout_marginTop="24dp"
        app:layout_constraintEnd_toEndOf="parent"
        app:layout_constraintHorizontal_bias="0.494"
        app:layout_constraintStart_toStartOf="parent"
        app:layout_constraintTop_toBottomOf="@+id/textView"
        app:srcCompat="@drawable/androidlover" />

    <Button                                              //执行更新操作的命令按钮
        android:id="@+id/button"
        android:layout_width="158dp"
        android:layout_height="69dp"
        android:layout_marginTop="24dp"
        android:text="开始课程"
        android:textSize="24sp"
        app:layout_constraintEnd_toEndOf="parent"
        app:layout_constraintHorizontal_bias="0.498"
        app:layout_constraintStart_toStartOf="parent"
        app:layout_constraintTop_toBottomOf="@+id/imageView" />

</androidx.constraintlayout.widget.ConstraintLayout>
```

其中，<TextView.../>就是用来显示课程数信息的视图元素，而<Button.../>是"开始课程"按钮，是供用户点击操作的视图元素。由此可见，View 层是用户与应用程序直接交互的媒介，通过 View 层接受用户的操作动作，将最终运行结果以可视化的方式反馈给用户。

（3）Controller 层。

Android 中一般由 Activity 来充当 Controller。Controller 一方面接收来自 View 的事件，另一方面通知 Model 处理数据。本工程中只有一个 Activity，也就是主程序类 MainActivity，我们对其稍加改动，增加它与视图及模型交互控制的功能。

修改后的 MainActivity.java 代码如下：

```java
package com.easybooks.helloworld;
……                                                     //导入控件代码
public class MainActivity extends AppCompatActivity {
    private TextView textView;
    private ImageView imageView;
    private Button button;
    private CourseModel courseModel;                    // (a)

    @Override
    protected void onCreate(Bundle savedInstanceState) {
        super.onCreate(savedInstanceState);
        setContentView(R.layout.activity_main);
        findViews();
```

```
        courseModel = new CourseModel();                          // (a)

        Toolbar toolbar = findViewById(R.id.toolbar);
        ……                                                         //系统生成的其他代码
        button.setOnClickListener(new View.OnClickListener() {    // (b) 接收来自 View 的事件
            @Override
            public void onClick(View view) {
                //……                                              //注释掉原来的功能代码
                courseModel.addCourse(MainActivity.this);         // (c) 通知 Model 处理数据
            }
        });
    }

    private void findViews() { ... }

    public void updateUI(int num) {                               //更新 UI
        textView.setText("您开始了  " + Integer.toString(num) + " 门课程");
    }
    ……                                                            //系统生成的其他代码
}
```

其中：

（a）**private CourseModel courseModel;** 和 **courseModel = new CourseModel();**：由于本例控制器 MainActivity 需要与后台模型 CourseModel 交互，所以要将 CourseModel 声明为 MainActivity 的私有成员，并在 onCreate()方法初始化时实例化该模型的一个类对象，用于接下来程序中的操作。

（b）**button.setOnClickListener(new View.OnClickListener(){ ... });**：从这句可以看出，用来监听按钮点击事件的监听器实质上是 View（即 Android 视图类）的一个接口。View 是 Android 系统最为重要的基类，所有在界面上可见的控件都是 View 的子类，控件通过其上绑定的监听器接收视图传进来的事件动作。所以，监听器实际上就是 View 层与 Controller 层传递信息的桥梁。

（c）**courseModel.addCourse(MainActivity.this);**：调用 addCourse()方法向模型对象 CourseModel 传递消息，通知模型处理数据，对开课数加 1。

2.1.2 控件概述

1. 控件的表达

在 Android Studio 中，用 XML 文件中的 UI 元素来描述界面控件，其表达形式自然是 XML 的标签，有如下两种写法。

（1）规范表达：
```
<控件名
     控件属性
     ……>
</控件名>
```
（2）简单表达：
```
<控件名
     控件属性
     ……/>
```
这两种形式是等价的，用户可根据需要灵活地使用这两种写法。

Android 支持的控件有基本控件、布局控件和高级控件。基本控件是几乎所有 App 界面都会用到的最基础的 UI 元素；布局控件一般作为容器使用，"盛放"其他控件来对界面布局进行整体组织规划；高级控件则用于开发有复杂丰富效果的用户界面。

各类主要控件的 XML 标签及其功能作用如表 2.1 所示。

表 2.1　控件的 XML 标签

类　别	控　件	XML 标签	功 能 作 用
基本控件	文本视图	<TextView.../>	显示文本、图标
	文本编辑框	<EditText.../>	输入文本，当中可显示图标
	按钮	<Button.../>	响应点击，执行用户操作
	图像按钮	<ImageButton.../>	显示带图片背景的按钮
	图像视图	<ImageView.../>	显示图片
	复选框	<CheckBox.../>	向用户提供选项，可多选
	单选按钮	<RadioButton.../>	向用户提供选项，单选互斥
	单选按钮组	<RadioGroup.../>	容器，将多个相关的单选按钮组合在一起
	下拉框	<Spinner.../>	提供下拉列表选项（浮动、可带图标）
	自动完成文本视图	<AutoCompleteTextView.../>	制作搜索框自动选择关键词功能
	日期选择器	<DatePicker.../>	选择日期
	时间选择器	<TimePicker.../>	选择时间
布局控件	线性布局	<LinearLayout...>...</LinearLayout>	按水平或垂直方向顺序排列控件
	相对布局	<RelativeLayout...>...</RelativeLayout>	按相对位置关系来确定其中控件位置
	表格布局	<TableLayout...> <TableRow>...</TableRow> <TableRow>...</TableRow> ... </TableLayout>	将控件添加到指定行和列的表格中
	网格布局	<GridLayout...>...</GridLayout>	将控件添加到网格中，一个控件可跨多行多列
	绝对布局	<AbsoluteLayout...>...</AbsoluteLayout>	指定当中每一个控件的绝对位置坐标
	板块布局	<FrameLayout...>...</FrameLayout>	将多个控件重叠放置于屏幕左上角
高级控件	标签栏	<TabHost... android:id="@android:id/tabhost"...> <RelativeLayout...> <FrameLayout android:id="@android:id/tabcontent".../> <TabWidget android:id="@android:id/tabs".../> ... </RelativeLayout> </TabHost>	切换显示 App 底部的多个标签页
	翻页视图	<androidx.viewpager.widget.ViewPager.../>	手机相册照片翻页，与单选按钮/组结合可实现广告轮播条功能
	循环视图	<androidx.recyclerview.widget.RecyclerView.../>	功能十分强大的控件，可实现 App 频道栏、商品展示等多种复杂功能

续表

类别	控件	XML 标签	功能作用
高级控件	列表视图	\<ListView.../\>	结合多层线性布局嵌套图像和文本视图制作的列表项，可实现丰富的产品列表展示效果
	网格视图	\<GridView.../\>	功能类同列表视图，只不过换了以网格的形式展示产品，其网格项若是结合自定义的文本视图和图像视图，再加上线性布局的嵌套使用，可呈现出更加完美的视觉效果
	Tab 导航栏	\<TabHost...\> \<LinearLayout...\> \<TabWidget android:id="@android:id/tabs".../\> \<FrameLayout android:id="@android:id/tabcontent"...\> \<FrameLayout...\>...\</FrameLayout\> \</FrameLayout\> \</LinearLayout\> \</TabHost\>	多用于实现 App 页面顶部的类别标签

表 2.1 中，大多数控件的标签都是单一的 XML 元素，如\<TextView.../\>（文本视图）、\<Button.../\>（按钮）、\<LinearLayout...\>...\</LinearLayout\>（线性布局）等，但也有少数控件（主要是高级控件）的标签是由一系列 XML 元素嵌套组合而成的，例如，标签栏和 Tab 导航栏的标签就是由一系列基本控件和布局控件的标签元素按照一定方式结合在一起构成的，其组合方式是 Android 系统内部定义的，用户不可擅自修改，只能在既有结构的基础上定制开发。

2. 控件的属性

控件的属性直接在其标签内赋值，格式如下：

```
<控件名
    android:id="@+id/<名称>"
    android:<属性名 1>="<值 1>"
    android:<属性名 2>="<值 2>"
    ……
    android:<属性名 n>="<值 n>"
</控件名>
```

每个控件可供赋值设置的属性和类型都不同，这里列举几种最为通用的属性。

（1）指定控件标识：android:id 属性。

例如：

android:id="@+id/TextView1"：表示新建标识 id 为 TextView1 的 UI 元素。

android:id="@android: id/TextView1"：表示不是新添加的资源，或属于 Android 框架的资源，必须添加其 Android 包的命名空间。

指定控件 id 属性是为了在其后引用该控件，当前界面的所有控件 id 属性的值不能相同。

（2）指定控件的大小：android:layout_width 属性设置宽度；android:layout_height 属性设置高度。

例如：

android:layout_width="wrap_content"：控件宽度只要能够包含所显示的字符串即可。

android:layout_width="match_parent"：控件宽度匹配父控件的宽度。

android:layout_height="240dp"：控件高度为 240 个像素。

（3）指定控件显示的字符串：android:text 属性。在运行过程中，可以通过 setText()去修改该属性值，以改变显示内容。

例如：

android:text="abc 汉字"：显示"abc 汉字"。

android:text="@string/text1"：显示字符串资源文件（strings.xml）中"text1"标识符指定的内容。

（4）指定控件显示字符的大小：android:textSize 属性。

例如：

android:textSize="18sp"：设置字符大小为 18sp。

（5）指定控件位置：新版 Android Studio 支持约束布局，即在设计模式并不明确定义控件在界面上的绝对位置，仅指明它与其他控件的相对位置关系，这么做可以在运行时由程序根据手机屏幕尺寸和方位自动调整界面以达到最佳的显示效果。

例如：

app:layout_constraintTop_toBottomOf="@+id/EditText1"：表示将该控件的顶部上边沿紧贴于 EditText1 控件的底部。

同样地，还有指定将控件底部、右部、左部贴于另一控件顶部、左部和右部的属性。

（6）指定控件对齐方式。

例如：

android:layout_alignLeft="@+id/Button1"：该控件与 Button1 控件左对齐。

同样地，也有与指定控件上对齐、下对齐和右对齐的属性。

例如：

android:layout_alignParentLeft="true"：将该控件的左部与其父控件的左部对齐。

另外，还有指定将控件的右部、上部、下部与其父控件相应的部位对齐的属性。

3. 控件的事件

控件常用公共事件除了鼠标点击事件，还有按键事件、触摸事件等。在这些事件发生时，Android 用户界面框架调用控件的事件处理方法对事件进行处理。在 MVC 模型中，控制器根据事件类型的不同，将其传递给控件不同的事件处理方法。例如，点击事件传递给 onClick()方法；按键事件传递给 onKey()方法；触击事件传递给 onTouch()方法。

实际开发中常用的控件事件如表 2.2 所示。

表 2.2　常用的控件事件

事件名	监听器	接收方法	应用场合
点击	View.OnClickListener	public void onClick(View v)	点击控件时
长按	View.OnLongClickListener	public boolean onLongClick(View v)	手指按在控件上停留达一定时间
复选框选项变化	CompoundButton.OnCheckedChangeListener	public void onCheckedChanged(CompoundButton v, boolean checked)	增加/取消选择新的复选框
单选按钮变化	RadioGroup.OnCheckedChangeListener	public void onCheckedChanged(RadioGroup radioGroup, int i)	点选更改按钮组中选项
编辑框内容变化	TextWatcher	public void afterTextChanged(Editable text)	往文本编辑框中输入、删除或更改内容后

续表

事件名	监听器	接收方法	应用场合
下拉框选择	AdapterView.OnItemSelectedListener	public void onItemSelected(AdapterView<?> arg0, View arg1, int arg2, long arg3)	用户从下拉框中选中某个选项时
日期改变	DatePicker.OnDateChangedListener	public void onDateChanged(DatePicker dpview, int year, int month, int day)	从日期选择器选择新的日期
时间改变	TimePicker.OnTimeChangedListener	public void onTimeChanged(TimePicker tpview, int hour, int minute)	从时间选择器选择新的时间
按键	View.OnKeyListener	public boolean onKey(View v, int keyCode, KeyEvent event)	用户按下某个键时
触击手机屏	View.OnTouchListener	public boolean onTouch(View v, MotionEvent event)	用户手指触击或在屏幕上滑动时

事件通常由与控件绑定的监听器接收，监听器是 Android 视图（或某些特殊类）上的接口实现，编程开发时，只需重写接口中的方法，当事件发生时就会自动传递给对应的接收方法进行处理。后面我们将结合具体的实例来看这些事件的应用。

4. 控件的类

Android 中的每一种控件都对应一个类，这个类可以是 Android 内置的，也可以由第三方提供。第三方提供的类必须由用户先安装和导入类库后，才能在程序中使用。

各主要控件对应的类如表 2.3 所示。

表 2.3 控件对应的类

类 别	控 件	对 应 类
基本控件	文本视图	android.widget.TextView
	文本编辑框	android.widget.EditText
	按钮	android.widget.Button
	图像按钮	android.widget.ImageButton
	图像视图	android.widget.ImageView
	复选框	android.widget.CheckBox
	单选按钮	android.widget.RadioButton
	单选按钮组	android.widget.RadioGroup
	下拉框	android.widget.Spinner
	自动完成文本视图	android.widget.AutoCompleteTextView
	日期选择器	android.widget.DatePicker
	时间选择器	android.widget.TimePicker
布局控件	线性布局	android.widget.LinearLayout
	相对布局	android.widget.RelativeLayout
	表格布局	android.widget.TableLayout
	网格布局	android.widget.GridLayout
	绝对布局	android.widget.AbsoluteLayout
	板块布局	android.widget.FrameLayout

类别	控件	对应类
高级控件	标签栏	android.app.TabActivity
	翻页视图	androidx.viewpager.widget.ViewPager
	循环视图	androidx.recyclerview.widget.RecyclerView
	列表视图	android.widget.ListView
	网格视图	android.widget.GridView
	Tab 导航栏	android.widget.TabHost

本章我们仅介绍基本控件，布局控件的应用将在第 3 章介绍；而高级控件多为复合控件，使用较为复杂，将在第 4 章专门展开详解。

2.2 基本的界面控件

2.2.1 文本视图：TextView

TextView 既可用于显示文本字符，也可用于图片的显示，还可以做出滚动字幕特效。
TextView 控件比较重要的属性和方法介绍如下。
属性：
- **android:background**：设置文字背景。
- **android:ellipsize**：设置当文字过长时内容该如何显示。有如下选项值：
  start——开头显示省略号；
  end——结尾显示省略号；
  middle——中间显示省略号；
  marquee——以"跑马灯"（横向滚动字幕）的方式显示。
- **android:gravity**：设置文字对齐方式。
- **android:singleLine**：设为单行文本模式。
- **android:lines**：多行显示模式设置显示行数。
- **android:maxLines**：设置文本的最大行数。
- **android:hint**：设置默认显示的提示信息文字。

方法：
- **setCompoundDrawables()**：设置图片在文本视图中的显示方位。
- **setFocusable()**：设置文本视图的焦点使能状态，为"true"能获取焦点，为"false"则无法获得焦点。
- **setFocusableInTouchMode()**：设置文本视图在触摸模式下的焦点使能状态。

【例 2.1】 设计显示图书销售信息，用滚动字幕介绍，同时显示书的封面图片，效果如图 2.3 所示。

图 2.3 TextView 显示效果

1. 准备

创建 Android Studio 工程 TextViewDemo，在工程的 drawable 目录下放入一个 192×192 的图书封面

图片 androidbook.png。

2. 定义文本视图

在 content_main.xml 中定义文本视图，代码如下：

```
<TextView                                                    //显示静态文本的视图
    android:id="@+id/myTitleTextView"
    android:layout_width="0dp"
    android:layout_height="wrap_content"
    android:layout_marginTop="140dp"
    android:gravity="center_horizontal"
    android:text="Android 实用教程（第 2 版）"               //静态文本内容
    android:textSize="24sp"
    android:textStyle="bold"
    .../>

<TextView                                                    //显示图片的文本视图
    android:id="@+id/myImageTextView"
    android:layout_width="403dp"
    android:layout_height="249dp"
    ...
    app:layout_constraintTop_toBottomOf="@+id/myTitleTextView" />

<TextView                                                    //显示滚动字幕的文本视图
    android:id="@+id/myMoveTextView"
    android:layout_width="372dp"
    android:layout_height="25dp"
    android:background="?attr/colorButtonNormal"            //字幕文字背景加浅灰色底
    android:ellipsize="marquee"                             //设置为滚动效果
    android:gravity="bottom|center"        //文字对齐方式为底部居中，设为底部是为了给图片显示留出空间
    android:singleLine="true"                               //单行滚动文本
    android:text="本书介绍 Android 的最新功能及其实例典型用法"
    android:textColor="#E91E63"
    android:textSize="18sp"
    android:textStyle="bold"
    ...
    app:layout_constraintTop_toBottomOf="@+id/myImageTextView" />
```

3. 设置图片方位并开启滚动字幕

实现思路：

（1）onCreate()方法初始化时加载预先准备的资源图片，显示在文本视图文字上方。

（2）在文本视图上绑定点击事件监听器 View.OnClickListener。

（3）当发生点击事件时，根据视图传来的 id 判断，如果是该文本视图的 id，则根据启动字幕滚动的控制变量的状态来执行字幕的滚动与停止。

在 MainActivity.java 中实现功能，代码如下：

```
public class MainActivity extends AppCompatActivity implements View.OnClickListener {
    //定义控件对象、程序变量
    private TextView myImageTextView;                       //定义图片文本视图对象
    private TextView myMoveTextView;                        //定义滚动字幕文本视图对象
    private boolean start = false;                          //定义控制启动字幕滚动的变量
```

```java
        private Drawable myDrawable;                                    //定义图书封面图片资源对象

    //创建事件代码
    @Override
    protected void onCreate(Bundle savedInstanceState) {
        super.onCreate(savedInstanceState);
        setContentView(R.layout.activity_main);
        myImageTextView = findViewById(R.id.myImageTextView);
        myMoveTextView = findViewById(R.id.myMoveTextView);
        myMoveTextView.setOnClickListener(this);                        //在文本视图上绑定点击事件监听器
        myDrawable = getResources().getDrawable(R.drawable.androidbook);
                                                                        //(a)
        myDrawable.setBounds(0, 0, myDrawable.getMinimumWidth(), myDrawable.getMinimumHeight());
                                                                        //设置图片的位置和大小
        myImageTextView.setCompoundDrawables(null, myDrawable, null, null);
                                                                        //(b)
        ......
    }
     ......
    //点击事件代码
    @Override
    public void onClick(View v) {
        if (v.getId() == R.id.myMoveTextView) {                         //(c)
            if (!start) {                                               //如果 start= false（尚未启动）
                myMoveTextView.setFocusable(true);
                myMoveTextView.setFocusableInTouchMode(true);           //启动滚动
            } else {
                myMoveTextView.setFocusable(false);
                myMoveTextView.setFocusableInTouchMode(false);          //停止滚动
            }
            start = !start;                                             //控制变量取反
        }
    }
}
```

其中：

（a）**myDrawable = getResources().getDrawable(R.drawable.androidbook);**：用 getDrawable()方法获取图片资源的对象引用，这个方法要带一个参数，它是 drawable 目录下图书封面图片资源的 id。在 Android 中所有资源的 id 都是一个整型数，在程序中以字符串的形式加以引用。

（b）**myImageTextView.setCompoundDrawables(null, myDrawable, null, null);**：文本视图的 setCompoundDrawables()方法用来设置图片在其中的显示方位，该方法带有 4 个参数，使用时只有一个参数被赋值，其余 3 个皆为 null。根据被赋值的参数所在的位置来决定图片显示在视图中的什么位置，具体如下：

setCompoundDrawables(myDrawable, null, null, null)：左边；
setCompoundDrawables(null, myDrawable, null, null)：上边；
setCompoundDrawables(null, null, myDrawable, null)：右边；
setCompoundDrawables(null, null, null, myDrawable)：下边。

（c）**if (v.getId() == R.id.myMoveTextView)**：View 的 getId()方法可获取点击事件发生时接收它的

UI 组件的 id 标识，由于可接收点击事件的 UI 组件有很多，在编程时一般都是采用这种方式来得知究竟是哪一个组件被点击了。

2.2.2 按钮和图像按钮：Button/ImageButton

通常 Button 按钮控件上只显示文字，如果要想在按钮上显示图像，可使用 ImageButton 控件，但 ImageButton 只能显示图像却不能同时显示文字。若是想要同时显示文字配图标的按钮，仍然需要通过对 Button 的设置来实现。

ImageButton 控件比较重要的属性和方法介绍如下。

属性：
- **android:background**：设置图像按钮背景色。
- **app:srcCompat**：设置图像按钮上的图片。
- **android:adjustViewBounds**：设置图像的边框自动适应按钮的边界显示。

方法：
- **setImageResource()**：设置图像按钮上的图片资源。

【例 2.2】 制作配有文字和图片的按钮，要求点击按钮时能改变上面的文字和图片内容，同时还能够变换图片的方位及按钮背景色。

本例需要用到 Android 的状态列表图形技术，不仅要预先准备好一系列的图片资源，还要在工程 drawable 目录下创建一个用于切换界面上 UI 元素状态的 XML 文件，在其中编写样式选择的配置项。

1. 准备

创建 Android Studio 工程 ButtonDemo，在工程的 drawable 目录下放入如表 2.4 所示处理过的图片文件。

表 2.4 制作图文按钮需要用到的图片资源

文件名	尺寸	缩略图	作用
icon_1.png	48×48		初始 Button 上的图标
icon_2.png	48×48		点击 Button 后其上显示的图标
bg_1.png	160×50		初始 Button 的背景
bg_2.png	160×50		在 Button 上按下鼠标呈现的背景
image_1.png	144×144		初始 ImageButton 上的图片
image_2.png	144×144		点击 ImageButton 后显示的图片

在 drawable 目录下创建 button_style.xml，用于设置当用户按下 Button 时其背景色外观的变化，内容如下：

```xml
<?xml version="1.0" encoding="utf-8"?>
<selector xmlns:android="http://schemas.android.com/apk/res/android">
    <item android:state_pressed="true" android:drawable="@drawable/bg_2" />   //按钮被按下时背景为 bg_2
    <item android:drawable="@drawable/bg_1" />                                //默认状态下的背景为 bg_1
</selector>
```

2. 设计按钮外观

在 content_main.xml 中设计按钮外观，代码如下：

```xml
<Button                                          //这个按钮上有文字和图标,点击可变换图标及其位置
    android:id="@+id/myButton"
    android:layout_width="260dp"
    android:layout_height="75dp"
    android:layout_marginTop="68dp"
    android:background="@drawable/button_style"  // (a)
    android:text="确   定"
    android:textSize="36sp"
    .../>

<ImageButton                                     //这是个图片按钮,只能显示图片,点击更换图片
    android:id="@+id/myImageButton"
    android:layout_width="144dp"
    android:layout_height="144dp"
    android:background="#FFFFFF"                 // (b)
    ...
    app:srcCompat="@drawable/image_1" />         //初始显示的图片为 image_1
```

其中:

(a) **android:background="@drawable/button_style"**:设置背景属性值为预定义在 drawable 目录下的样式文件 button_style.xml,这样就可以实现在运行时由用户按下按钮使其外观产生变化的动态效果,按钮的背景由浅绿变成浅蓝色,如图 2.4 所示。

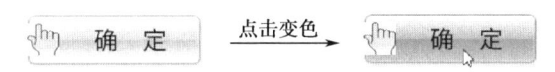

图 2.4 按钮点击变色

(b) **android:background="#FFFFFF"**:设置 ImageButton 的背景为#FFFFFF(白色),是为了更清楚地显示图片。

3. 改变按钮文字图片及变换图标方位

实现思路:

(1)在 onCreate()方法中初始加载 Button 的图片为"手形"图标,位于文字左边;ImageButton 上的图片为"对钩",并在两个按钮上都绑定 View.OnClickListener 监听器。

(2)点击事件发生时,获取视图传来的 id,分别与两个按钮的 id 进行比较。

(3)如果为 Button 的 id,就重设按钮上的文字和图片方位。

(4)如果为 ImageButton 的 id,则改变其上的图片。

在 MainActivity.java 中实现功能,代码如下:

```java
public class MainActivity extends AppCompatActivity implements View.OnClickListener {
    private TextView myTextView;                 //文本视图(用于显示提示文字)
    private Button myButton;                     //按钮
    private Drawable myDrawable;                 //图片资源对象
    private ImageButton myImageButton;           //图片按钮

    @Override
    protected void onCreate(Bundle savedInstanceState) {
        super.onCreate(savedInstanceState);
        setContentView(R.layout.activity_main);
        myTextView = (TextView) findViewById(R.id.myTextView);
```

```
        myButton = (Button) findViewById(R.id.myButton);
        myDrawable = getResources().getDrawable(R.drawable.icon_1);        //获取图片资源的对象引用
        myDrawable.setBounds(0, 0, myDrawable.getMinimumWidth(), myDrawable.getMinimumHeight());
                                                                           //设置图片大小
        myButton.setCompoundDrawables(myDrawable, null, null, null);       //（a）
        myButton.setOnClickListener(this);
        myImageButton = (ImageButton) findViewById(R.id.myImageButton);
        myImageButton.setOnClickListener(this);                            //图像按钮上绑定点击事件监听器
        ...
    }
    ...
    @Override
    public void onClick(View v) {
        switch (v.getId()) {              //根据接收到点击事件的视图 id 判断是哪个按钮被点击
            case R.id.myButton:                                            // Button 被点击
                myTextView.setText("按下了 Button 按钮");
                //改变按钮上文字及图片方位
                myButton.setText("OK");                                    //修改按钮文字
                myDrawable = getResources().getDrawable(R.drawable.icon_2);
                                                                           //获取新图片资源的引用
                myDrawable.setBounds(0, 0, myDrawable.getMinimumWidth(), myDrawable.getMinimumHeight());
                                                                           //设置图片大小
                myButton.setCompoundDrawables(null, null, myDrawable, null);
                                                                           //（a）将图片方位改到文字右边
                return;
            case R.id.myImageButton:                                       // ImageButton 被点击
                //改变按钮上的图片
                myTextView.setText("按下了 ImageButton 按钮");
                myImageButton.setImageResource(R.drawable.image_2);        //（b）修改其上的图片
                return;
        }
    }
}
```

其中：

（a）**myButton.setCompoundDrawables(myDrawable, null, null, null);** 和 **myButton.setCompoundDrawables(null, null, myDrawable, null);**：对比【例 2.1】的代码，我们发现，在文本视图（TextView）和按钮（Button）上都既可以显示文本又可以显示图片，且都是通过 Drawable 获取图片资源的引用后，再用 setCompoundDrawables()方法来设置图片要显示的方位，编程的流程逻辑代码完全一样，这是因为 Android 中的 TextView 和 Button 这两个控件在**本质上**是一样的，它们都继承派生自同一个父类（View 视图类），所以有着相似的功能和行为特性。在处理点击事件时重新设置按钮上的图标及方位，释放鼠标后按钮上的文字变为"OK"，图标则由"手形"变成"对钩"，且位置移到了文字的右边，如图 2.5 所示。

（b）**myImageButton.setImageResource(R.drawable.image_2);**：ImageButton 的 setImageResource()方法可以设置图像按钮上的图片，将标识该图片资源的整型 id 作为参数传给该方法即可。运行程序时点击 ImageButton，其上图片由"对钩"变成"纽扣"，显示效果如图 2.6 所示。

第 2 章　Android 用户界面   45

图 2.5　改变按钮文字图片及变换图标方位　　　图 2.6　点击改变图像按钮上的图片

2.2.3　图像视图：ImageView

ImageView 既可以显示现成的资源图片，也可以显示 TextView 中文本的截图。ImageButton 具有与之相同的性质。

ImageView 控件的属性和方法同 ImageButton，见 **2.2.2** 小节。

【例 2.3】　设计 ImageView 和 ImageButton 初始都显示各自预置的资源图片，点击 ImageButton 向 TextView 中随机添加图书条目（可上下滚动浏览）；长按 TextView 对其中的文本截图，将所截取的图片实时显示在 ImageView 和 ImageButton 上。整个演示过程的效果如图 2.7 所示。

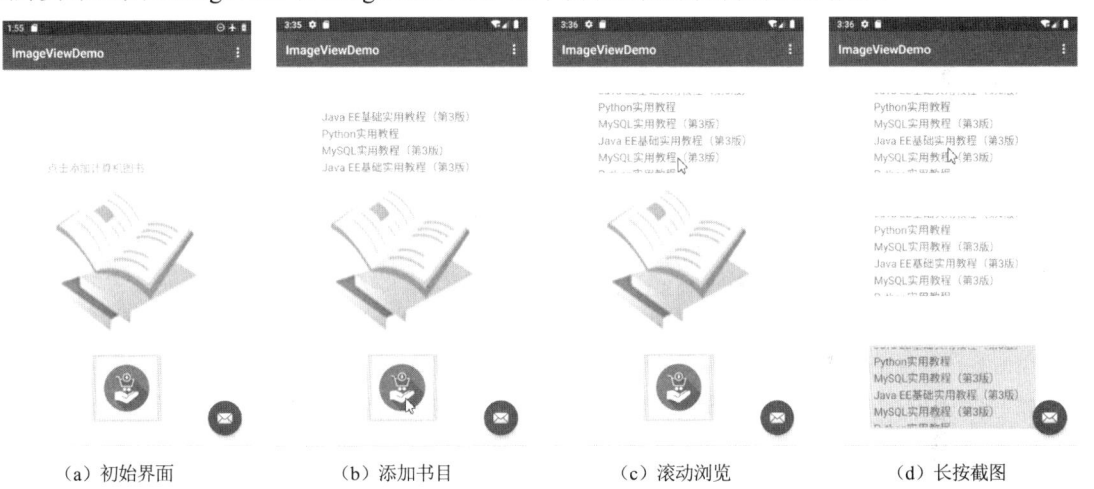

（a）初始界面　　　　（b）添加书目　　　　（c）滚动浏览　　　　（d）长按截图

图 2.7　ImageView 和 ImageButton 显示图片及截图演示过程的效果

1. 准备

创建 Android Studio 工程 ImageViewDemo，在工程的 drawable 目录下放入 books.png（ImageView 显示，100×100）和 cart.png（ImageButton 显示，100×100）。

2. 设计界面

在 content_main.xml 中设计界面，代码如下：

```
<TextView                                      //这个文本视图用来显示图书条目
    android:id="@+id/myTextView"
    android:layout_width="262dp"
    android:layout_height="127dp"
    android:gravity="bottom|left"              //图书条目的对齐方式为底部居左，即从下方逐条加入
    android:hint="点击添加计算机图书"
    android:lines="6"                          //设置为多行显示
    android:maxLines="6"                       //（a）
    android:textSize="18sp"
    .../>
```

```
<ImageView                                      //定义图像视图
    android:id="@+id/myImageView"
    android:layout_width="262dp"
    android:layout_height="234dp"
    android:layout_marginTop="16dp"
    ...
    app:srcCompat="@drawable/books" />          //初始显示一摞书的图片

<ImageButton                                    //定义图像按钮
    android:id="@+id/myImageButton"
    android:layout_width="wrap_content"         //（b）
    android:layout_height="wrap_content"        //（b）
    android:adjustViewBounds="true"             //图片边框自适应显示
    ...
    app:srcCompat="@drawable/cart" />           //初始显示一幅购物车的图标
```

其中：

（a）**android:maxLines="6"**：设置文本视图的最大行数为 6 行，实际运行时，当加入内容超过 6 行后，文本视图形状大小不会随加入内容的多少而变化，反正它显示的内容就是 6 行。

（b）**android:layout_width="wrap_content"** 和 **android:layout_height="wrap_content"**：这里设置成 "wrap_content" 是为了在稍后显示截图的时候能让 ImageButton 自适应截图的尺寸。

3. 添加图书条目滚动浏览及截图

实现思路：

（1）在 MainActivity 中定义全局变量数组，用于预先存储书名信息。

（2）在 onCreate()方法中初始打开文本视图的绘图缓存，将文本视图与长按监听器绑定，图像按钮与点击事件监听器绑定。

（3）点击事件发生时，从变量数组中随机取一个书名添加进文本视图。

（4）长按事件发生时，从绘图缓存获取当前文本视图的截图，分别设于图像视图和图像按钮上显示。

（5）开一个线程专门负责定时重置绘图缓存的内容。

在 MainActivity.java 中实现功能，代码如下：

```
public class MainActivity extends AppCompatActivity implements View.OnClickListener,
View.OnLongClickListener {                                  //（a）
    private TextView myTextView;                //文本视图（显示图书条目，执行截图操作）
    private ImageView myImageView;              //图像视图（显示图片和截图）
    private ImageButton myImageButton;          //图像按钮（点击添加书目，显示图片和截图）
    private String[] myBookName = {"Qt 5 开发及实例（第 4 版）", "Python 实用教程", "MySQL 实用教程（第
3 版）", "Java EE 基础实用教程（第 3 版）"};     //字符串数组（提供随机显示的图书名称）

    @Override
    protected void onCreate(Bundle savedInstanceState) {
        super.onCreate(savedInstanceState);
        setContentView(R.layout.activity_main);
        myTextView = findViewById(R.id.myTextView);
        myTextView.setMovementMethod(new ScrollingMovementMethod());
                                                //设置文本视图可滚动查看
        myTextView.setDrawingCacheEnabled(true);        //打开绘图缓存
        myTextView.setOnLongClickListener(this);        //文本视图绑定长按监听器用于截图操作
        myImageView = findViewById(R.id.myImageView);
```

```java
            myImageButton = findViewById(R.id.myImageButton);
            myImageButton.setOnClickListener(this);              //图像按钮绑定点击事件监听器（添加书目）
            ...
        }
        ...
        @Override
        public boolean onLongClick(View v) {
            if (v.getId() == R.id.myTextView) {
                Bitmap map = myTextView.getDrawingCache();       //（b）获取绘图缓存的内容（即截取的图片）
                myImageView.setImageBitmap(map);                 //将图片设置于图像视图上
                myImageButton.setImageBitmap(map);               //将图片设置于图像按钮上
                myHnd.postDelayed(myDrawCache, 100);             //（c）
            }
            return true;                                         // onLongClick 方法必须返回一个布尔值
        }

        @Override
        public void onClick(View v) {
            if (v.getId() == R.id.myImageButton) {
                int index = (int) (Math.random() * 10) % 4;      //生成随机图书名称文本的索引
                String bookInf = String.format("%s\n%s", myTextView.getText(), myBookName[index]);
                                                                 //加入新书目
                myTextView.setText(bookInf);                     //更新文本视图
            }
        }

        private Handler myHnd = new Handler();                   //定义任务处理句柄
        private Runnable myDrawCache = new Runnable() {          //（d）启用线程
            @Override
            public void run() {                                  //重置绘图缓存，为下一次截图做好准备
                myTextView.setDrawingCacheEnabled(false);
                myTextView.setDrawingCacheEnabled(true);
            }
        };
    }
```

其中：

（a）**public class MainActivity extends AppCompatActivity implements View.OnClickListener, View.OnLongClickListener**：由于本例需要响应用户"点击"和"长按"这两种类型的事件，所以主程序类 MainActivity 要同时实现 OnClickListener、OnLongClickListener 这两种监听器接口。

（b）**Bitmap map = myTextView.getDrawingCache();**：文本视图控件本身内置有一个绘图缓存，里面保存的是视图上当前所显示内容的图片（当显示内容有变化时，视图会自动刷新绘图缓存），所谓"截图"其实也就是将绘图缓存中的图片设置到其他控件视图上加以显示。在使用文本视图的绘图缓存之前，先要用 setDrawingCacheEnabled(true)语句打开绘图缓存，然后在需要时通过 getDrawingCache() 方法获取里面的缓存图片。

（c）**myHnd.postDelayed(myDrawCache, 100);**：截图完毕后不能马上重置绘图缓存，因为画面渲染需要时间，这里延迟 100 毫秒就是为了等待图片渲染完成。

（d）**private Runnable myDrawCache = new Runnable()**：由于绘图缓存的重置操作是耗时的，要另开一个线程来实现，不能放在 UI 线程里。

Android 实用教程（第 2 版）（含视频分析）

从上面这个例子可见，图像按钮 ImageButton 与图像视图 ImageView 一样都可用于显示图片和截图，它们具有相同的特性，这是因为它们都是由同一个父类派生而来的。在 Android 的类体系中，ImageButton 与 ImageView 的亲缘关系比之与 Button 按钮还要来得近，甚至可以认为 ImageButton 就是一种具备 Button 性质的特殊的 ImageView。

2.2.4 复选框：CheckBox

CheckBox 控件称为复选框，它有两种状态：在非选择状态，控件显示为一个方框；而在选择状态，控件显示方框中包含对钩。界面上可以同时放置多个 CheckBox 控件，每个 CheckBox 控件的状态之间互不影响，一般用于表示可以选择多个选项的栏目。

CheckBox 控件比较重要的属性有：
- **android:drawableLeft**：在复选框左侧添加小图标。
- **android:drawablePadding**：设置复选框文字与方框之间的间距。

【例 2.4】 设计复选框应用程序，初始界面如图 2.8（a）所示；点击"南京"和"苏州"复选框，显示如图 2.8（b）所示。

图 2.8 复选框应用程序界面

1. 准备

创建 Android Studio 工程 CheckBoxDemo，在工程的 drawable 目录下放入 checked.png（复选框选中的外观，72×72）和 uncheck.png（复选框未选中的外观，72×72）。

这里，我们还是用状态列表图形技术来美化复选框的外观。在 drawable 目录下创建 checkbox_style.xml，用于设置当用户选中和取消勾选时复选框外观的变化，内容如下：

```
<?xml version="1.0" encoding="utf-8"?>
<selector xmlns:android="http://schemas.android.com/apk/res/android">
    <item android:state_checked="true" android:drawable="@drawable/checked" />   //选中外观为 checked
    <item android:drawable="@drawable/uncheck" />                                 //未选外观为 uncheck
</selector>
```

2. 定义复选框

在 content_main.xml 中定义复选框，下面仅列出其中一个复选框的标签元素代码，如下：
```
<CheckBox
    android:id="@+id/myCheckBox_nj"
```

界面完整代码

```
        android:layout_width="wrap_content"
        android:layout_height="wrap_content"
        android:layout_marginTop="60dp"
        android:button="@null"                              //（a）
        android:drawableLeft="@drawable/checkbox_style"     //引用状态列表图形的样式文件设置复选框的外观
        android:drawablePadding="@dimen/fab_margin"         //（b）
        android:text="南京"
        android:textSize="36sp"
        .../>
```
//其他两个复选框与第一个设置属性基本相同，id 分别为 myCheckBox_sz、myCheckBox_wx，代码略

其中：

（a）**android:button="@null"**：在使用 drawableLeft 设置外观时，button 属性必须设为"@null"。

（b）**android:drawablePadding="@dimen/fab_margin"**：设置复选框文字与方框之间的间隔。这里使用 android:drawableLeft 配合 android:drawablePadding 属性共同决定复选框的外观，可以使复选框文字与方框间的间距恰到好处、美观大方。

3. 根据复选框选择状态显示信息

实现思路：

（1）在 onCreate()方法中初始化一个控件数组，成员为 3 个复选框的引用；同时创建一个布尔数组与控件数组成员一一对应，记录每个复选框的选择状态。

（2）复选框选项变化事件发生时，根据从视图中获取的 id 检查并保存对应复选框的状态，然后遍历输出用户选择信息。

在 MainActivity.java 中实现功能，代码如下：

```java
public class MainActivity extends AppCompatActivity implements CompoundButton.OnCheckedChangeListener {
                                                            //（a）
    private CheckBox myCheckBox_nj;                         //"南京"复选框
    private CheckBox myCheckBox_sz;                         //"苏州"复选框
    private CheckBox myCheckBox_wx;                         //"无锡"复选框
    private CheckBox[] myCheckBoxes;                        //控件数组，存放复选框控件对象的引用
    private Boolean[] myChecked;                            //（b）
    private TextView myTextView;                            //文本视图（用于显示用户选择信息）
    private String myString;                                //记录文本

    @Override
    protected void onCreate(Bundle savedInstanceState) {
        super.onCreate(savedInstanceState);
        setContentView(R.layout.activity_main);
        myCheckBox_nj = findViewById(R.id.myCheckBox_nj);
        myCheckBox_sz = findViewById(R.id.myCheckBox_sz);
        myCheckBox_wx = findViewById(R.id.myCheckBox_wx);
        myCheckBoxes = new CheckBox[3];                     //创建复选框控件数组（必不可少！）
        //控件数组赋值
        myCheckBoxes[0] = myCheckBox_nj;
        myCheckBoxes[1] = myCheckBox_sz;
        myCheckBoxes[2] = myCheckBox_wx;
        myChecked = new Boolean[3];                         //创建布尔数组（必不可少！）
        for (int i = 0; i < 3; i++) {                       //初始化
            myChecked[i] = false;                           //初始默认所有复选框为不勾选状态
```

```
                myCheckBoxes[i].setOnCheckedChangeListener(this);    //复选框控件绑定监听类
            }
            myTextView = findViewById(R.id.myTextView);
            myString = myTextView.getText().toString();              //记录标头信息
            ...
        }
        ...
        @Override
        public void onCheckedChanged(CompoundButton v, boolean checked) {   // (c)
            switch (v.getId()) {                                            // (d)
                case R.id.myCheckBox_nj:
                    myChecked[0] = checked ? true : false;           //保存第一个复选框的状态
                    break;
                case R.id.myCheckBox_sz:
                    myChecked[1] = checked ? true : false;           //保存第二个复选框的状态
                    break;
                case R.id.myCheckBox_wx:
                    myChecked[2] = checked ? true : false;           //保存第三个复选框的状态
                    break;
            }
            String s = "";                                           //记录当前选择内容
            for (int i = 0; i < 3; i++) {                            //遍历得到当前的选择状态信息
                s += myChecked[i] ? myCheckBoxes[i].getText() + "   " : "";
            }
            myTextView.setText(String.format("%s\n%s", myString, s));//与标头拼接后显示在文本视图上
        }
    }
```

其中：

（a）**public class MainActivity extends AppCompatActivity implements CompoundButton. OnCheckedChangeListener**：复选框类继承自 CompoundButton（复合按钮），OnCheckedChangeListener 是它特有的监听器，主程序类通过实现该监听器接口来处理复选框选择状态的改变。复选框也可以用 OnClickListener 监视和获取选择状态信息，但此种情况下还要通过 CheckBox 的 isChecked 方法判断复选框是否被选中，而用 OnCheckedChangeListener 则可直接从 onCheckedChanged 方法的传入参数得到复选框的选择状态，故一般建议还是使用复选框原生的 OnCheckedChangeListener 监听器为佳。

（b）**private Boolean[] myChecked;**：定义布尔数组，记录对应复选框的选中状态，其中 true 为选中，false 为取消。

（c）**public void onCheckedChanged(CompoundButton v, boolean checked)**：onCheckedChanged() 方法传入两个参数，前一个表示接收选择状态改变的复选框控件的视图，后一个是当前它的选择状态。

（d）**switch (v.getId())**：CompoundButton 也可通过其获取控件标识 id，这一点与之前用的 View 作用类同，根本原因在于 CompoundButton 类实际上也是从 View 派生而来的，继承了所有视图类公共的特性。

2.2.5 单选按钮及其容器：RadioButton 和 RadioGroup

RadioButton 控件也有两种状态：在非选择状态，控件显示为一个空心小圆圈；在选择状态，该小圆圈中会包含一个点。一般将多个 RadioButton 置于 RadioGroup 控件中组成单选按钮组，RadioGroup 是 RadioButton 的容器，它在程序运行时是不可见的。在同一个 RadioGroup 中，用户仅能够选择其中

的一个 RadioButton。

RadioButton 控件的属性用法与 **CheckBox** 的基本相同，见 **2.2.4** 小节。

【例 2.5】 设计单选按钮应用程序，选择"江苏省 GDP 排名**第一**的城市"（只有一个），提供"南京""苏州""无锡"三个城市让用户选择。初始界面如图 2.9（a）所示；点击"苏州"单选按钮，显示如图 2.9（b）所示。

图 2.9 单选按钮应用程序界面

1. 准备

创建 Android Studio 工程 RadioButtonDemo，在工程的 drawable 目录下放入 on.png（单选按钮选中的外观，分辨率 56×56）和 off.png（单选按钮未选中的外观，分辨率 56×56）。同样使用状态列表图形技术，在 drawable 目录下创建 radiobutton_style.xml，设置当用户选中时单选按钮外观的变化，内容如下：

```xml
<?xml version="1.0" encoding="utf-8"?>
<selector xmlns:android="http://schemas.android.com/apk/res/android">
    <item android:state_checked="true" android:drawable="@drawable/on" />   //选中外观为 on
    <item android:drawable="@drawable/off" />                                 //未选外观为 off
</selector>
```

2. 定义单选按钮

在 content_main.xml 中定义单选按钮，下面仅列出其中一个单选按钮的标签元素代码，如下：

```xml
<RadioGroup                                                             // （a）
    android:id="@+id/myRadioGroup"
    ...>
    ......
    <RadioButton
        android:id="@+id/myRadioButton_sz"
        android:layout_width="match_parent"
        android:layout_height="wrap_content"
        android:button="@null"                              //使用 drawableLeft 时此项必须设为"@null"
        android:drawableLeft="@drawable/radiobutton_style"  //引用状态列表样式文件设置单选按钮外观
        android:drawablePadding="@dimen/fab_margin"         // （b）
        android:paddingTop="@dimen/fab_margin"              // （c）
        android:text="苏州"
        android:textSize="36sp" />
```

界面完整代码

```
        ……
        //其他两个单选按钮与之设置属性基本相同，id 分别为 myRadioButton_nj、myRadioButton_wx，略
</RadioGroup>
```

其中：

（a）**<RadioGroup...></RadioGroup>**：单选按钮必须定义在它所属单选按钮组的 RadioGroup 容器标签之内。

（b）**android:drawablePadding="@dimen/fab_margin"**：设置 drawablePadding 属性是为了使文字与按钮之间保持恰当的间距。

（c）**android:paddingTop="@dimen/fab_margin"**：设置 paddingTop 属性是为了使该单选按钮与上一个单选按钮之间保持一定的距离间隔。

3. 根据单选按钮选中状态显示信息

实现思路：

直接在 RadioGroup 上注册监听器，将 RadioButton 按钮全都交予 RadioGroup 管理，单选按钮变化事件发生时可直接从监听器接口 onCheckedChanged()方法中获知被选中的单选按钮。

在 MainActivity.java 中实现功能，代码如下：

```java
public class MainActivity extends AppCompatActivity {
    private TextView myTextView;                        //文本视图（用于显示用户选择信息）
    private String myString;                            //记录文本标头

    @Override
    protected void onCreate(Bundle savedInstanceState) {
        super.onCreate(savedInstanceState);
        setContentView(R.layout.activity_main);
        myTextView = findViewById(R.id.myTextView);
        myString = myTextView.getText().toString();
        RadioGroup myRadioGroup = findViewById(R.id.myRadioGroup);
        myRadioGroup.setOnCheckedChangeListener(new RadioGroup.OnCheckedChangeListener() {
                                                                                    // (a)
            @Override
            public void onCheckedChanged(RadioGroup radioGroup, int i) {            // (b)
                if (i == R.id.myRadioButton_nj) {
                    myTextView.setText(myString + "南京");
                } else if (i == R.id.myRadioButton_sz) {
                    myTextView.setText(myString + "苏州");
                } else if (i == R.id.myRadioButton_wx) {
                    myTextView.setText(myString + "无锡");
                }
            }
        });
        ...
    }
    ...
}
```

其中：

（a）**myRadioGroup.setOnCheckedChangeListener(new RadioGroup.OnCheckedChangeListener() { ... });**：在 RadioGroup 上绑定注册的 OnCheckedChangeListener 监听器，这个监听器虽然与之前使用

的 CheckBox 上的监听器同名,但却是属于不同的类,故只能直接在 RadioGroup 上注册监听绑定实现,而**不能**在主程序类 MainActivity 上实现这个接口。当然,用户也可以分别在每个 RadioButton 上绑定 OnCheckedChangeListener 或者 OnClickListener 来实现相同的功能,但那样做会烦琐很多,使得代码极为冗长,故建议最好还是通过 RadioGroup 原生的监听器接口来获取和操作其中的 RadioButton。

(b) **public void onCheckedChanged(RadioGroup radioGroup, int i)**:i 是用户当前选中的单选按钮的 id。

2.2.6 文本编辑框:EditText

EditText 用于接收用户输入,与文本视图和按钮的特性一样,在 EditText 里面也可以同时显示文字和图片。

EditText 控件的属性、方法与 **TextView** 的基本相同(见 **2.2.1** 小节),但它比 **TextView** 多了编辑功能,故增加了如下几个特有的属性:

- **android:ems**:设定需要编辑的字符宽度。
- **android:inputType**:设置编辑框输入内容的类型。
- **android:textCursorDrawable**:设置光标颜色。

下面通过制作一个手机 App 里十分常见的"余额提现"界面来演示 EditText 的应用。

【例 2.6】 设计"余额提现"界面,如图 2.10 所示。

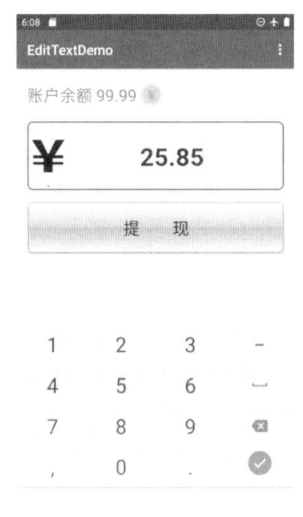

图 2.10 "余额提现"界面

这个界面的顶部是一个 TextView,用来显示当前账户的余额;中间 EditText 接收用户输入要提现的金额,其前方的"¥"符号为一图片;下方"提现"按钮根据用户输入是否符合要求来动态置可用性。当用户将焦点置于 EditText 中准备输入时,界面下方会自动弹出软键盘;而当输入完成,点击界面其他任一处地方时,软键盘又会收起隐藏。

初始启动程序时,EditText 是不带边框的,用户点击 EditText 获得焦点,其上会出现激活背景框,如图 2.11 所示。

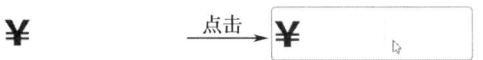

图 2.11 EditText 获得焦点后出现激活背景框

当输入内容不符合要求（为 0 或超账户余额）时，"提现"按钮自动置为不可用，如图 2.12 所示。

当输入完成点击界面其他任一处地方时，软键盘隐藏，同时 EditText 上的背景框消失，如图 2.13 所示。

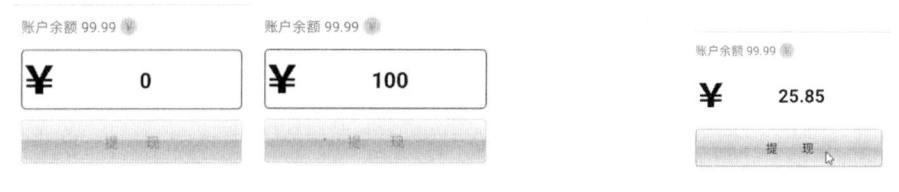

图 2.12　输入内容不符合要求时"提现"按钮自动置为不可用　　　　图 2.13　输入完成

1. 准备

（1）创建 Android Studio 工程 EditTextDemo，在工程的 drawable 目录下放入 yuan_1.png（TextView 中的账户余额"¥"图片，32×32）、yuan_2.png（EditText 中的前导单位"¥"图片，48×48）、edit_bg.png（EditText 输入激活背景框，128×75）、button_bg1.png（"提现"按钮不可用，160×50）和 button_bg2.png（"提现"按钮可用，160×50）。

（2）本例状态列表图形需要在 drawable 目录下准备两个文件。

① editext_style.xml（控制当 EditText 获得焦点时的激活背景框的显隐），内容如下：

```xml
<?xml version="1.0" encoding="utf-8"?>
<selector xmlns:android="http://schemas.android.com/apk/res/android">
    <item android:state_focused="true" android:drawable="@drawable/editext_bg" />   //当获得焦点时显示
</selector>
```

② button_style.xml（控制"提现"按钮的可用性外观），内容如下：

```xml
<?xml version="1.0" encoding="utf-8"?>
<selector xmlns:android="http://schemas.android.com/apk/res/android">
    <item android:state_enabled="true" android:drawable="@drawable/button_bg2" />   //可用，背景为浅绿色
    <item android:drawable="@drawable/button_bg1" />   //不可用，背景为灰色
</selector>
```

（3）为避免控件图片边缘显示由于图片尺寸的自适应拉伸而变得模糊，在 Android Studio 中一般都会将控件背景图处理成 .9 格式的 PNG 图片，本例对 EditText 和 Button 的背景图片都进行了这种处理。具体操作如下：

① 在 Android Studio 开发环境中，展开工程目录树，右击 res\drawable 目录下需要处理的图片文件，从弹出的快捷菜单中选择"Create 9-Patch file..."命令，弹出对话框，单击"OK"按钮，如图 2.14 所示。

② 确定之后，在 res\drawable 目录下就生成了对应图片的 .9 格式 PNG 图，双击打开编辑器对其进行处理，如图 2.15 所示。用鼠标单击图片上下左右四个边框，其上会出现黑色线，拖曳调整黑线的长度，位于此长度覆盖范围内的部分在拉伸尺寸时其每个像素都会同步地缩放，而未覆盖的区域（如四个边角）像素比例保持不变，这就有效地避免了因拉伸导致的图片边缘外观变模糊的现象。

经过这种处理的图片，无论怎样缩放都不会改变其清晰度，尤其边角处的分辨率始终维持不变，就达到了我们想要的效果。经 .9 格式 PNG 处理的按钮其边缘要清晰得多。

> ◉◉ **注意：**
> 在制作好 .9 格式 PNG 图片后，要记得将 drawable 目录下的原图删掉，否则界面 XML 文件中将无法引用 .9 图片。引用时**不需要写** .9 后缀，仍旧用原图名即可。

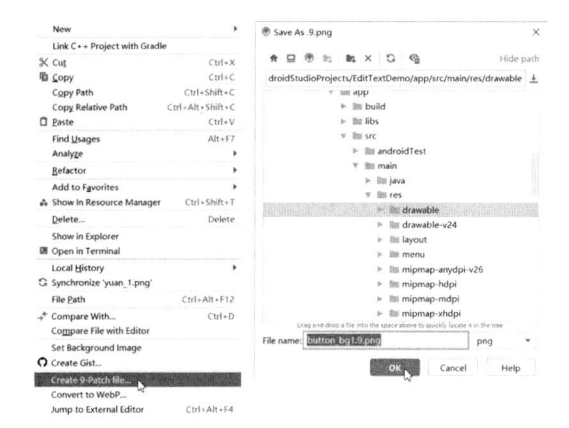

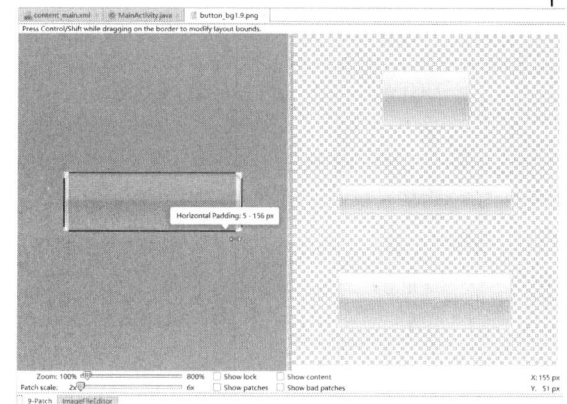

图2.14 另存为.9格式PNG图片 　　　　图2.15 限制原图拉伸时需要缩放的范围

2. 设计界面

在content_main.xml中设计界面，代码如下：

```
<LinearLayout                                              //（a）所有控件都放入这个线性布局
    android:id="@+id/myLinearLayout"
    android:layout_width="match_parent"
    android:layout_height="match_parent"
    android:orientation="vertical"
    tools:layout_editor_absoluteX="196dp"
    tools:layout_editor_absoluteY="462dp">

    <TextView                                              //显示账户余额的文本视图
        android:id="@+id/myTextView"
        android:layout_width="200dp"
        android:layout_height="75dp"
        android:layout_marginLeft="@dimen/fab_margin"      //与屏幕左边留一定间距，更美观
        android:gravity="center_vertical"                  //设置账户余额文本纵向居中显示
        android:textSize="24sp"
        .../>

    <EditText                                              //文本编辑框控件
        android:id="@+id/myEditText"
        android:layout_width="match_parent"
        android:layout_height="100dp"
        android:layout_marginLeft="@dimen/fab_margin"      //与屏幕左右两端均留一定间距（为了美观）
        android:layout_marginRight="@dimen/fab_margin"
        android:background="@drawable/editext_style"       //激活时显示绿色输入框
        android:ems="10"
        android:gravity="center|center_vertical"           //输入金额居中（为了醒目）
        android:inputType="numberDecimal"                  //设置只能输入数字（可以含小数点）
        android:textCursorDrawable="@android:color/black"  //设置光标颜色为黑色
        android:textSize="36sp"                            //设置为大号字体、加粗（为了醒目）
        android:textStyle="bold" />

    <Button                                                //"提现"按钮
        android:id="@+id/myButton"
```

```
            android:layout_width="match_parent"
            android:layout_height="75dp"
            android:layout_marginLeft="@dimen/fab_margin"
            android:layout_marginTop="@dimen/fab_margin"
            android:layout_marginRight="@dimen/fab_margin"
            android:background="@drawable/button_style"        //可用时背景为浅亮绿色，不可用时为浅灰色
            android:enabled="false"                             //初始不可用
            android:text="提    现"
            android:textSize="24sp" />
</LinearLayout>
```

其中：

(a) **\<LinearLayout>\</LinearLayout>**：这里将所有控件都放入一个 LinearLayout 内，是为了让界面的任何区域都能接收并响应用户点击事件，来控制软键盘的开关和 EditText 背景框的显隐状态。

3. 功能实现

实现思路：

（1）在初始化 onCreate()方法中设定文本视图和编辑框的外观，为编辑框绑定 TextWatcher 监听器，为屏幕线性布局绑定点击事件监听器。

（2）编辑框内容改变后，通过 TextWatcher 监听器的 afterTextChanged()方法获取用户输入的金额数值，判断是否合法，置"提现"按钮的可用性。

（3）线性布局（屏幕）被用户点击后，从 Android 系统服务中获取 InputMethodManager（输入方法管理器），控制软键盘的显隐。

在 MainActivity.java 中实现功能，代码如下：

```
public class MainActivity extends AppCompatActivity implements View.OnClickListener, TextWatcher {
                                                                    //（a）
    private LinearLayout myLinearLayout;                            //线性布局容器（用于接收用户点击）
    private TextView myTextView;                                    //文本视图（显示账户余额）
    private EditText myEditText;                                    //文本编辑框
    private Button myButton;                                        //"提现"按钮
    private Drawable myDrawable;                                    //图片资源引用对象
    private double total = 99.99;                                   //（b）

    @Override
    protected void onCreate(Bundle savedInstanceState) {
        super.onCreate(savedInstanceState);
        setContentView(R.layout.activity_main);
        myLinearLayout = findViewById(R.id.myLinearLayout);
        myLinearLayout.setOnClickListener(this);                    //（c）
        myTextView = findViewById(R.id.myTextView);
        myTextView.setText("账户余额 " + total);                     //初始显示账户余额
        myEditText = findViewById(R.id.myEditText);
        //在文本视图中显示人民币"¥"图标
        myDrawable = getResources().getDrawable(R.drawable.yuan_1);
        myDrawable.setBounds(0, 0, myDrawable.getMinimumWidth(), myDrawable.getMinimumHeight());
        myTextView.setCompoundDrawables(null, null, myDrawable, null);  //设置显示在金额右边
        //在文本编辑框中显示人民币"¥"单位标识
        myDrawable = getResources().getDrawable(R.drawable.yuan_2);
        myDrawable.setBounds(0, 0, myDrawable.getMinimumWidth(), myDrawable.getMinimumHeight());
```

```
            myEditText.setCompoundDrawables(myDrawable, null, null, null);      //顶着行首显示
            myEditText.addTextChangedListener(this);                             //（a）
            myButton = findViewById(R.id.myButton);
            ...
        }
        ...
        @Override
        public void onClick(View v) {
            if (v.getId() == R.id.myLinearLayout) {                              //点击 LinearLayout 时执行的处理代码
                InputMethodManager manager = (InputMethodManager) getSystemService
(Context.INPUT_METHOD_SERVICE);                                                 //（d）
                manager.hideSoftInputFromWindow(v.getWindowToken(), 0);          //隐藏软键盘
                //如果用户已经输入了符合要求的金额，则取消文本编辑框的焦点
                if (myEditText.getText().length() > 0 && myButton.isEnabled())
                    myEditText.setFocusable(false);
            }
        }

        @Override
        public void beforeTextChanged(CharSequence s, int start, int count, int after) {
                                                                                 //必须重写的方法
        }

        @Override
        public void onTextChanged(CharSequence s, int start, int before, int count) {
                                                                                 //必须重写的方法
        }

        @Override
        public void afterTextChanged(Editable text) {                            //在文本编辑框内容改变后触发
            String editStr = text.toString();                                    //获取文本字符串（用户输入的金额值）
            //如果输入内容为空或用户删除了金额，"提现"按钮置为不可用
            if (editStr.length() <= 0) {
                myButton.setEnabled(false);
                return;
            }
            Double withdraw = Double.parseDouble(editStr);                       //获取用户输入的金额数值
            //要进行判断，金额必须大于 0 且小于账户余额才允许提现
            if (withdraw > 0 && withdraw <= total) myButton.setEnabled(true);
            else myButton.setEnabled(false);                                     //否则置"提现"按钮为不可用
        }
    }
```

其中：

（a）**public class MainActivity extends AppCompatActivity implements View.OnClickListener, TextWatcher{ ... } myEditText.addTextChangedListener(this);**：本例用到 TextWatcher 监听器来监测文本编辑框中内容的变化，这个监听器比较特殊，它并**不属于** Android 的视图（View）系统，而是位于 android.text 中。程序里我们通过主程序类来实现这个监听器接口，然后在文本编辑框上调用 addTextChangedListener()方法绑定监听接口，在文本内容变化时就会触发动作，根据文本内容（金额数值）的改变来实时地响应用户输入。TextWatcher 接口中一共有 3 个方法：beforeTextChanged()、onTextChanged()和 afterTextChanged()，分别对应用来监听文本内容变化前、变化过程中及变化后的事件，

本例用到的仅仅是最后一个 afterTextChanged()方法（在文本变化后触发事件），但是另外两个方法即使不用也要在主程序类中声明重写实现，并以@Override 加以注解，这是 Java 面向对象编程的规范，必须遵守。

（b）**private double total = 99.99;**：声明这个实数型的私有全局变量 total 用来保持账户中总的余额，便于后面的程序中进行比较以判断用户输入的合法性。

（c）**myLinearLayout.setOnClickListener(this);**：在 LinearLayout 布局容器上绑定点击事件监听类，以感知用户的点击动作。

（d）**InputMethodManager manager = (InputMethodManager) getSystemService(Context. INPUT_METHOD_SERVICE);**：本例是用 InputMethodManager（输入方法管理器）来控制软键盘开关的，在 Android 中这个管理器作为一个系统服务提供给用户使用，编程时通过 getSystemService()方法从上下文获取（以常量名 INPUT_METHOD_SERVICE 引用）。

2.2.7 下拉框：Spinner

Spinner 是从多个选项中选择一个选项的控件，类似于桌面应用程序的组合框（ComboBox），但没有组合框的下拉菜单，而是使用浮动菜单为用户提供选项。

Spinner 控件比较重要的属性和方法介绍如下。

属性：
- **android:background**：设置下拉框背景样式。
- **android:spinnerMode**：设置下拉框的显示模式。

方法：
- **setPrompt()**：设置选项顶部提示文字。
- **setAdapter()**：将下拉框与指定适配器绑定。
- **setSelection()**：设置当前默认选项。

【例 2.7】 在页面上设计下拉框，包含"学士""硕士""博士"选项，每个选项同时带有图标。运行程序时，下拉框中的初始项为"硕士"，如图 2.16（a）所示；点击后出现一个浮动菜单，内有"学士""硕士""博士"三个选项，如图 2.16（b）所示。

（a）初始项　　　　　　　　　（b）浮动菜单

图 2.16　下拉框应用程序界面

1. 准备

创建 Android Studio 工程 SpinnerDemo，在工程的 drawable 目录下放入 bachelor.png（学士，120×120）、master.png（硕士，120×120）、doctor.png（博士，120×120）。

2. 定义下拉框

（1）在 content_main.xml 中定义下拉框，代码如下：

```
<Spinner
    android:id="@+id/mySpinner"
    android:layout_width="335dp"
    android:layout_height="wrap_content"
    android:background="@android:drawable/spinner_background"          //（a）
    android:spinnerMode="dialog"                                        //（b）
    .../>
```

其中：

（a）**android:background="@android:drawable/spinner_background"**：这里将下拉框的背景设置为 Android 系统内置的 spinner_background 样式，这种样式外观上会带边框，显得更加醒目。

（b）**android:spinnerMode="dialog"**：在 Android 中，下拉框控件有多种显示模式，由 spinnerMode 属性指定，这里设为 dialog 表示以浮动菜单的样式显示下拉框中的选项（此为当前手机 App 界面普遍通用的模式）；若设为 dropdown，则是以通常计算机桌面应用程序下拉列表的样式显示，但这在手机 App 上并不常见。

（2）在工程的 layout 目录下再创建一个名为 myspinner_item.xml（名字可任取，程序代码中引用一致即可）的文件，用于进一步设计 Spinner 中选项的具体样式，内容如下：

```
<LinearLayout xmlns:android="http://schemas.android.com/apk/res/android"
    android:layout_width="match_parent"           //宽度匹配下拉框的宽度
    android:layout_height="wrap_content"          //高度与内容相适应
    android:orientation="horizontal" >            //下拉框中每一个选项的内容都放在一个水平线性布局中

    <ImageView                                    //显示选项所对应的图片
        android:id="@+id/degree_icon"
        android:layout_width="0dp"
        android:layout_height="50dp"
        android:layout_marginTop="@dimen/fab_margin"      //设置图片与选项顶部的间距
        android:layout_marginBottom="@dimen/fab_margin"   //设置图片与选项底部的间距
        android:layout_weight="1"
        android:gravity="center" />                       //居中

    <TextView                                     //显示选项的文本
        android:id="@+id/degree_name"
        android:layout_width="0dp"
        android:layout_height="match_parent"              //高度与选项相匹配
        android:layout_marginTop="@dimen/fab_margin"
        android:layout_marginBottom="@dimen/fab_margin"
        android:layout_weight="3"
        android:gravity="center"                          //居中
        android:textColor="#ff0000"                       //红色
        android:textSize="24sp" />                        //字号
</LinearLayout>
```

今后大家会看到，对于像 Spinner 这类选项列表型的视图控件，在编程的时候除了在 content_main.xml 中定义控件元素标签外，还需要另外创建一个选项样式文件（在本例中也就是上面的这个 myspinner_item.xml），在其中具体设计每个选项的样式布局。

3. 使用适配器实现选择功能

实现思路：

（1）下拉框选项文本预先存储在一个字符串数组中，选项图片的 id 引用则存放在另一个整型数组中。

（2）在 onCreate()方法中初始化时构造出一个 List<Map<String, Object>>列表结构，以键值对映射的形式保存每个选项的文字和图片。

（3）下拉框绑定到一个简单适配器 SimpleAdapter 并初始化。

（4）当选择事件发生时，从 onItemSelected()方法参数中得到用户选中项的索引，根据索引检索数组得到该项所对应的学历名称，再由文本视图将其显示出来。

在 MainActivity.java 中定义和使用适配器，实现选择功能，代码如下：

```java
public class MainActivity extends AppCompatActivity implements AdapterView.OnItemSelectedListener {
                                                                    //（a）
    private Spinner mySpinner;                              //下拉框控件变量
    private TextView myTextView;                            //文本视图控件变量
    private String[] degreeName = {"学士", "硕士", "博士"};    //字符串数组存储选项的文本内容
    private int[] degreeIcon = {R.drawable.bachelor, R.drawable.master, R.drawable.doctor};
                                                            //整型数组存储图片资源 id
    private String myDegree;                                //提示文本

    @Override
    protected void onCreate(Bundle savedInstanceState) {
        super.onCreate(savedInstanceState);
        setContentView(R.layout.activity_main);
        myTextView = findViewById(R.id.myTextView);
        myDegree = myTextView.getText().toString();         //初始提示文本"您的学历是："
        List<Map<String, Object>> degree_list = new ArrayList<Map<String, Object>>();
                                                            //（b）
        for (int i = 0; i < degreeName.length; i++) {
            Map<String, Object> item = new HashMap<String, Object>();
                                                            //Map 项中存放"名称-图片"键值对
            item.put("name", degreeName[i]);                //名称
            item.put("icon", degreeIcon[i]);                //图片
            degree_list.add(item);                          //添加进 List 结构
        }
        //创建一个简单适配器
        SimpleAdapter adapter = new SimpleAdapter(this, degree_list, R.layout.myspinner_item, new String[]{"name", "icon"}, new int[]{R.id.degree_name, R.id.degree_icon});
        adapter.setDropDownViewResource(R.layout.myspinner_item);   //设置适配器列表项的样式
        mySpinner = (Spinner) findViewById(R.id.mySpinner);
        mySpinner.setPrompt("请选择学历");
        mySpinner.setAdapter(adapter);                      //（c）
        mySpinner.setSelection(1);                          //默认选中第 2 项（索引从 0 开始）
        mySpinner.setOnItemSelectedListener(this);          //设置选择动作监听类
        ...
```

```
        }
        ...
        @Override
        public void onItemSelected(AdapterView<?> arg0, View arg1, int arg2, long arg3) {
            String degree = degreeName[arg2];                   //arg2 参数是用户选中项的索引
            myTextView.setText(myDegree + degree);              //显示用户选中的学历名称
        }

        @Override
        public void onNothingSelected(AdapterView<?> arg0) {
        }
}
```

其中：

（a）**public class MainActivity extends AppCompatActivity implements AdapterView.OnItem SelectedListener**：这里用主程序类实现 AdapterView（适配器视图）的 OnItemSelectedListener（列表项选择监听）接口，然后由下拉框控件注册该监听类实现对用户选择动作的响应，在 onItemSelected()方法中完成当用户选择某个选项时需要进行的处理。该监听接口中还有另一个 onNothingSelected()（未选任何项）方法，在一般编程中不会用，与之前 TextWatcher 监听器中的方法一样，虽没有实现，但也必须写出它的声明和方法体。

（b）**List<Map<String, Object>> degree_list = new ArrayList<Map<String, Object>>();**：创建一个 Map 类型对象的 List（队列）数据结构，存储学历名称与图片的映射信息。以后大家还会学到，其他几种与 Spinner 类似的选项列表型视图也都普遍使用这样一种数据结构来存储列表项的数据信息。

（c）**mySpinner.setAdapter(adapter);**：用下拉框控件的 setAdapter()方法将适配器与之绑定，在使用适配器绑定界面控件和底层数据后，应用程序就不再需要时刻监视底层数据的变化，而用户界面显示的内容与底层数据始终保持一致。

2.2.8 自动完成文本视图：AutoCompleteTextView

在我们用手机上网打开网页，在搜索栏中只要输入个别字词就能看到系统自动匹配的一些关联词组的下拉列表，这个效果就是用 Android 的自动完成文本视图 AutoCompleteTextView 实现的。

AutoCompleteTextView 控件比较重要的属性和方法介绍如下。

属性：
- **android:background**：设置搜索框的背景样式。
- **android:completionHint**：设置关键词下拉列表底部的提示文字。
- **android:completionThreshold**：设置当用户输入几个字符时显示下拉完成列表。
- **android:hint**：搜索框初始的提示文字。

方法：
- **setCompoundDrawables()**：设置搜索框头部显示的图标。
- **setAdapter()**：设置自动完成文本视图所要绑定的适配器。

下面来设计一个自动搜索框演示它的用法。

【例 2.8】 设计一个搜索框，输入"中"，列表会自动匹配出"中国""中国梦"等正能量词语，如图 2.17 所示。

图 2.17 搜索框界面

1. 准备

创建 Android Studio 工程 AutoCompleteTextViewDemo，在工程的 drawable 目录下放入 search.png（放大镜图标，48×48），另放一个 editext_bg.9.png 及其状态列表图形切换的样式文件 editext_style.xml（用【例 2.6】中现成的资源）。

2. 设计搜索框

（1）在 content_main.xml 中设计搜索框的外观，代码如下：

```
<AutoCompleteTextView
    android:id="@+id/myAutoCompleteTextView"
    android:layout_width="318dp"
    android:layout_height="55dp"
    android:background="@drawable/editext_style"    //背景设为样式文件，使搜索框在获得焦点时显示边框
    android:completionHint="关联主题"                //运行时出现在关键词列表底部的提示文字
    android:completionThreshold="1"                 //设定当用户输入 1 个字符时就开始显出下拉完成列表
    android:hint="请输入关键词"                      //初始的提示文本
    android:textSize="24sp"
    .../>
```

（2）在工程的 layout 目录下再创建一个名为 keyword_item.xml（名字可任取，程序代码中引用一致即可）的文件，在其中定义搜索框下拉完成列表中出现词组的文本样式，内容如下：

```
<TextView xmlns:android="http://schemas.android.com/apk/res/android"
    android:id="@+id/myKeyword"
    android:layout_width="match_parent"
    android:layout_height="40dp"
    android:gravity="left"                  //文本靠左对齐显示
    android:singleLine="true"
    android:textColor="#4CAF50"             //文本颜色为浅绿色
    android:textSize="24sp" />
```

3. 自动完成功能实现

实现思路：

（1）用一个字符串数组预先存放搜索提示关键词文本。

（2）在 onCreate()方法中初始化将 AutoCompleteTextView 与 ArrayAdapter 适配器绑定即可。

在 MainActivity.java 中实现自动完成功能，代码如下：

```
public class MainActivity extends AppCompatActivity {
    private Drawable myDrawable;                    //用于获取和显示放大镜图标
```

```
        private AutoCompleteTextView myAutoCompleteTextView;        //自动完成文本视图
        private String[] keyWords = {"中国", "中国梦", "中国女排", "中国机长", "中华民族伟大复兴", "中华人民
共和国成立70周年"};                                                   //自动完成的词组文本集

        @Override
        protected void onCreate(Bundle savedInstanceState) {
            super.onCreate(savedInstanceState);
            setContentView(R.layout.activity_main);
            myAutoCompleteTextView = findViewById(R.id.myAutoCompleteTextView);
            myDrawable = getResources().getDrawable(R.drawable.search);    //获取放大镜图片资源
            myDrawable.setBounds(0, 0, myDrawable.getMinimumWidth(), myDrawable.getMinimumHeight());
            myAutoCompleteTextView.setCompoundDrawables(myDrawable, null, null, null);
                                                                //在搜索框头部显示放大镜
            ArrayAdapter<String> adapter = new ArrayAdapter<String>(this, R.layout.keyword_item, keyWords);
                                                                //（a）
            myAutoCompleteTextView.setAdapter(adapter);         //（b）
            ...
        }
        ...
}
```

其中：

（a）**ArrayAdapter<String> adapter = new ArrayAdapter<String>(this, R.layout.keyword_item, keyWords);**：创建 ArrayAdapter 适配器，将 keyword_item.xml 样式文件与要显示的词组文本集（保存在数组中）关联起来。

（b）**myAutoCompleteTextView.setAdapter(adapter);**：将自动完成文本视图与适配器绑定。

可见，AutoCompleteTextView 实现自动完成功能的实质与 Spinner 其实一样，都是通过**适配器**做到的，而搜索功能则是 AutoCompleteTextView 自身内置的特性。运行程序试验会发现，只有输入"中"才会显示下拉完成列表，而输入其他内容则不会出现列表，故 AutoCompleteTextView 实现搜索的能力范围受限于使用它的程序内部所预置存储的信息（本例中也就是字符串数组 keyWords 中的内容）。如果想让 AutoCompleteTextView 具备更强大的匹配功能，可以考虑专门定义一个复杂的数据结构并封装成独立的类，甚至使用后台数据库来存储更多的数据。

AutoCompleteTextView 可以像 Spinner 一样绑定适配器，但 TextView、EditText 和 Button 皆不可以，从这个角度上看，AutoCompleteTextView 似乎与 Spinner（而非 TextView）的亲缘关系更近。

2.2.9　日期时间选择器：DatePicker/TimePicker

DatePicker/TimePicker 可以提供用户选择日期和时间，它们各有两种基本的样式：一种是直接布局在界面上；另一种是以日期时间选择对话框的形式出现，结合日历组件 Calendar 可实现灵活的日期时间选择功能。

DatePicker 控件比较重要的属性和方法介绍如下。

属性：
- **style**：设置日期选择器的样式。
- **android:fadingEdge**：设置滑动操作时边框渐变的方向。

方法：
- **setOnDateChangedListener()**：设置绑定日期改变监听类。

TimePicker 控件比较重要的属性和方法介绍如下。

属性：
- **android:timePickerMode**：设置时间选择器的样式。

方法：
- **setOnTimeChangedListener()**：设置绑定时间改变监听类。

下面还是通过实例来了解它们的应用。

【例 2.9】 设计一个手机闹钟时间设定程序，DatePicker/TimePicker 均直接布局在界面上，用户可以上下滑动来设定日期和时间，如图 2.18 所示，Switch 开关可控制闹钟的开启与关闭。

图 2.18 手机闹钟时间设定程序界面

1. 准备

创建 Android Studio 工程 DateTimePickerDemo，在工程的 drawable 目录下放入图片 on.png 和 off.png（闹钟开关 CheckBox 的背景，尺寸均为 91×46）。另创建样式控制文件 switch_style.xml，内容如下：

```
<?xml version="1.0" encoding="utf-8"?>
<selector xmlns:android="http://schemas.android.com/apk/res/android">
    <item android:state_checked="true" android:drawable="@drawable/on" />    //呈现开背景 on
    <item android:drawable="@drawable/off" />                                //呈现关背景 off
</selector>
```

2. 设计界面

Android Studio 的工具面板中并未提供现成的 DatePicker/TimePicker 控件，故需要用户自己在 content_main.xml 中直接编写代码来定义：

```
<DatePicker                                                                  //定义日期选择器
    android:id="@+id/myDatePicker"
    style="@android:style/Widget.DatePicker"                                  // (a)
    android:layout_width="354dp"
    android:layout_height="134dp"
    android:fadingEdge="horizontal|vertical"
    .../>

<TimePicker                                                                  //定义时间选择器
    android:id="@+id/myTimePicker"
    android:layout_width="263dp"
    android:layout_height="154dp"
```

```
        android:layout_marginTop="24dp"
        android:timePickerMode="spinner"                        // (b)
        .../>

    <TextView                                                    // (c)
        android:id="@+id/myTextView"
        android:layout_width="214dp"
        android:layout_height="60dp"
        android:layout_marginStart="12dp"
        android:layout_marginLeft="12dp"
        android:layout_marginTop="16dp"
        android:text="闹钟时间："
        android:textSize="18sp"
        .../>

    <CheckBox                               // (d) 复选框仿 Switch 的外观，模拟闹钟开关
        android:id="@+id/myCheckBox"
        android:layout_width="129dp"
        android:layout_height="65dp"
        android:layout_marginStart="16dp"
        android:layout_marginTop="12dp"
        android:background="@drawable/switch_style"  //背景以状态列表图形机制实现动态更换
        android:button="@null"                //必须设为 null 才能使上面背景的设置生效
        .../>
```

其中：

（a）**style="@android:style/Widget.DatePicker"**：设置日期选择器在界面上的显示样式，此处设为 Widget.DatePicker，显示为支持用户上下翻动的滑条外观。

（b）**android:timePickerMode="spinner"**：设置时间选择器在界面上的显示样式，此处设定为 spinner，同样也是支持用户上下翻动的滑条外观，以保持与日期选择器在外观风格上的一致。

（c）**<TextView.../>**：这个文本视图由事件触发，动态地显示用户当前所设定的日期和时间。

3. 闹钟功能实现

实现思路：

（1）在 onCreate()方法初始化时，在日期选择器上绑定日期改变监听类 DatePicker.OnDateChangedListener，在时间选择器上绑定时间改变监听类 TimePicker.OnTimeChangedListener。

（2）创建一个日历控件的对象实例，获取初始日期、时间显示。

（3）当用户选择日期、时间改变时，分别触发 onDateChanged()、onTimeChanged()方法，在其中获取更新的日期、时间加以显示。

在 MainActivity.java 中实现闹钟功能，代码如下：

```
public class MainActivity extends AppCompatActivity implements DatePicker.OnDateChangedListener,
TimePicker.OnTimeChangedListener, CompoundButton.OnCheckedChangeListener {    // (a)
    private TextView myTextView;                    //显示用户当前设定的日期、时间
    private DatePicker myDatePicker;                //日期选择器
    private TimePicker myTimePicker;                //时间选择器
    private CheckBox myCheckBox;                    //复选框（模拟闹钟开关）
    private String myString;                        //提示文本
    private String myDate;                          //用户当前设定的日期信息
    private String myTime;                          //用户当前设定的时间信息
```

```java
    @Override
    protected void onCreate(Bundle savedInstanceState) {
        super.onCreate(savedInstanceState);
        setContentView(R.layout.activity_main);
        myTextView = findViewById(R.id.myTextView);
        myDatePicker = findViewById(R.id.myDatePicker);                 //获取日期选择器控件引用
        myDatePicker.setOnDateChangedListener(this);                    //绑定日期改变监听类
        myString = myTextView.getText().toString() + "\n";
        Calendar calendar = Calendar.getInstance();                     //创建一个日历控件的对象实例
        myDate = calendar.get(Calendar.YEAR) + "年" + (calendar.get(Calendar.MONTH) + 1) + "月" + calendar.get(Calendar.DAY_OF_MONTH) + "日";
                                                                        //由日历控件获取初始日期
        myTimePicker = findViewById(R.id.myTimePicker);                 //获取时间选择器控件引用
        myTimePicker.setOnTimeChangedListener(this);                    //绑定时间改变监听类
        myTime = " " + calendar.get(Calendar.HOUR) + ":" + calendar.get(Calendar.MINUTE);
                                                                        //由日历控件获取初始时间
        myTextView.setText(myString + myDate + myTime);                 //显示初始的日期和时间
        myCheckBox = findViewById(R.id.myCheckBox);
        myCheckBox.setOnCheckedChangeListener(this);                    //绑定选择状态改变监听类
        ...
    }
    ...
    @Override
    public void onDateChanged(DatePicker dpview, int year, int month, int day) {
        myDate = year + "年" + (month + 1) + "月" + day + "日";         // (b)
        myTextView.setText(myString + myDate + myTime);                 //更新文本视图
    }

    @Override
    public void onTimeChanged(TimePicker tpview, int hour, int minute) {
        myTime = " " + hour + ":" + minute;                             //从传入参数中获取时间信息
        myTextView.setText(myString + myDate + myTime);                 //更新文本视图
    }

    @Override
    public void onCheckedChanged(CompoundButton v, boolean checked) {
        if (v.getId() == R.id.myCheckBox && checked) {
            Toast.makeText(MainActivity.this, "闹钟开启", Toast.LENGTH_LONG).show();
                                                                        //消息提示闹钟开启
        }
    }
}
```

其中：

（a）**public class MainActivity extends AppCompatActivity implements DatePicker.OnDate Changed Listener, TimePicker.OnTimeChangedListener, CompoundButton.OnCheckedChangeListener**：分别实现 DatePicker/TimePicker 的 OnDateChangedListener/OnTimeChangedListener 接口，是为了在日期、时间改变的时候第一时间做出响应处理。

（b）**myDate = year + "年" + (month + 1) + "月" + day + "日";**：这里从重写 onDateChanged() 方法的传入参数中获取日期信息，由于 Android 系统的月份是从 0 开始排序的，故实际的月份应当加上 1 才等于正确的当前月份值。

DatePicker 上的 setOnDateChangedListener()方法要求开发环境 Android 的 API 版本最低为 26，需要进行设置，如图 2.19 所示。

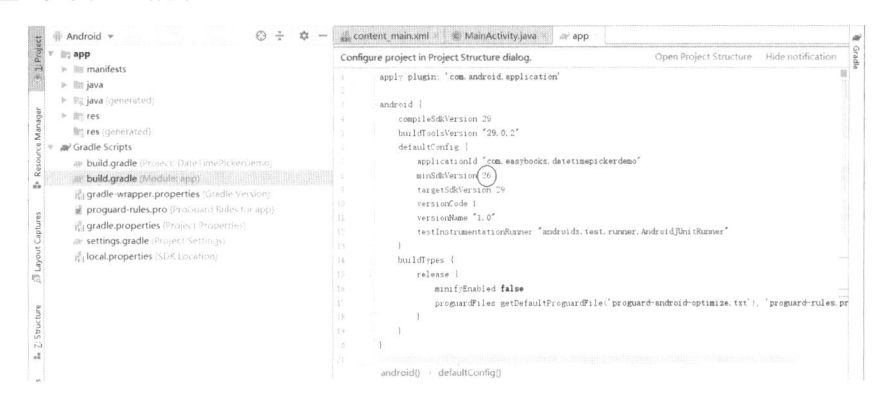

图 2.19 设置 Android API 的最低版本

展开工程的项目树，打开"Gradle Scripts"→"build.gradle"，将其中的 minSdkVersion 改为 26。今后遇到类似的问题都按照提示进行这样的修改处理即可。

2.3 界面事件

在 Android 系统中，除了种类丰富的界面控件，还存在多种界面事件（UI Event），如点击事件、触摸事件、焦点事件和菜单事件等，在这些事件发生时，Android 界面框架调用界面控件的事件处理方法对它们进行处理。

2.3.1 按键事件

在 MVC 模型中，控制器根据界面事件类型的不同，将其传递给界面控件不同的事件处理方法。例如，按键事件（KeyEvent）传递给 onKey()方法，触摸事件（TouchEvent）传递给 onTouch()方法。Android 系统对界面事件的传递和处理都遵循一定的规则：

（1）如果界面控件设置了事件监听器，则事件将优先传递给事件监听器；否则，会直接传递给界面控件的其他事件处理方法。

（2）如果该事件监听器处理方法的返回值为 true，表示该事件已经处理完毕，不需要其他处理方法参与处理，这样事件就不会再继续进行传递；否则，会传递给其他的事件处理方法。

例如，假设 EditText 控件已经设置了按键事件监听器 OnKeyListener。当用户按下键盘上的某个键时，控制器将产生 KeyEvent 按键事件。因为 EditText 已设置了按键事件监听器，所以按键事件先传递到该监听器的事件处理方法 onKey()中。如果 onKey()方法返回 false，事件将继续传递，EditText 就可以捕获该事件，将按键的内容显示出来；但如果 onKey()方法返回 true，将阻止事件继续传递，这样 EditText 就不能捕获按键事件，也不能显示按键的内容。

Android 界面框架支持对按键事件的监听，并能够将事件的详细信息传递给处理方法。为了处理控件的按键事件，需要先设置按键事件的监听器并重写 onKey()方法，示例代码如下：

```
entryText.setOnKeyListener(new OnKeyListener(){         //（a）
    @Override
    public boolean onKey(View view, int keyCode, KeyEvent keyEvent) {   //（b）
        //事件处理的代码……
        return true; //or false;                        //（c）
```

```
    }
});
```
其中：

（a）**entryText.setOnKeyListener(new OnKeyListener(){ ... });**：设置控件的按键事件监听器。

（b）**public boolean onKey(View view, int keyCode, KeyEvent keyEvent)**：第一个参数 view 表示产生按键事件的界面控件；第二个参数 keyCode 表示按键代码；第三个参数 keyEvent 则包含了事件的详细信息，如按键的重复次数、硬件编码和按键标志等。

（c）**return true; //or false;**：onKey()方法的返回值，返回 true，将阻止事件传递；返回 false，允许继续传递事件。

【例 2.10】 测试按键事件的处理及显示事件详细信息。

1. 设计界面

创建 Android Studio 工程 KeyEventDemo，设计测试程序的界面，如图 2.20 所示。

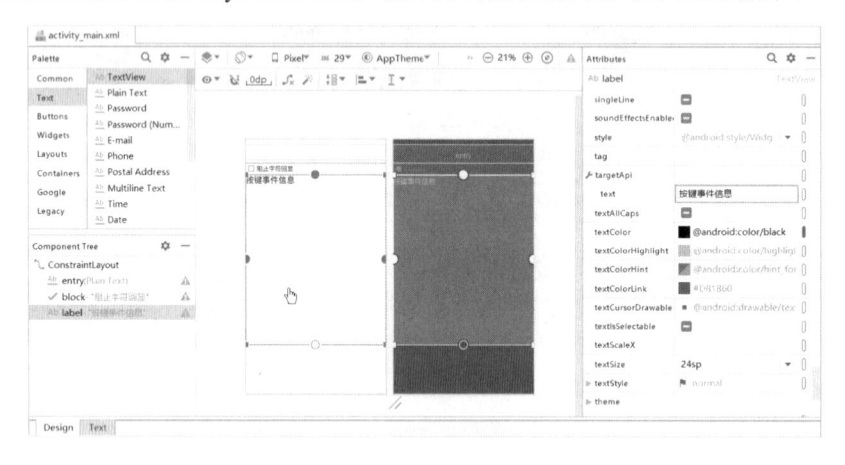

图 2.20 按键事件测试程序界面

界面顶部的 EditText 是字符输入区；中间 CheckBox 用来控制 onKey()方法的返回值；下方的 TextView 用来显示按键事件的详细信息，包括按键动作、按键代码、按键字符、Unicode 编码、重复次数、功能键状态、硬件编码和按键标志。

2. 处理事件

在 EditText 中，每当键盘任何一个键被按下或释放时都会引发按键事件。但为了使 EditText 能够处理按键事件，还需要使用 setOnKeyListener()方法在其上设置按键事件监听器，并在 onKey()方法中编写事件的处理代码，内容如下：

```
myentry.setOnKeyListener(new View.OnKeyListener() {
    @Override
    public boolean onKey(View v, int keyCode, KeyEvent event) {
        int metaState = event.getMetaState();                                    //（a）
        int unicodeChar = event.getUnicodeChar();                                //（b）
        String msg = "";
        msg += "按键动作：" + String.valueOf(event.getAction()) + "\n";          //（c）
        msg += "按键代码：" + String.valueOf(keyCode) + "\n";
        msg += "按键字符：" + (char) unicodeChar + "\n";                         //（b）
        msg += "UNICODE：" + String.valueOf(unicodeChar) + "\n";
        msg += "重复次数：" + String.valueOf(event.getRepeatCount()) + "\n";     //（d）
```

```
            msg += "功能键状态：" + String.valueOf(metaState) + "\n";         // (a)
            msg += "硬件编码：" + String.valueOf(event.getScanCode()) + "\n";    // (e)
            msg += "按键标志：" + String.valueOf(event.getFlags()) + "\n";       // (f)
            mylabel.setText(msg);
            if (myblock.isChecked()) return true;
            return false;
        }
    });
```

其中：

（a）**int metaState = event.getMetaState(); msg += "功能键状态：" + String.valueOf(metaState) + "\n";**：用来获取功能键状态。功能键包括左 Alt 键、右 Alt 键和 Shift 键，当这 3 个键被按下时，metaState 值（功能键代码）分别为 18、34 和 65；若没有功能键被按下，metaState 值都为 0。

（b）**int unicodeChar = event.getUnicodeChar();** 和 **msg += "按键字符：" + (char)unicodeChar + "\n";**：获取按键的 Unicode 值，并将其转换为字符显示在 TextView 中。

（c）**msg += "按键动作：" + String.valueOf(event.getAction()) + "\n";**：获取按键动作，0 表示按下，1 表示释放。

（d）**msg += "重复次数：" + String.valueOf(event.getRepeatCount()) + "\n";**：获取按键的重复次数，当按键被长时间按下时，就会产生这个属性值。

（e）**msg += "硬件编码：" + String.valueOf(event.getScanCode()) + "\n";**：获取按键的硬件编码，不同硬件设备的按键硬件编码都不相同，此值一般用于调试。

（f）**msg += "按键标志：" + String.valueOf(event.getFlags()) + "\n";**：获取按键事件的标识符。

3．运行测试

运行程序，按下键盘上的"A"键，下方显示出该按键事件的详细信息，同时顶部的字符输入区也会显示出'a'字符，如图 2.21（a）所示；在勾选"阻止字符回显"复选框后，按下键盘上的"B"键，下方显示按键事件的信息，但字符输入区却不会同时显示'b'字符，如图 2.21（b）所示。实际进行 App 开发时往往需要用这些信息来做出判断，以便进行进一步的处理或者提示用户。

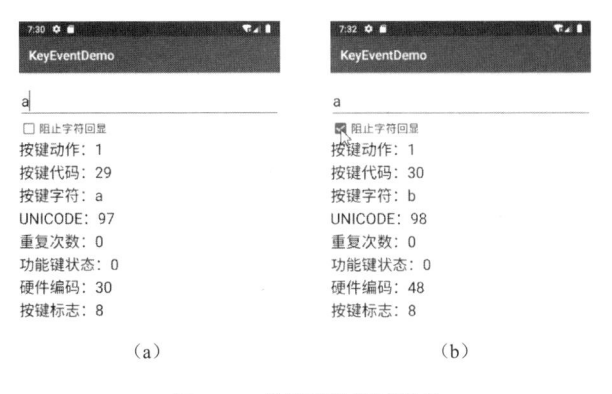

(a)　　　　　　　　(b)

图 2.21　按键事件测试效果

2.3.2　触摸事件

Android 界面框架支持对触摸事件的监听，并能将触摸事件的详细信息传递给处理方法。为了处理控件的触摸事件，需要先设置该事件的监听器并重写 onTouch() 方法，示例代码如下：

```
touchView.setOnTouchListener(new View.OnTouchListener(){                       // (a)
    @Override
    public boolean onTouch(View view, MotionEvent event) {                     // (b)
        //事件处理的代码……
        return true; //or false;                                               // (c)
    }
});
```

其中：

（a）**touchView.setOnTouchListener(new View.OnTouchListener(){ ... });**：设置控件的触摸事件监听器。

（b）**public boolean onTouch(View view, MotionEvent event)**：第一个参数 view 表示产生触摸事件的界面控件；第二个参数 event 是触摸事件的详细信息，如产生时间、坐标和触点压力等。

（c）**return true; //or false;**：onTouch()方法的返回值。

【例 2.11】 测试触摸事件的处理及显示事件详细信息。

1. 设计界面

创建 Android Studio 工程 TouchEventDemo，设计测试程序的界面如图 2.22 所示，界面实现源码请扫描右边二维码查看。

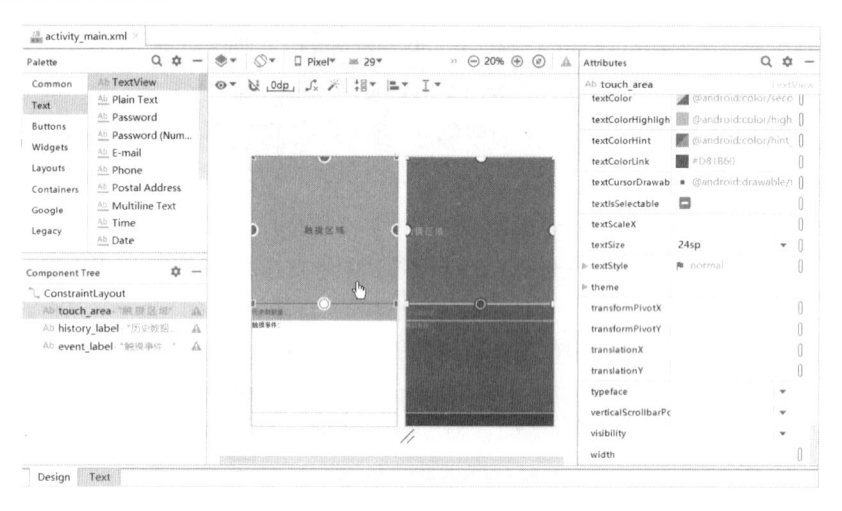

图 2.22 触摸事件测试程序界面

界面中包含 3 个 TextView：第一个（上方浅色区域）是可以接收触摸事件的区域；第二个（中间灰色条）用来显示触摸事件的历史数据量；第三个（底部区域）用来显示触摸事件的详细信息，包括类型、相对坐标、绝对坐标、触点压力和尺寸。

2. 处理事件

当用户手指触碰到手机屏、在屏幕上移动或离开时，分别会引发 ACTION_DOWN、ACTION_UP 和 ACTION_MOVE 触摸事件。而无论是哪种类型的触摸事件，都会调用 onTouch()方法来进行处理，具体类型包含在 MotionEvent 参数中，可以通过 getAction()方法获取，然后根据不同类型进行不一样的处理。为了让第一个 TextView 能够处理触摸事件，需使用 setOnTouchListener()方法在其上设置触摸事件监听器，并在 onTouch()方法中编写事件的处理代码，内容如下：

```
touchView.setOnTouchListener(new View.OnTouchListener() {
    @Override
```

触摸事件代码

```
        public boolean onTouch(View v, MotionEvent event) {
            int action = event.getAction();
            switch (action) {
                case MotionEvent.ACTION_DOWN:
                    display("触击屏幕", event);
                    break;
                case MotionEvent.ACTION_UP:
                    processHistory(event);
                    display("离开屏幕", event);
                    break;
                case MotionEvent.ACTION_MOVE:
                    display("在屏幕上移动", event);
                    break;
            }
            return true;
        }
    });
```

一般情况下,如果用户将手指放在手机屏上(但不移动),然后抬起手指,会先后产生 ACTION_DOWN 和 ACTION_UP 两个触摸事件。但如果用户在屏幕上移动手指,然后再抬起,则会产生这样的事件序列:ACTION_DOWN→ACTION_MOVE→ACTION_MOVE→ACTION_MOVE→……→ACTION_UP。

上面的代码中,调用了 display()和 processHistory()两个自定义的方法,下面分别具体说明。

(1) display()方法。该方法将 MotionEvent 参数中的事件信息提取出来显示在用户界面上,不仅有触摸事件的类型信息,还有触点的坐标、压力、尺寸等信息,实现代码如下:

```
private void display(String eventType, MotionEvent event) {
    int x = (int) event.getX();                                    // (a)
    int y = (int) event.getY();                                    // (a)
    int rawX = (int) event.getRawX();                              // (b)
    int rawY = (int) event.getRawY();                              // (b)
    float pressure = event.getPressure();                          // (c)
    float size = event.getSize();                                  // (d)

    String msg = "";
    msg += "事件类型: " + eventType + "\n";
    msg += "相对坐标: (" + String.valueOf(x) + ", " + String.valueOf(y) + ")\n";
    msg += "绝对坐标: (" + String.valueOf(rawX) + ", " + String.valueOf(rawY) + ")\n";
    msg += "触点压力: " + String.valueOf(pressure) + "\n";
    msg += "触点尺寸: " + String.valueOf(size) + "\n";
    eventView.setText(msg);
}
```

其中:

(a) **int x = (int) event.getX();和 int y = (int) event.getY();**:获取的是触点相对于父界面元素的坐标信息。

(b) **int rawX = (int) event.getRawX();和 int rawY = (int) event.getRawY();**:获取的是触点绝对坐标的坐标信息。

(c) **float pressure = event.getPressure();**:获取触点压力。触点压力是一个介于 0 和 1 之间的浮点数,表示用户手指对手机屏施加压力的大小,接近 0 表示压力较小,接近 1 表示压力较大。

(d) **float size = event.getSize();**:获取触点尺寸。触点尺寸是指用户手指与手机屏接触面积的大小,也是一个介于 0 和 1 之间的浮点数，接近 0 表示接触面较小，接近 1 表示接触面较大。

（2）processHistory()方法。当下的智能手机基本上都是以屏幕触摸来接收和响应用户的操作的，如果 Android 界面框架不能产生足够多的触摸事件，App 就不能很精确地描绘用户的动作；相反地，如果 Android 界面框架产生了过多的触摸事件，虽然能满足精度要求，但却降低了 App 的运行效率。为了解决这个矛盾，Android 采用了"打包"的策略：在触点移动速度较快时会产生大量的数据，每隔一段时间便会产生一个 ACTION_MOVE 事件，在这个事件中，除了有当前触点的相关信息外，还包含了这段时间间隔内触点轨迹的历史数据，这样既能够保持精度，又不至于产生过多的触摸事件。通常情况下，在 ACTION_MOVE 的事件处理方法中，都是先处理历史数据，然后再处理当前数据。本程序中，用自定义的 processHistory()方法实现对历史数据的处理，代码如下：

```
private void processHistory(MotionEvent event) {
    int historySize = event.getHistorySize();                    //获取历史数据数量
    //下面开始循环处理历史数据
    for (int i = 0; i < historySize; i++) {
        long time = event.getHistoricalEventTime(i);             //历史事件发生时间
        //获取历史事件的相对坐标
        float x = event.getHistoricalX(i);
        float y = event.getHistoricalY(i);
        float pressure = event.getHistoricalPressure(i);         //历史事件触点压力
        float size = event.getHistoricalSize(i);                 //历史事件触点尺寸
        //处理过程……
    }
    historyView.setText("历史数据量：" + historySize);             //界面显示历史数据量
}
```

3. 运行测试

运行程序，用鼠标在模拟器屏幕上点击（模拟指尖触击），然后拖曳鼠标（模拟手指在屏幕上移动），最后释放鼠标（模拟手指离开触摸屏）的过程，界面下方显示出其间发生的事件类型及详细信息，如图 2.23 所示。

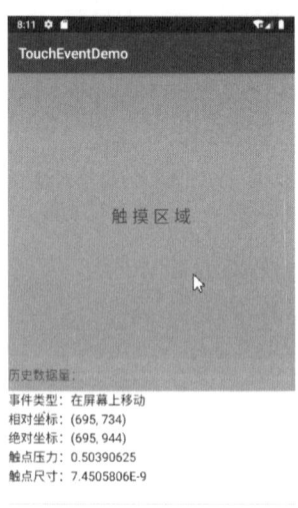

图 2.23 触摸事件测试效果

第3章 界面布局与活动页

3.1 界面布局

界面布局（Layout）是对用户界面结构的描述，它定义了界面中各个控件之间的相对位置关系。在 Android Studio 中设计界面布局有两种方式：使用系统默认的约束布局（ConstraintLayout）和自定义布局。

3.1.1 约束布局：ConstraintLayout

约束布局在设计一开始并不限定控件在界面上的绝对位置，而是由用户通过鼠标拖曳锚定控件上的约束点来指定该控件相对于整个页面或其他控件的方位，这么做极大地方便了用户对 Android 程序界面的可视化操作。

1. 约束布局库

自 Android Studio 3.x 起，创建工程的主设计文件 content_main.xml 默认采用的就是约束布局。在设计模式下，选中 HelloWorld 文本视图，在其四周出现 4 个约束点（实心蓝点），每个点都以约束线（弹簧箭头线）锚定到页面的四个边上，如图 3.1 所示。这表示给文本视图上下左右四个方向添加了到页边的约束，这样运行时文本视图的位置就确定下来，只能显示在页面中央。

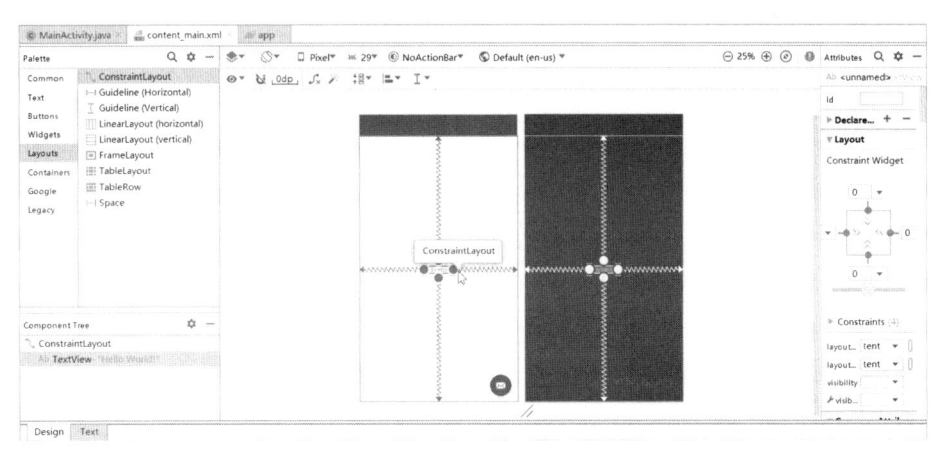

图 3.1 HelloWorld 文本视图默认的约束

单击左下"Text"选项卡切换到代码视图，可以看到，在 HelloWorld 文本视图控件的周边以及控件元素标签的尾部自动生成了约束布局代码，内容如下（加粗处）：

```
<?xml version="1.0" encoding="utf-8"?>
<androidx.constraintlayout.widget.ConstraintLayout
    xmlns:android="http://schemas.android.com/apk/res/android"......>
    <TextView
        android:id="@+id/textView"
```

```
            ...
            android:text="Hello World!"
            app:layout_constraintBottom_toBottomOf="parent"
            app:layout_constraintLeft_toLeftOf="parent"
            app:layout_constraintRight_toRightOf="parent"
            app:layout_constraintTop_toTopOf="parent" />
</androidx.constraintlayout.widget.ConstraintLayout>
```

同时，在工程 app\build.gradle 文件中，自动添加依赖并实现了 Android 官方提供的约束布局库，内容如下：

```
dependencies {
    ...
    implementation 'androidx.constraintlayout:constraintlayout:1.1.3'
    ...
}
```

2. 设计约束布局

当用户拖曳一个控件（比如按钮）到设计界面上时，初始这个控件四周的约束点都是空心的，此时尚未给该控件添加约束，也就是说，它在界面上的显示位置是不确定的（运行时一律默认显示在界面的左上角）。要想设定控件的位置，就必须给它添加约束。下面介绍几种基本的约束情形。

（1）约束一个控件在页面中的位置。通过鼠标点选约束点然后拖曳至页边锚定的方式来约束该控件在页面中的位置。例如，若要使按钮在页面上居中，分别点选按钮左右两边的约束点拖曳至页的左右两边锚定即可，如图 3.2 所示。

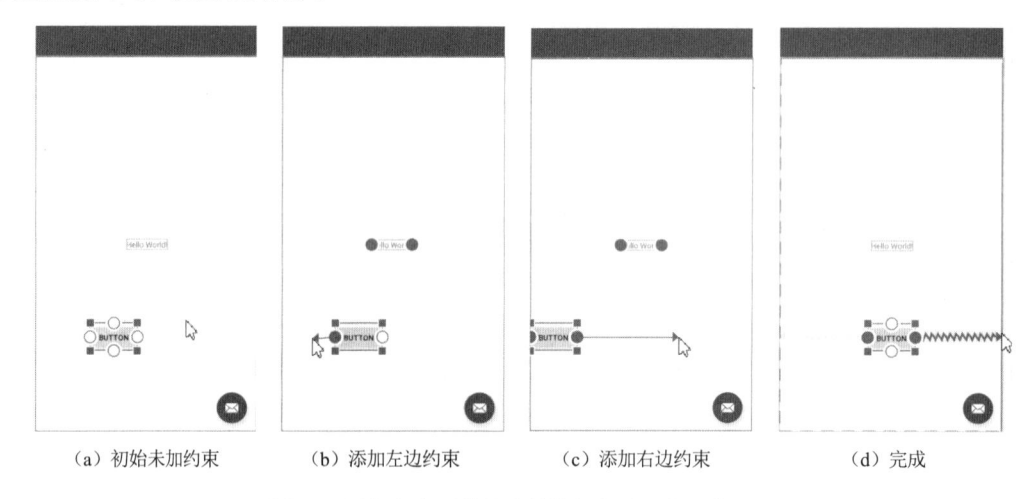

(a) 初始未加约束　　(b) 添加左边约束　　(c) 添加右边约束　　(d) 完成

图 3.2　添加左右两边约束使按钮在页面上居中显示

（2）约束一个控件相对于另一个控件的位置。只要点选控件上的约束点，用鼠标拖曳锚定至另一控件的目标位置约束点即可。例如，要想让按钮控件紧贴文本视图的底部下边沿显示，只需选中按钮上边的约束点拖曳锚定至文本视图下边的约束点就可以了，具体操作如图 3.3 所示。

若要删除已添加的约束，只需选中约束点，在右键快捷菜单中单击 "Clear Constraints of Selection" 命令即可，如图 3.3（c）所示。

(a) 选中按钮上边的约束点拖曳　　　(b) 完成　　　(c) 删除已添加的约束

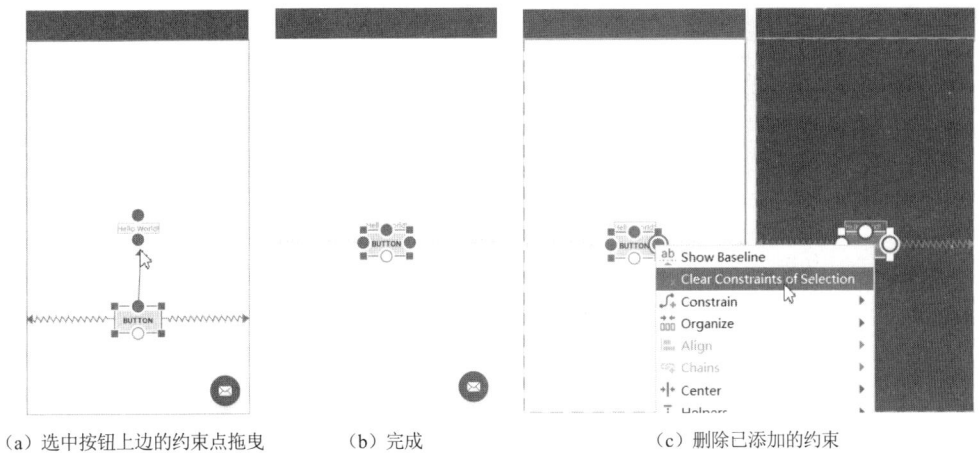

图 3.3　添加约束使按钮紧贴文本视图的底部下边沿显示及删除约束

（3）约束两个（或多个）控件在页面上对齐。在约束布局下对齐控件需要使用准线（Guideline），首先单击通知栏中的 Guidelines 图标，选择 "Add Vertical Guideline" 可以添加一个垂直方向上的准线。准线默认使用 dp（像素）尺定位，但我们想用更方便的百分比尺，只要点一下准线上端的箭头就可以切换到百分比尺，然后用鼠标拖曳准线调整到 50% 的位置，如图 3.4 所示。

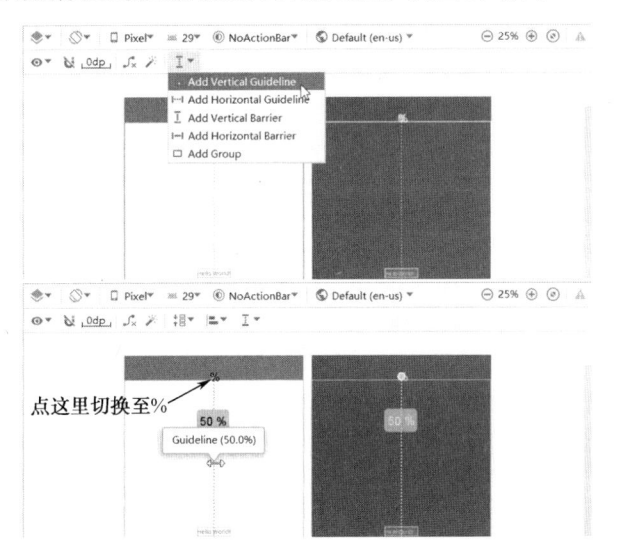

图 3.4　添加和调整准线（Guideline）

然后，往界面上放两个按钮，对第一个按钮添加相对页面底部及到准线的约束，如图 3.5（a）所示；对第二个按钮添加到准线及到第一个按钮底部的约束，如图 3.5（b）所示。再进行一些细微的拖曳以调整间距，这样就实现了两个按钮在水平方向上对齐且在页面上水平居中。同理，使用水平方向上的准线可实现按钮在垂直方向上对齐及在页面上垂直居中的效果，读者可以自己操作试试看。

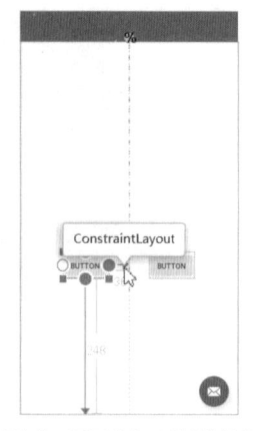

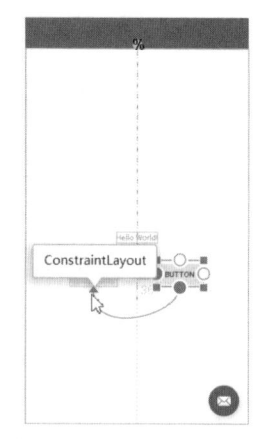

(a) 添加相对页面底部及到准线的约束　　　　(b) 添加到准线及到第一个按钮底部的约束

图 3.5　添加约束使两个按钮在水平方向上对齐且在页面上水平居中

3.1.2　自定义布局

除了使用系统默认提供的约束布局，以可视化的方式进行拖曳设计外，用户也可以用传统的布局（Layouts）组件结合编写 XML 代码的方式来自定义布局。自定义布局首先要创建界面布局文件（.xml），然后在其中定义所需类型的布局元素标签或拖曳放置进布局类组件。

1. 定义布局的操作

（1）创建界面布局文件。用户创建的界面布局文件与系统默认生成的主设计文件 content_main.xml 一样，都位于工程 layout 目录下，用于定义设计用户界面。创建界面布局文件有如下两种操作方式：

① 右击工程 layout 目录，在弹出的快捷菜单中选择"New"→"Layout resource file"命令，出现"New Resource File"对话框，如图 3.6 所示。在"File name"栏输入要创建的界面布局文件名，在"Root element"栏输入布局类型名，单击"OK"按钮，系统生成界面布局文件并自动保存到工程 layout 目录下。

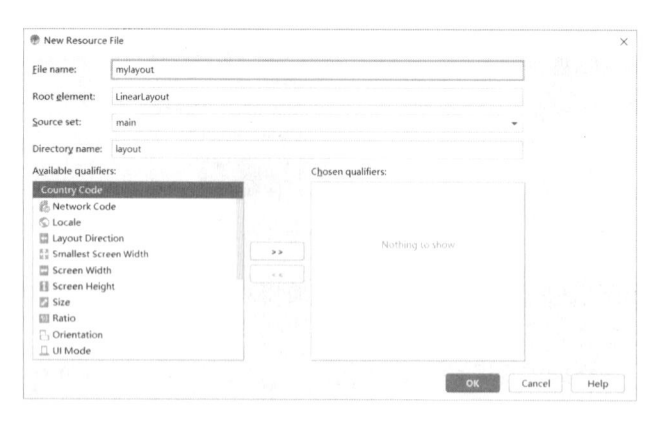

图 3.6　"New Resource File"对话框

② 右击工程 layout 目录，在弹出的快捷菜单中选择"New"→"XML"→"Layout XML File"命令，出现"New Android Component"对话框，如图 3.7 所示。在"Layout File Name"栏填写界面布局文件名，在"Root Tag"栏输入布局类型名，单击"Finish"按钮。

（2）布局的嵌套。系统在创建界面布局文件时仅仅确定了整个页面的大布局（称为根元素），除了可以直接在该布局下放置控件外，还可以在其中再安排新的布局，也就是说布局可以嵌套使用。在"Design"选项卡（设计视图）下，工具箱面板（Palette）的"Layouts"下包含了 Android 系统所支持的布局组件，如图 3.8 所示。用户可根据需要选用特定类型的组件拖曳至界面上已有的布局中作为其子布局，在子布局中再放入控件，如此可设计出更为丰富复杂的界面来。当然，也可以直接在 XML 文件中编写子元素标签代码来定义嵌套的布局。

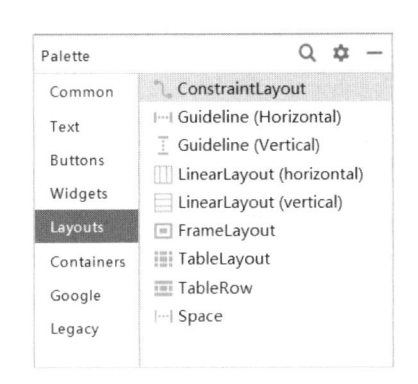

图 3.7 "New Android Component" 对话框　　图 3.8 Android 系统所支持的布局组件

Android 支持多种类型的布局，包括 LinearLayout（线性布局）、RelativeLayout（相对布局）、TableLayout（表格布局）、GridLayout（网格布局）、AbsoluteLayout（绝对布局）和 FrameLayout（板块布局），下面分别进行介绍。

2. 线性布局

线性布局是最常用的布局，在线性布局中所有的控件都沿水平或垂直方向顺序排列。该布局类型对应在 Palette 中有两项：

（1）LinearLayout（Horizontal）：属性 android:orientation="horizontal"，包含的控件水平排列，每列仅有一个控件，如图 3.9（a）所示。

（2）LinearLayout（Vertical）：属性 android:orientation="vertical"，包含的控件垂直排列，每行仅有一个控件，如图 3.9（b）所示。

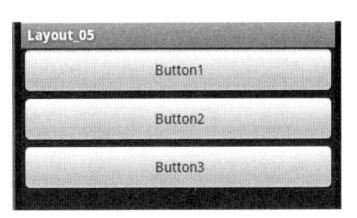

(a) 水平排列（Horizontal）　　　　(b) 垂直排列（Vertical）

图 3.9 两种形式的线性布局

下面是两种形式线性布局嵌套使用的代码：

```
<LinearLayout xmlns:android="http://schemas.android.com/apk/res/android"
    android:layout_width="match_parent"
    android:layout_height="match_parent"
    android:orientation="horizontal" >                    //外层是水平线性布局
    <Button
        android:layout_width="wrap_content"
        android:layout_height="wrap_content"
        android:text="BUTTON1"
        />
    <LinearLayout
        android:orientation="vertical"                    //内层是垂直线性布局
        android:layout_width="fill_parent"
        android:layout_height="fill_parent">
        <Button
            android:layout_width="wrap_content"
            android:layout_height="wrap_content"
            android:text="BUTTON2"   />
        <Button
            android:layout_width="wrap_content"
            android:layout_height="wrap_content"
            android:text="BUTTON3"   />
    </LinearLayout>
</LinearLayout>
```

显示的效果如图 3.10 所示。

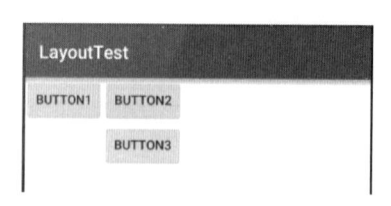

图 3.10 两个线性布局嵌套

3. 相对布局

相对布局是一种非常灵活的布局方式，它通过指定当前控件与特定 id 值控件的相对位置关系，来确定界面中所有控件的方位。这一点与 Android 系统默认的约束布局是相似的，都能够最大限度地保证在各种屏幕类型的手机上正确显示界面，只是相对布局通常需要手工编写代码来定义控件间的相对关系。例如，一段典型的使用相对布局定义界面的代码如下：

```
<RelativeLayout xmlns:android="http://schemas.android.com/apk/res/android"    //定义一个相对布局
    android:layout_width="fill_parent"                                          //布局宽度（填满页宽）
    android:layout_height="fill_parent">                                        //布局高度（填满页高）
    <Button
        ...
        android:text="Button1"
        android:id="@+id/btn1"
        />
    <Button
        android:layout_width="fill_parent"                                      //按钮宽度（填满布局）
        android:layout_height="wrap_content"
```

```xml
            android:text="Button2"
            android:id="@+id/btn2"
            android:layout_below="@id/btn1"                    // (a)
            />
    <Button
            ...
            android:text="Button3"
            android:id="@+id/btn3"
            android:layout_below="@id/btn2"
            android:layout_alignRight="@id/btn2"                // (b)
            />
    <Button
            ...
            android:text="Button4"
            android:id="@+id/btn4"
            android:layout_below="@id/btn3"
            android:layout_alignParentRight="true"              // (c)
            />
    <Button
            ...
            android:text="Button5"
            android:id="@+id/btn5"
            android:layout_below="@id/btn4"
            android:layout_centerHorizontal="true"              // (d)
            />
</RelativeLayout>
```

其中：

(a) **android:layout_below="@id/btn1"**：这里的属性 android:layout_below 设定了 Button2 在 Button1 的下方，而 Button1 的位置是默认的（在界面左上角）。常用的定义控件之间位置关系的属性如下：

android:layout_above——将控件放在指定 id 控件的上方。
android:layout_below——将控件放在指定 id 控件的下方。
android:layout_toLeftOf——将控件放在指定 id 控件的左边。
android:layout_toRightOf——将控件放在指定 id 控件的右边。

(b) **android:layout_alignRight="@id/btn2"**：属性 android:layout_alignRight 设定 Button3 对齐 Button2 右边缘。常用的定义控件对齐方式的属性如下：

android:layout_alignBaseline——将该控件与指定 id 控件的中心线对齐。
android:layout_alignTop——将该控件与指定 id 控件的顶部对齐。
android:layout_alignBottom——将该控件与指定 id 控件的底部对齐。
android:layout_alignLeft——将该控件与指定 id 控件的左边缘对齐。
android:layout_alignRight——将该控件与指定 id 控件的右边缘对齐。

(c) **android:layout_alignParentRight="true"**：属性 android:layout_alignParentRight 设定 Button4 对齐父组件（相对布局）的右边缘。常用的定义当前控件与父组件对齐方式的属性（属性值="true"生效）如下：

android:layout_alignParentTop——该控件与父组件的顶部对齐。
android:layout_alignParentBottom——该控件与父组件的底部对齐。
android:layout_alignParentLeft——该控件与父组件的左边缘对齐。
android:layout_alignParentRight——该控件与父组件的右边缘对齐。

(d) **android:layout_centerHorizontal="true"**：属性 android:layout_centerHorizontal 设定 Button5

在水平方向的中央位置。常用的设定控件放置位置的属性（属性值="true"生效）如下：
android:layout_centerHorizontal——将该控件放置在水平方向中央的位置。
android:layout_centerVertical——将该控件放置在垂直方向中央的位置。
android:layout_centerInParent——将该控件放置在父组件的水平中央及垂直中央的位置。
相对布局的显示效果如图 3.11 所示。

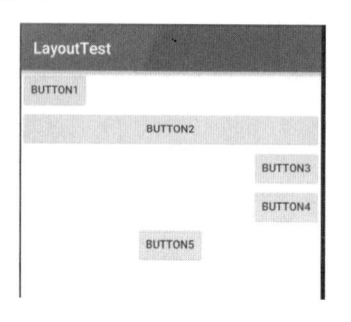

图 3.11　相对布局的显示效果

4. 表格布局与网格布局

（1）表格布局。表格布局在 Palette 中的对应项为 TableLayout，它将屏幕划分为表格，通过指定行和列将控件添加到表格中。表格布局用<TableRow>…</TableRow>来表示一行，有多少行就添加多少个行标签对。有多少列就看最多的一个行中添加了多少个控件，直到把屏幕占满，超出屏幕的就不再显示。若直接在 TableLayout 中添加控件，默认会占据一整行。

表格布局的常用属性如下：

- **shrinkColumns**：当 TableRow 里面的控件填满布局时，指定列自动延伸以填充可用部分；而当 TableRow 里面的控件尚未填满布局时，该属性不起作用。
- **strechColumns**：指定列对空白部分进行填充。
- **collapseColumns**：隐藏指定的列。
- **layout_column**：设置组件显示指定列。
- **layout_span**：设置组件显示占用的列数。

> ◎◎ 注意：
> 列号都是从 0 开始，收缩、拉伸或隐藏可以指定多个列，列之间必须用逗号隔开，例如，android:stretchColumns = "1,2"。也可以用"*"代替所有列。

几种典型的表格布局界面的效果如图 3.12 所示。

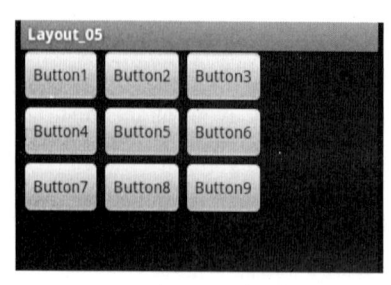

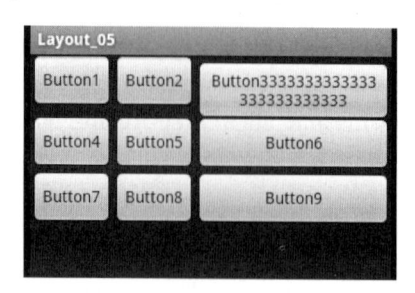

(a) 一个 3 行 3 列表格　　　　　　(b) 指定第 3 列填充可用部分

图 3.12　几种典型的表格布局界面的效果

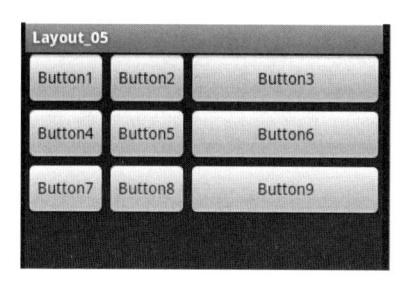

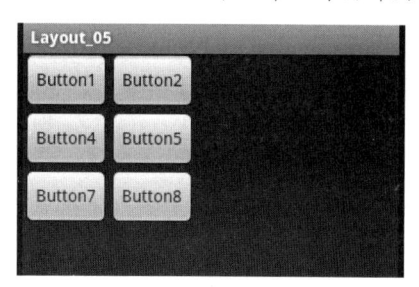

（c）指定第3列填充空白部分　　　　（d）指定第3列隐藏

图3.12　几种典型的表格布局界面的效果（续）

其中：

图3.12（b）中，android:shrinkColumns="2"。

图3.12（c）中，android:strechColumns="2"。

图3.12（d）中，android:collapseColumns="2"。

下面举个例子，有如下表格布局的代码：

```xml
<TableLayout xmlns:android="http://schemas.android.com/apk/res/android"
    android:layout_width="fill_parent"
    android:layout_height="fill_parent">
    <TableRow>
        <Button android:text="Button1"
            android:layout_span="3"/>         //（a）占用3列
        <Button android:text="Button2"/>
        <Button android:text="Button3"/>
    </TableRow>
    <TableRow>
        <Button android:text="Button4"
            android:layout_column="2"/>       //（a）显示在第3列
        <Button android:text="Button5"
            android:layout_column="0"/>       //（b）显示在第1列
        <Button android:text="Button6"/>
    </TableRow>
    <TableRow>
        <Button android:text="Button7"/>
        <Button android:text="Button8"/>
        <Button android:text="Button9"/>
    </TableRow>
</TableLayout>
```

其中：

（a）**android:layout_span="3"和 android:layout_column="2"**：Button1被设置成占用了3列，Button4被设置显示在第3列，实际运行的显示效果如图3.13所示。

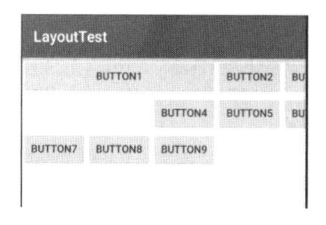

图3.13　表格布局的显示效果

（b）android:layout_column="0"：细心的读者会发现，上述代码中指定了 Button5 显示在第 1 列，而运行的结果却没有按照设定值显示。这是因为在表格布局中，TableRow 一行里的控件都会自动放在前一控件的右侧，依次排列，当某个控件已经确定了所在列，其后面的控件就无法再自由地设定位置。

（2）网格布局。与表格布局类似，网格布局是将屏幕划分为许多网格，界面控件可随意摆放在这些网格中，一个控件可占用多个网格（可以跨多行也可以跨多列），相比表格布局将控件限定在一个表格行上（不能跨越多行）的方式要灵活得多。网格布局的边界对用户是不可见的，在界面设计器中可以看到虚线网格，但在 Android 仿真器的运行结果中是看不到的。网格布局的块可以根据界面控件动态划分。

网格布局的常用属性如下：
- **android:rowCount**：设置网格布局行数。
- **android:columnCount**：设置网格布局列数。
- **android:layout_row**：设置组件位于第几行。
- **android:layout_column**：设置组件位于第几列。
- **android:layout_rowSpan**：设置组件纵向纵跨几行。
- **android:layout_columnSpan**：设置组件横向横跨几列。

与其他布局不同，网格布局可以不为控件设置 layout_width 和 layout_height 属性，因为控件的宽和高已经由其所在的行和列决定了。当然，也可以设成 wrap_content。

在实际的 App 开发中，网格布局的效果往往用 Android 的某些高级视图控件（如 GridView、RecyclerView 等）来实现。单纯的网格布局用得并不多，这里就不举例了。

5. 其他布局

（1）绝对布局。绝对布局（AbsoluteLayout）通过指定每一个界面控件的绝对坐标来定位设计界面。在这种布局方式下，屏幕上每一个控件都必须指定坐标（X，Y），坐标原点（0，0）在屏幕的左上角。绝对布局是一种**不推荐**使用的布局类型，因为用绝对位置确定的控件，虽然在特定目标手机上非常完美，但 Android 系统无法根据屏幕尺寸的改变对界面上控件的位置做出调整，导致在其他不同类型的手机上，界面会变得混乱，故实际开发中基本不用。

（2）板块布局。板块布局（FrameLayout）是最简单的界面布局，因为板块布局在新加入控件的时候都会将其放置在屏幕的左上角，即使在此布局中添加多个控件，后一个也总会将前一个覆盖，除非这后一个控件是透明的。板块布局多用在基于板块（Fragment）的 Android 程序界面设计中，第 4 章还会进一步涉及。

3.1.3 界面布局的应用

【例 3.1】 综合运用前述界面布局的知识，设计一个用户登录界面。

（1）创建 Android Studio 工程，工程名为 LoginPage，类型选择为"Empty Activity"（空活动页）。

（2）直接在 activity_main.xml 中设计界面，设计模式下的视图效果如图 3.14 所示。

本例的程序界面使用到了前面介绍的相对布局、网格布局和线性布局。activity_main.xml 文件中的代码如下：

图 3.14 登录界面在设计模式下的视图效果

```xml
<?xml version="1.0" encoding="utf-8"?>
<RelativeLayout xmlns:android="http://schemas.android.com/apk/res/android"    //最外层是相对布局
    android:layout_width="match_parent" android:layout_height="match_parent">

    <GridLayout                                                                //（a）里面是网格布局
        android:layout_width="match_parent"
        android:layout_height="match_parent"
        android:layout_alignParentTop="true"
        android:layout_centerHorizontal="true"
        android:useDefaultMargins="true"
        android:rowCount="5"
        android:columnCount="2">

        <TimePicker                                                            //（b）
            android:layout_width="wrap_content"
            android:layout_height="99dp"
            android:id="@+id/myTimePicker"
            android:layout_columnSpan="2"/>

        <TextView                                                              //（c）
            android:layout_width="wrap_content"
            android:layout_height="wrap_content"
            android:text="用户名："
            android:id="@+id/myLabelName"
            android:textStyle="bold"
            android:textSize="@dimen/abc_text_size_display_1_material"
            android:layout_row="1"
            android:layout_column="0" />

        <EditText                                                              //（d）
            android:layout_width="wrap_content"
            android:layout_height="wrap_content"
            android:inputType="textPersonName"
            android:ems="10"
```

```xml
        android:id="@+id/myTextName"
        android:textSize="@dimen/abc_text_size_display_1_material"
        android:layout_row="1"
        android:layout_column="1" />

    <TextView                                                              // (e)
        android:layout_width="wrap_content"
        android:layout_height="wrap_content"
        android:text="密    码："
        android:id="@+id/myLabelPwd"
        android:textStyle="bold"
        android:textSize="@dimen/abc_text_size_display_1_material"
        android:layout_row="2"
        android:layout_column="0" />

    <EditText                                                              // (e)
        android:layout_width="wrap_content"
        android:layout_height="wrap_content"
        android:inputType="textPassword"
        android:ems="10"
        android:id="@+id/myTextPwd"
        android:textSize="@dimen/abc_text_size_display_1_material"
        android:layout_row="2"
        android:layout_column="1" />

    <LinearLayout                                                          // (f) 嵌套一个线性布局
        android:orientation="horizontal"
        android:layout_width="match_parent"
        android:layout_height="wrap_content"
        android:layout_columnSpan="2"
        android:weightSum="1"                                              // (f)
        android:gravity="center">

        <Button
            android:layout_width="wrap_content"
            android:layout_height="wrap_content"
            android:text="登录"
            android:id="@+id/myButtonOk"
            android:textSize="@dimen/abc_text_size_display_1_material"
            android:textStyle="bold"
            android:layout_weight="0.5"/>                                  // (f)

        <Button
            android:layout_width="wrap_content"
            android:layout_height="wrap_content"
            android:text="重置"
            android:id="@+id/myButtonCancel"
            android:textSize="@dimen/abc_text_size_display_1_material"
            android:textStyle="bold"
            android:layout_weight="0.5"/>                                  // (f)
    </LinearLayout>
```

```
        <DatePicker                                              // (g)
            android:layout_width="wrap_content"
            android:layout_height="wrap_content"
            android:id="@+id/myDatePicker"
            android:visibility="invisible"                       // (g)
            android:layout_columnSpan="2"/>

    </GridLayout>
</RelativeLayout>
```

其中：

（a）**<GridLayout... android:rowCount="5" android:columnCount="2">**：因为登录界面需要多行多列，而每一行包含的列不相同，大小不一致，所以创建一个网格布局。界面内容包含 4 行 2 列，但为了界面下面留空行，所以这里设置 5 行 2 列。

（b）**<TimePicker...android:layout_columnSpan="2"/>**：网格第 1 行放时间控件显示时间，占满一行，所以需要合并 2 列。

（c）**<TextView...android:text="用户名："... android:layout_row="1" android:layout_column="0" />**：网格第 1 行第 1 列显示"用户名："，放 TextView 控件。

（d）**<EditText... android:layout_row="1" android:layout_column="1" />**：网格第 2 行第 2 列需要用户输入用户名，放 EditText 控件。

（e）**<TextView...android:text="密　码："...android:layout_row="2" android:layout_column="0" />和<EditText...android:inputType="textPassword"...android:layout_row="2"android:layout_column="1" />**：网格第 3 行与第 2 行类似，但用户输入密码时不能显示密码字符本身，所以需要设置 EditText 控件的属性 android:inputType="textPassword"。

（f）**<LinearLayout... android:weightSum="1" android:gravity="center"> <Button...android:text="　登　录　"...android:layout_weight="0.5"/> <Button...android:text="重置"...android:layout_weight="0.5"/>**：网格第 4 行需要放置 2 个命令按钮，希望分别控制占用宽度。所以需要在本行嵌套加入一个水平的线性布局。在线性布局<LinearLayout...>中设置属性 android:weightSum="1"，同时在"登录"和"重置"命令按钮<Button...>中设置属性 android:layout_weight="0.5"，通过调整 layout_weight 属性值可以改变命令按钮的宽度。

（g）**<DatePicker...android:visibility="invisible" android:layout_columnSpan="2"/>**：网格第 4 行后面放置日期控件，在后续章节使用本实例时要用到获取日期，因为不需要显示出来，所以这里设置属性 android:visibility="invisible"。

登录界面的运行效果如图 3.15 所示。

图 3.15　登录界面的运行效果

3.2　Activity 活动页

手机 App 是由很多页面构成的，在 Android 程序中，页面以一个个 Activity（活动）的形式存在，故又称作"活动页"。之前我们设计的程序都只有一个页面（MainActivity），稍复杂点的程序往往要涉及多个页面。本章接下来将学习多页面 Android 程序的运作原理及开发。

3.2.1 Activity 概述

Activity 是 Android 组件中最基本也是最为常用的四大组件（Activity、Service、ContentProvider、BroadcastReceiver）之一。Activity 中的所有操作都与用户密切相关，是一个负责与用户交互的组件，可以通过 setContentView(View)设定页面要显示的视图。

一个多页面的 Android 应用程序包含多个 Activity，所有的 Activity 都必须在工程的 AndroidManifest.xml 文件中注册，但包含：

   `<action android:name="android.intent.action.MAIN" />`

项的 Activity 为启动页面。

> **注意：**
> 只有在 AndroidManifest.xml 中注册了全部（不论是否启动页面的）Activity，程序才能够正常启动运行，否则界面会出现闪退。

一个 Activity 可以通过调用 startActivity()方法启动其他 Activity 进行相关的操作，也可以用 startActivityForResult()方法请求其他 Activity 返回处理结果数据。当启动其他的 Activity 时，当前 Activity 将会停止，新的 Activity 被压入栈中，同时获取焦点，这时用户就可在新的 Activity 上进行操作了。因为栈遵循先进后出的原则，当用户按返回键时，当前的这个 Activity 被销毁，前一个 Activity 重新恢复。

3.2.2 页面间的数据交互

在手机 App 页面跳转、切换或返回时，常常需要将当前页面上操作产生的数据信息传递给下一个页面，在 Android 中，这种数据交互是通过 Intent 消息传递机制实现的。

1. Intent 原理

Intent 是 Android 中通用的数据传递技术，不仅仅用于 Activity，也同样适用于其他类型的组件（如 Service 和 BroadcastReceiver 等，本书后续章节将会介绍）之间的数据传递。Intent 负责对应用中一次操作的动作、动作涉及的数据、附加数据等进行描述。Android 则根据此 Intent 的描述，找到对应的组件，然后将 Intent 传递给该组件，完成对组件的调用。通过 Intent，应用程序可以向 Android 表达某种请求（意愿），Android 则会根据请求的内容选择适当的组件来完成请求。

Intent 一旦被发出，Android 就会准确找到相匹配的一个或多个组件（可以是 Activity、Service 或者 BroadcastReceiver 等）做出响应。但由于调用方法接口的不同，不同类型的 Intent 消息不会出现重叠，由 startActivity()方法传递的消息只会发给 Activity，而决不会发送给 Service 或 BroadcastReceiver。

（1）Intent 的构成。要在不同的 Activity 之间传递数据，就要在 Intent 中包含相应的内容，通常 Intent 中都包含以下基本属性：

- **Action**：用来指明要实施的具体动作是什么。
- **Data**：即执行动作要操作的数据，Android 采用指向数据的一个 URI 来表示。
- **Category**：表示一个类别字符串，包含了有关处理该 Intent 组件的类别信息。
- **Type**：用于显式指定 Intent 的数据类型 MIME。
- **Component**：指定 Intent 的目标组件的类名称。通常 Android 会根据 Intent 中包含的其他属性（如 Action、Data/Type、Category 等）的信息进行查找，最终找到一个与之匹配的目标组件。
- **Extras**：是其他所有附加信息的集合，为组件提供扩展信息。

以上各属性有机结合在一起形成了一种语言，使 Android 系统能够理解诸如"查看某联系人的详细信息"之类的短语。

（2）Intent 的解析。解析 Intent 主要是通过查找已注册在 AndroidManifest.xml 中的所有<intent-filter>及其中定义的 Intent，最终找到匹配的 Intent。Android 是通过 Intent 的 Action、Type 和 Category 这三个属性来进行判断的，方法如下：

① 如果 Intent 指定了 Action，则目标组件的 IntentFilter 的 Action 列表中必须包含这个 Action，否则不能匹配。

② 如果 Intent 没有提供 Type，系统将从 Data 中得到数据类型。和 Action 一样，目标组件的数据类型列表中也必须包含 Intent 的数据类型，否则不能匹配。

③ 如果 Intent 指定了一个或多个 Category，这些类别必须全部出现在组件的类别列表中。

（3）Intent 的两种形式。

① **显式 Intent**。

显式 Intent 是通过调用 setComponent(ComponentName) 或者 setClass(Context,Class) 指定了 Component 属性的 Intent。通过明确指定具体的组件类，来通知应用启动对应的组件。由于显式 Intent 指定了具体的组件对象，因此不需要设置其他意图过滤对象。

② **隐式 Intent**。

隐式 Intent 是没有指定 Component 属性的 Intent，即没有明确指定组件名，系统根据隐式意图中设置的动作（Action）、类别（Category）、数据（Data）等来匹配最合适的组件，故这种 Intent 需要包含足够的信息，让系统能够根据这些信息在所有可用组件中找出满足条件的组件。当一个 App 要激活另一个 App 中的 Activity 时，只能使用隐式 Intent，根据 Activity 配置的 Intent 过滤器创建一个 Intent，让其中的各项参数的值都跟过滤器匹配，这样就可以激活其他 App 中的 Activity。所以，隐式 Intent 通常都是跨 App 使用的。

2. 数据传递

（1）Activity 间的数据传递。

一个 Android 应用程序可能包含多个 Activity，要从一个 Activity 切换到另一个 Activity，必须通过 Intent，因为 Intent 存储着切换时所需的重要信息，如图 3.16 所示为 Activity01 通过 Intent 切换到 Activity02 的过程。

如果想要在两个 Activity 切换时携带额外数据，可以将数据存储在 Bundle 内。Bundle 依附在 Intent 上，是一个专门用来存储附加数据的对象，如图 3.17 所示。

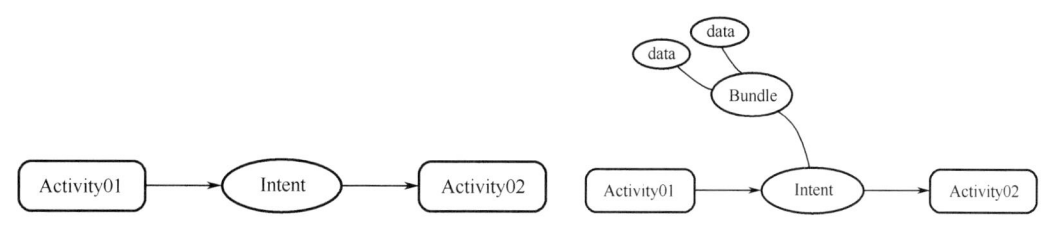

图 3.16　Activity01 通过 Intent 切换到 Activity02　　　图 3.17　Bundle 的作用

具体包括以下几种情形：

① 直接向 Intent 对象中传入键值对，相当于 Intent 对象具有 Map 键值对功能。

② 定义一个 Bundle 对象，在该对象中加入键值对，然后将该对象加入 Intent 中。

③ 向 Intent 中添加 ArrayList 集合对象。

④ Intent 传递 Object 对象，被传递对象实现 Parcelable（或者 Serializable）接口。

(2) Activity 退出时的返回结果：

① 通过 startActivityForResult 方式启动一个 Activity。

② 新 Activity 中设定 setResult 方法，通过该方法可以传递 responseCode（响应码）和 Intent 对象。

③ 在 MainActivity 中重写 onActivityResult()方法，新的 Activity 一旦退出，就会执行该方法从而得到返回结果。

3. 应用举例

【例 3.2】 在【例 3.1】用户登录界面的基础上增加一个欢迎页面，应用 Intent 将登录界面上输入的用户名和密码传递到欢迎页面，由欢迎页面验证密码是否正确并做出响应，返回登录页后提示响应结果信息。

程序启动后，首先出现登录页面，输入用户名和密码，如图 3.18（a）所示。点击"登录"按钮，跳转到欢迎页面，如果密码正确，显示欢迎信息及登录时间，如图 3.18（b）所示。

图 3.18　密码正确，显示欢迎信息及登录时间

如果密码不正确，提示"密码错！"，如图 3.19（a）所示。点击"返回"按钮回到登录页面，密码输入框自动清空，显示此前输入的密码并提示重新输入，如图 3.19（b）所示。

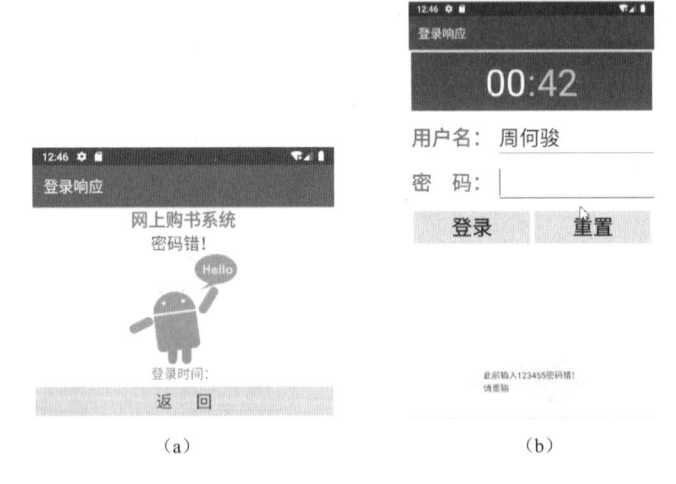

图 3.19　密码错误，显示此前输入的密码并提示重新输入

（1）增加欢迎页面。打开工程 LoginPage，添加设计一个欢迎页面 welcome.xml，如图 3.20 所示。页面很简单，用 TextView 显示欢迎信息和登录时间，用 ImageView 显示一个"Android 机器人招手 Hello"的图片，图片资源 androidwelcomer.gif 放在工程的 drawable 目录下，代码略。

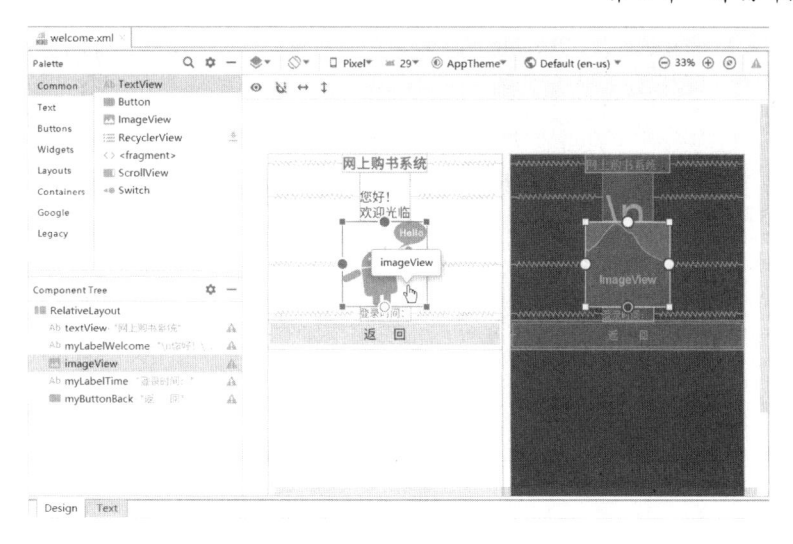

图 3.20 添加设计一个欢迎页面 welcome.xml

（2）数据传递功能实现。

实现思路：

本例数据的传递在 MainActivity（登录页）与 WelcomeActivity（欢迎页）之间进行，大致的流程如下：

① 运行 App，初始启动 MainActivity。

② 在 MainActivity 的页面上点击"登录"按钮，执行 onLoginClick()事件方法代码，将用户输入的用户名和密码信息数据绑定在 Bundle 上，通过 Intent 传递给 WelcomeActivity。

③ WelcomeActivity 在 showWelcome()方法中解析 Intent 消息，得到 Bundle 中的用户名和密码信息并进行验证，验证的结果通过 resultCode 返回给 MainActivity。

④ 当用户点击"返回"按钮重新回到登录页面时，MainActivity 中的 onActivityResult()方法会被自动执行，在其中对 WelcomeActivity 的返回码 resultCode 进行处理，界面上给出消息通知提示。

MainActivity.java 实现登录页面功能，代码如下：

```java
public class MainActivity extends AppCompatActivity {
    private EditText myName;              //"用户名"输入框
    private EditText myPwd;               //"密码"输入框
    private Button myOk;                  //"登录"按钮
    private TimePicker myTime;            //时间控件
    private DatePicker myDate;            //日期控件

    @Override
    protected void onCreate(Bundle savedInstanceState) {
        super.onCreate(savedInstanceState);
        setContentView(R.layout.activity_main);
        findViews();
    }

    private void findViews() {
        myName = findViewById(R.id.myTextName);
        myPwd = findViewById(R.id.myTextPwd);
        myOk = findViewById(R.id.myButtonOk);
```

```
                myTime = findViewById(R.id.myTimePicker);
                myDate = findViewById(R.id.myDatePicker);
                myTime.setIs24HourView(true);                           //时间采用 24 小时制
            }

            public void onLoginClick(View view) {
                String name = myName.getText().toString();              //获得用户名
                String pass = myPwd.getText().toString();               //获得密码
                int year, month, day, hour, minute;
                year = myDate.getYear();                                //获得日期中的年
                month = myDate.getMonth() + 1;                          //获得日期中的月
                day = myDate.getDayOfMonth();                           //获得日期中的日
                hour = myTime.getCurrentHour();                         //获得时间中的时
                minute = myTime.getCurrentMinute();                     //获得时间中的分
                //（a）定义 Intent 数据传递机制
                Intent intent = new Intent(this, WelcomeActivity.class);
                Bundle bundle = new Bundle();                           //定义 Bundle 对象 bundle
                bundle.putString("name", name);                         //字符串 name 内容绑定到 bundle
                bundle.putString("pass", pass);
                bundle.putInt("year", year);
                bundle.putInt("month", month);
                bundle.putInt("day", day);
                bundle.putInt("hour", hour);
                bundle.putInt("minute", minute);
                intent.putExtras(bundle);                               //bundle 放入 Intent 传输机制
                //（b）启动欢迎页面的 Activity
                //startActivity(intent);
                startActivityForResult(intent, 200);
            }

            protected void onActivityResult(int requestCode, int resultCode, Intent data) {    //（c）
                if (resultCode == 101) Toast.makeText(this, "重新登录", Toast.LENGTH_LONG).show();
                else if (resultCode == 404) {
                    Toast.makeText(this, "此前输入" + data.getStringExtra("pass") + "密码错！\n 请重输",
Toast.LENGTH_LONG).show();
                    myPwd.setText("");                                  //清空密码框
                }
            }
        }
```

其中：

（a）一个完整的 Intent 机制程序代码段必须包括以下流程：

① 获得页面交互信息（数据）。

② **Intent intent = new Intent(this, WelcomeActivity.class);**：定义数据传递机制。这里定义 Intent 对象 intent，用于传输到 WelcomeActivity.class 指定的页面。

③ **Bundle bundle = new Bundle();**：定义数据的载体。这里定义 Bundle 对象 bundle，用于绑定要传递的数据。

④ **bundle.putString("name", name);...bundle.putInt("minute", minute);**：将数据绑定在 Bundle 对象上，使用 putXXX(参数1,参数2)方法，这里 XXX 表示要绑定的数据类型，第 1 个参数为数据名称，第 2 个参数为数据内容（变量名）。

⑤ **intent.putExtras(bundle);**：将 Bundle 放入 Intent 的传输机制，然后启动页面间的数据传输。
（b）启动新页面有两种方式：
startActivity(Intent)：不要求新 Activity 退出时返回值。
startActivityForResult(Intent, 请求码)：要求新 Activity 在退出时向本页面返回处理的结果信息。本例由于在返回时需要提示用户密码正误，故使用的是第二种方式。
（c）**protected void onActivityResult(int requestCode, int resultCode, Intent data)**：当重新调用本页面时，该方法自动执行，本页面 Activity 从其传入参数中获知新页面的处理结果。其中：
- **requestCode**：带回请求码。
- **resultCode**：带回响应码。
- **data**：带回数据。

接下来，创建 WelcomeActivity.java 来实现欢迎页面的功能，代码如下：

```java
public class WelcomeActivity extends AppCompatActivity {
    private TextView myWelcome;                        //文本视图（显示欢迎信息）
    private TextView myTime;                           //文本视图（显示登录时间）
    private String name, pass;                         //存储登录页传来的用户名和密码
    private int resultCode;                            //响应码：登录成功 101；错误 404

    @Override
    protected void onCreate(Bundle savedInstanceState) {
        super.onCreate(savedInstanceState);
        setContentView(R.layout.welcome);              //设置显示页面为 welcome.xml
        findViews();
        showWelcome();
    }

    private void findViews() {
        myWelcome = findViewById(R.id.myLabelWelcome);
        myTime = findViewById(R.id.myLabelTime);
    }

    private void showWelcome() {
        Bundle bundle = getIntent().getExtras();       //定义 Bundle 对象 bundle
        name = bundle.getString("name");               //从 bundle 中得到 name 内容（输入的用户名）
        pass = bundle.getString("pass");               //从 bundle 中得到 pass 内容（输入的密码）
        if (pass.equals("123456")) {                   //验证密码
            myWelcome.setText("\n" + name + " 您好！\n 欢迎光临");
            int year, month, day, hour, minute;        //显示欢迎信息
            year = bundle.getInt("year");
            month = bundle.getInt("month");
            day = bundle.getInt("day");
            hour = bundle.getInt("hour");
            minute = bundle.getInt("minute");
            myTime.setText("登录时间：" + year + "-" + month + "-" + day + " " + hour + ":" + minute);
            resultCode = 101;                          //显示登录时间，返回成功码 101
        } else {
            myWelcome.setText("密码错！");
            resultCode = 404;                          //返回错误码 404
        }
```

```
        //这里定义的 Intent 机制是为了向登录页面返回响应码及处理的结果
        Intent data = new Intent(this, MainActivity.class);
        data.putExtras(bundle);
        setResult(resultCode, data);
    }

    public void onBackClick(View view) {
        finish();                              //结束当前 Activity，回到前一个 Activity
    }
}
```

（3）注册 Activity，设定启动顺序。在 AndroidManifest.xml 中，将新添加的欢迎页面的 Activity（WelcomeActivity）注册其中，系统已经默认设定了登录页面的 Activity（MainActivity）作为启动项。

```
<application
    android:allowBackup="true"
    android:icon="@mipmap/ic_launcher"
    android:label="@string/app_name"
    android:roundIcon="@mipmap/ic_launcher_round"
    android:supportsRtl="true"
    android:theme="@style/AppTheme">
    <activity android:name=".MainActivity">
        <intent-filter>
            <action android:name="android.intent.action.MAIN" />
            <category android:name="android.intent.category.LAUNCHER" />
        </intent-filter>
    </activity>
    <activity android:name=".WelcomeActivity"></activity>
</application>
```

其中，MainActivity 中 "<action android:name="android.intent.action.MAIN" />" 就表示 MainActivity 为启动页面。

3.2.3 页面生命周期

1. 生命周期的基本概念

从 MVC（Model-View-Controller）设计模式的角度来说，Android 应用程序的一页其实是由 Layout XML 文件（View）与 Activity（Controller）组成的，其中的 Activity 扮演着控制页面流程的角色，是一个页面最核心的部分。

Activity 控制的页面从产生到结束，会经历 7 个阶段，这 7 个阶段就是一个页面的生命周期。Android 为了方便开发者能够轻易地指定页面在每个阶段要执行什么程序，在 Activity 类中预定义了与生命周期各阶段相对应的 7 个方法，如表 3.1 所示。开发者通过重写这些方法，就可以在页面生命周期的任一个阶段加入自己想要执行的程序功能。

表 3.1 与生命周期各阶段对应的方法

方法	说明	对应阶段
onCreate()	当 Activity 第一次被创建时调用，用来加载必要的数据到内存中，完成初始化，方便后续使用	生成（加载）
onStart()	当 Activity 画面即将显示前会调用	启动

续表

方　法	说　明	对应阶段
onResume()	当每次 Activity 画面要显示时调用。如果画面显示时所要使用的特定功能很耗电，不想太早打开，可以放在这里打开	恢复
onPause()	当前的 Activity 画面无法完全显示时调用。应该在此阶段释放较耗电的资源	暂停（休眠）
onStop()	当前 Activity 画面被其他 Activity 画面完全取代时调用	停止
onRestart()	当 Activity 从 onStop()状态要恢复到 onStart()状态时调用	重启
onDestroy()	Activity 要结束之前调用。建议此阶段释放所有尚未释放的资源	销毁

说明：

（1）表 3.1 中的每一个方法都代表 Activity 生命周期的一个阶段。开发者可以自行创建一个类继承 Activity 类，并按照自己的需要重写对应的方法。Activity 执行到该阶段，就会自动调用被重写的方法。

（2）重写的方法内部都必须调用父类对应的方法，例如重写 onStart()时必须加上 super.onStart()。

（3）Activity 的执行程序可否被强制停止并删除：一般而言，Activity 页面从加载到显示阶段（onCreate()→onStart()→onResume()→页面显示）都是重要阶段，因为此时用户正等待画面打开，如果程序被删除，画面将无法显示，用户会不满意。而当 Activity 页面处于休眠至结束阶段（onPause()→onStop→onDestroy()→页面结束）时，不仅程序已停止，连画面也无显示，这时候可以强制删除 Activity 程序，对用户影响不大。

2. 生命周期的进程调度

现在的智能手机基本上都具备多任务能力，例如，使用手机听音乐的同时还可以浏览网页、聊微信，因而所耗费的系统内存往往很大。为了兼顾所有应用程序的正常运行与内存的有效利用，Android 有自己的一套程序管理模式，它实质上就是通过对各程序中 Activity 生命周期的管理来实现的，如图 3.21 所示。

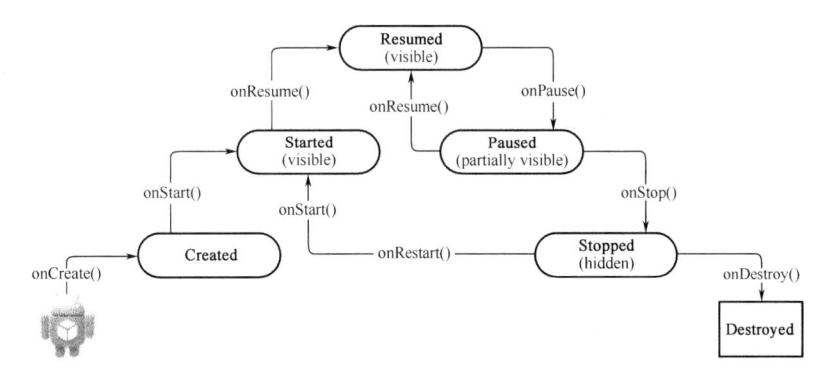

图 3.21　Activity 生命周期的管理

从图 3.21 可见，7 个方法虽然决定了一个 Activity 的完整生命周期，但是 Activity 却不一定会运行所有方法，而有些方法运行还不止一次。一个 Activity 究竟会经历哪些方法，是由 Android 系统按照用户操作与系统资源使用情况加以控制的。将 Activity 经历过的方法结合起来，其实代表的就是一个完整的程序流程，也可以称作进程（process）。进程由 Android 系统调度，只要内存不足，Android 系统可以随时按照重要程度决定要终止的进程。各类进程的重要程度从高到低的顺序排列如下：

- **前台进程**：用户正在使用的进程，而且该进程的画面正显示在手机屏上。这类进程几乎不会被终止，除非现有内存所剩无几且没有其他进程可以终止，前台进程才可能被迫终止。

当 Activity 的 onCreate()、onStart()、onResume()方法被调用时，该 Activity 就进入前台进程。

- **可视进程**：虽然不是前台进程，但用户仍然可以看到该进程所显示的画面。例如按电源键会开启键盘锁，此时仍可看到主画面的背景图，解开键盘锁后仍会回到该进程。进入这个进程时，Activity 的 onPause()方法会被调用。
- **服务进程**：就是 Android 的 Service，与前两类进程都属于 Activity 有所不同，该类进程在启动后会始终保持运行状态，以持续对用户提供服务。
- **后台进程**：这类进程的画面既没有显示，对用户也无直接的影响，进入这个进程，Activity 的 onStop()方法会被调用。
- **空进程**：当后台进程被终止，会将所占用的内存空间释放，该进程就会变成空进程，但其 Activity 仍然存在（只要 onDestroy()方法未被调用，Activity 就不会被删除）。Android 系统保留空进程的目的在于，在需要的时候可以快速将其恢复成前台进程，而无须再重新生成 Activity。

在 Android 系统中，往往同时运行着以上五类进程，在内存吃紧时，越不重要的进程，越容易被终止掉。

3. 生命周期观察示例

【例 3.3】 在【例 3.2】的基础上，程序代码中添加生命周期监测功能，再次运行程序，从 Android Studio 开发环境界面下方的日志输出中观察程序中各个页面的生命周期过程。

（1）登录页面添加生命周期监测代码。

打开工程 LoginPage，在 MainActivity.java 中添加代码（加粗处），内容如下：

```
...
import android.util.Log;                                    //日志输出命名空间

public class MainActivity extends AppCompatActivity {
    ...
    private final static String TAG = "登录页面（MainActivity）";    // (a) 标记页面的名称

    @Override
    protected void onCreate(Bundle savedInstanceState) {
        super.onCreate(savedInstanceState);
        setContentView(R.layout.login);
        Log.i(TAG, "onCreate");                              //在页面生成（加载）阶段显示
        findViews();
    }
    private void findViews() { ... }
    public void onLoginClick(View view) { ... }

    //Activity 生命周期监测
    @Override
    protected void onStart() {                               //启动阶段
        super.onStart();                                     // (b)
        Log.i(TAG, "onStart");
    }

    @Override
```

```
            protected void onResume() {                    //恢复阶段
                super.onResume();                          // (b)
                Log.i(TAG, "onResume");
            }

            @Override
            protected void onPause() {                     //暂停（休眠）阶段
                super.onPause();                           // (b)
                Log.i(TAG, "onPause");
            }

            @Override
            protected void onStop() {                      //停止阶段
                super.onStop();                            // (b)
                Log.i(TAG, "onStop");
            }

            @Override
            protected void onRestart() {                   //重启阶段
                super.onRestart();                         // (b)
                Log.i(TAG, "onRestart");
            }

            @Override
            protected void onDestroy() {                   //销毁阶段
                super.onDestroy();                         // (b)
                Log.i(TAG, "onDestroy");
            }
        }
```

其中：

（a）**private final static String TAG = "登录页面（MainActivity）";**：我们在 Activity 类中声明一个 final static 类型的私有字符串常量，其内容就是本页面 Activity 的名称，这样在后面运行的时候借助 android.util.Log 打印出的日志文本，就可以一目了然地看出来是哪一个页面执行了操作。

（b）**super.onStart();**、**super.onResume();**、**super.onPause();**、**super.onStop();**、**super.onRestart();** 和 **super. onDestroy();**：之前已经讲过，用户重写的生命周期方法中必须首先调用其父类的对应方法。

（2）欢迎页面添加生命周期监测代码。

在 WelcomeActivity.java 中添加代码（加粗处），内容如下：

```
...
import android.util.Log;                                  //日志输出命名空间

public class WelcomeActivity extends AppCompatActivity {
    ...
    private final static String TAG = "欢迎页面（WelcomeActivity）";

    @Override
    protected void onCreate(Bundle savedInstanceState) {
        super.onCreate(savedInstanceState);
        setContentView(R.layout.welcome);
        Log.i(TAG, "onCreate");                           //页面生成（加载）阶段
```

```java
        findViews();
        showWelcome();
    }

    private void findViews() {    ...    }
    private void showWelcome() {    ...    }
    public void onBackClick(View view) {
        finish();
    }

    //Activity 生命周期监测
    @Override
    protected void onStart() {                              //启动阶段
        super.onStart();
        Log.i(TAG, "onStart");
    }

    @Override
    protected void onResume() {                             //恢复阶段
        super.onResume();
        Log.i(TAG, "onResume");
    }

    @Override
    protected void onPause() {                              //暂停(休眠)阶段
        super.onPause();
        Log.i(TAG, "onPause");
    }

    @Override
    protected void onStop() {                               //停止阶段
        super.onStop();
        Log.i(TAG, "onStop");
    }

    @Override
    protected void onRestart() {                            //重启阶段
        super.onRestart();
        Log.i(TAG, "onRestart");
    }

    @Override
    protected void onDestroy() {                            //销毁阶段
        super.onDestroy();
        Log.i(TAG, "onDestroy");
    }
}
```

(3) 运行程序,观察生命周期。

① 启动程序,显示登录页面,从输出日志中看到登录页面经过了生成(onCreate)、启动(onStart)和恢复(onResume)这 3 个阶段,如图 3.22 所示。

```
2019-12-05 09:04:39.542 11701-11701/com.easybooks.loginpage I/登录页面(MainActivity): onCreate
2019-12-05 09:04:39.564 11701-11701/com.easybooks.loginpage I/登录页面(MainActivity): onStart
2019-12-05 09:04:39.566 11701-11701/com.easybooks.loginpage I/登录页面(MainActivity): onResume
```

图 3.22　登录页面初始启动所经历的 3 个阶段

② 当用户点击"登录"按钮，跳转到欢迎页面，观察页面生命周期，如图 3.23 所示。

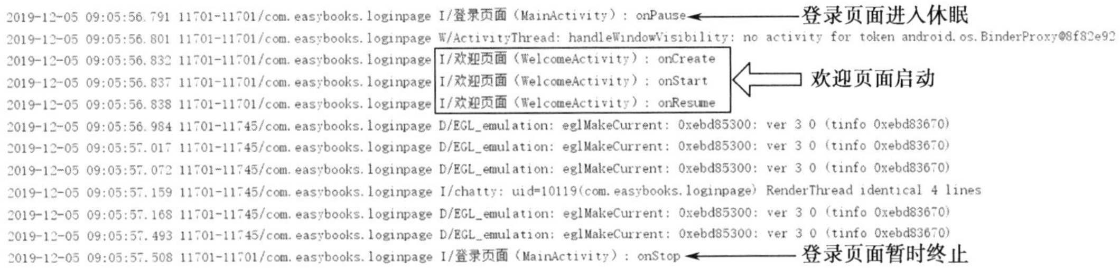

```
2019-12-05 09:05:56.791 11701-11701/com.easybooks.loginpage I/登录页面(MainActivity): onPause  ← 登录页面进入休眠
2019-12-05 09:05:56.801 11701-11701/com.easybooks.loginpage W/ActivityThread: handleWindowVisibility: no activity for token android.os.BinderProxy@8f82e92
2019-12-05 09:05:56.832 11701-11701/com.easybooks.loginpage I/欢迎页面(WelcomeActivity): onCreate
2019-12-05 09:05:56.837 11701-11701/com.easybooks.loginpage I/欢迎页面(WelcomeActivity): onStart   ← 欢迎页面启动
2019-12-05 09:05:56.838 11701-11701/com.easybooks.loginpage I/欢迎页面(WelcomeActivity): onResume
2019-12-05 09:05:56.984 11701-11745/com.easybooks.loginpage D/EGL_emulation: eglMakeCurrent: 0xebd85300: ver 3 0 (tinfo 0xebd83670)
2019-12-05 09:05:57.017 11701-11745/com.easybooks.loginpage D/EGL_emulation: eglMakeCurrent: 0xebd85300: ver 3 0 (tinfo 0xebd83670)
2019-12-05 09:05:57.072 11701-11745/com.easybooks.loginpage D/EGL_emulation: eglMakeCurrent: 0xebd85300: ver 3 0 (tinfo 0xebd83670)
2019-12-05 09:05:57.159 11701-11745/com.easybooks.loginpage I/chatty: uid=10119(com.easybooks.loginpage) RenderThread identical 4 lines
2019-12-05 09:05:57.168 11701-11745/com.easybooks.loginpage D/EGL_emulation: eglMakeCurrent: 0xebd85300: ver 3 0 (tinfo 0xebd83670)
2019-12-05 09:05:57.493 11701-11745/com.easybooks.loginpage D/EGL_emulation: eglMakeCurrent: 0xebd85300: ver 3 0 (tinfo 0xebd83670)
2019-12-05 09:05:57.508 11701-11701/com.easybooks.loginpage I/登录页面(MainActivity): onStop  ← 登录页面暂时终止
```

图 3.23　跳转到欢迎页面的生命周期

可以看到，登录页面先进入休眠（onPause），然后新的欢迎页面加载进来，也像登录页面启动时一样要经历 3 个阶段。接着登录页面进入终止（onStop）状态，进入终止状态的登录页，其正处于之前所述的"空进程"状态，不占内存，但它的 Activity 仍然存在，需要时仍可恢复。

③ 当用户点击"返回"按钮回到登录页面，登录页面又重新启动并恢复（onRestart→onStart→onResume），而欢迎页面则暂停、终止直至销毁（onPause→onStop→onDestroy），观察页面生命周期，如图 3.24 所示。

```
2019-12-05 09:07:09.039 11701-11701/com.easybooks.loginpage I/欢迎页面(WelcomeActivity): onPause  ← 欢迎页面暂停
2019-12-05 09:07:09.052 11701-11701/com.easybooks.loginpage I/登录页面(MainActivity): onRestart
2019-12-05 09:07:09.053 11701-11701/com.easybooks.loginpage I/登录页面(MainActivity): onStart    ← 登录页面恢复
2019-12-05 09:07:09.054 11701-11701/com.easybooks.loginpage I/登录页面(MainActivity): onResume
2019-12-05 09:07:09.137 11701-11745/com.easybooks.loginpage D/EGL_emulation: eglMakeCurrent: 0xebd85300: ver 3 0 (tinfo 0xebd83670)
2019-12-05 09:07:09.164 11701-11745/com.easybooks.loginpage D/EGL_emulation: eglMakeCurrent: 0xebd85300: ver 3 0 (tinfo 0xebd83670)
2019-12-05 09:07:09.282 11701-11745/com.easybooks.loginpage D/EGL_emulation: eglMakeCurrent: 0xebd85300: ver 3 0 (tinfo 0xebd83670)
2019-12-05 09:07:09.347 11701-11745/com.easybooks.loginpage I/chatty: uid=10119(com.easybooks.loginpage) RenderThread identical 2 lines
2019-12-05 09:07:09.647 11701-11745/com.easybooks.loginpage D/EGL_emulation: eglMakeCurrent: 0xebd85300: ver 3 0 (tinfo 0xebd83670)
2019-12-05 09:07:09.651 11701-11701/com.easybooks.loginpage I/欢迎页面(WelcomeActivity): onStop   ← 欢迎页面终止、销毁
2019-12-05 09:07:09.652 11701-11701/com.easybooks.loginpage I/欢迎页面(WelcomeActivity): onDestroy
```

图 3.24　登录页面重新启动并恢复，欢迎页面终止直至销毁

第 4 章　移动 App 高级界面开发技术

4.1　App 通用界面元素

在移动互联网和 5G 普及的今天，手机上的各种 App 界面丰富多彩，为人们的生活带来了极大的便利。学习 Android 开发的读者肯定想知道这些美轮美奂的 App 界面是如何制作出来的，本章就将在前面介绍 Android 基本知识的基础上，进一步讲解实际商用的 App 界面开发。

1. 主界面的构成

虽然各款手机应用的界面内容丰富，令人眼花缭乱，但若仔细观察你会发现，所有的界面几乎都是由一些类同的标准界面元素构成的。我们不妨来看一看京东、淘宝、口碑这几个著名互联网公司的 App 主界面，它们都包含了轮播条、频道栏、标签栏这几种通用的界面元素，如图 4.1 所示。

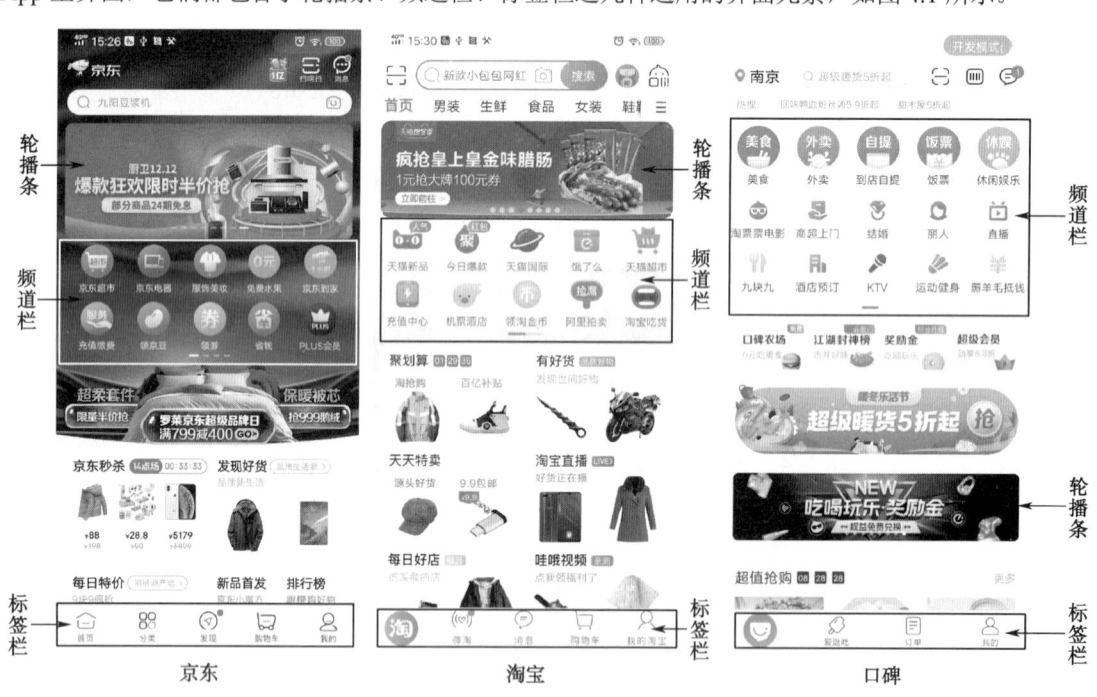

图 4.1　App 的通用界面元素示例

- **轮播条**：显示广告横幅，多幅广告图片会轮番循环滚动显示。
- **频道栏**：由一组带图片的文字栏目整齐排列所构成的网格状区域，作为系统各大主功能模块的入口。
- **标签栏**：一般位于界面底部，有多个带图标的选项按钮，点击可变色并切换至对应页面。

2. 内容的呈现形式

互联网电商的 App 都是用来向用户展示和推销商品的，对商品内容的呈现一般有列表、网格、类别标签这 3 种基本的形式。

（1）列表：以竖排顺序列表项的方式罗列出商品，其中的每个列表项又可包含丰富的内容，通常左边是商品图片，右边是商品的名称和价格等信息，如图 4.2（a）所示。

（2）网格：以网格状布局展示商品，每一件商品占据界面上的一块方形区域，其主体显示商品的大幅图片，图片底下再配上名称、价格等文字说明，如图 4.2（b）所示。

（3）类别标签：以标签选项页的形式切换显示不同类别的商品，如图 4.2（c）所示。这种形式可与前两种结合起来使用，其中每个选项标签下的内容又可以是一个列表或网格的形式，或者还可以是其他形式布局的页面。

（a）列表　　　　　　　　（b）网格　　　　　　　　（c）类别标签

图 4.2　内容的 3 种基本呈现形式

3. 易果鲜超市 App

本章以开发一个水果销售类 App——"易果鲜超市"的界面为例，来系统地演示 Android 商用 App 的诸多高级界面开发技术。这个案例 App 的主界面囊括了当前 App 开发所涉及的几乎所有常用界面元素：页面顶部是一个播放水果广告横幅的轮播条；下面两排水果图标排列成的区域是频道栏；再下面以列表形式展示了几种热销水果的条目；底部标签栏可切换至不同的页面。整个 App 主界面的整体效果如图 4.3 所示，下面我们先分别介绍各个界面基本元素的开发，最后再将它们整合在一起构成一个完整的 App。

图4.3 "易果鲜超市"主界面

4.2 界面元素开发

4.2.1 标签栏

【例4.1】 开发"易果鲜超市"App底部的标签栏，包含"首页""分类""推荐""购物车""我的"这5个标签按钮，如图4.4所示。当用户点击切换至不同标签页时，当前页面对应标签按钮的文字和图标均被着色突出显示，同时背景也上色。

图4.4 标签栏运行效果

创建Android Studio工程MyTabHost。下面先结合部分代码来介绍本例中用到的几项核心技术。

1. 核心技术

（1）状态列表图形。

要实现标签按钮的图标和背景随用户的点选状态而改变，我们很自然地想到，可以采用第2章讲的状态列表图形来实现。预先搜集（或制作）每个标签按钮在选中/未选状态下两个版本的图标和背景

（通常两个版本图片完全一样，只是选中版本的图片带色彩，而未选中版是黑白的），在 drawable 目录下创建用于控制图形状态的 XML 文件。例如，控制"首页"标签的文件 first_tab_selector.xml 内容为：

```xml
<selector xmlns:android="http://schemas.android.com/apk/res/android">
    <item android:state_selected="true" android:drawable="@drawable/first_tab_pressed" />
                                                //标签被选中时图标为 first_tab_pressed
    <item android:drawable="@drawable/first_tab_normal" />
                                                //标签未选中时图标为 first_tab_normal
</selector>
```

其他标签按钮的图标和背景切换机制与"首页"标签完全一样，在此不再重复罗列控制文件内容。

(2) 自定义控件风格。

本例标签栏上的 5 个标签按钮风格是一致的，选中/未选状态下文字都具有相同的字体、字号和颜色，倘若对每一个标签文本分别逐一设置这些属性，会很烦琐且造成代码冗余。为此，我们可以将这些公共属性抽取出来，定义成一个风格，再统一应用到每个控件元素上。Android 控件的自定义风格写在工程 values 目录下的 styles.xml 文件中，以"<style>…</style>"标注。这里我们定义一个名为"TabLabel"的风格元素如下：

```xml
<resources>
    ...
    <style name="TabLabel">
        <item name="android:layout_width">match_parent</item>      //宽匹配标签按钮尺寸
        <item name="android:layout_height">match_parent</item>     //高匹配标签按钮尺寸
        <item name="android:padding">5dp</item>                    //定义边距
        <item name="android:layout_gravity">center</item>          //布局居中
        <item name="android:gravity">center</item>                 //文字居中
        <item name="android:background">@drawable/bc_tab_selector</item>  //（a）
        <item name="android:textSize">10sp</item>                  //字号
        <item name="android:textStyle">normal</item>               //字体
        <item name="android:textColor">@drawable/text_tab_selector</item>  //（b）文字颜色
    </style>
</resources>
```

其中：

(a) **<item name="android:background">@drawable/bc_tab_selector</item>**：这里设置标签按钮的背景受 drawable 目录下的 bc_tab_selector.xml 文件控制，我们事先将两张候选的背景图置于 drawable 目录下，一张无色 bc_tab_normal.png 对应按钮未选中时的背景，另一张带颜色（浅绿色）bc_tab_pressed.png 则对应按钮选中时的背景。同样采用的是状态列表图形机制。

(b) **<item name="android:textColor">@drawable/text_tab_selector</item>**：标签按钮的文字颜色也是通过状态列表图形切换，但颜色的定义是在 values 目录下的 colors.xml 文件中。内容如下：

```xml
<resources>
    ...
    <color name="white">#ffffff</color>
    <color name="text_tab_pressed">#008B00</color>                 //深绿色（选中）
    <color name="text_tab_normal">#050505</color>                  //黑灰色（未选）
</resources>
```

在 colors.xml 中集中定义整个项目工程中用到的颜色，这是设计大型复杂 UI 界面色彩的普遍方式。

(3) 基于 TabActivity 的标签栏。

TabActivity 是 Android 中专用于实现标签栏的组件类，开发时只要继承该类就可以轻松实现带多个选项页的标签栏功能。TabActivity 对应前端的 UI 文件代码有着固定的框架，编程时用户必须在现成的框架中定义自己的界面。本例将 TabActivity 的代码框架写在 layout 目录下的 tab_host.xml 文件中，内容如下：

```xml
<TabHost xmlns:android="http://schemas.android.com/apk/res/android"    // (a)
    android:id="@android:id/tabhost"
    android:layout_width="match_parent"
    android:layout_height="match_parent" >

    <RelativeLayout                                                      // (a)
        android:layout_width="match_parent"
        android:layout_height="match_parent" >

        <FrameLayout                                                     // (a)
            android:id="@android:id/tabcontent"
            android:layout_width="match_parent"
            android:layout_height="match_parent"
            android:layout_marginBottom="@dimen/tabbar_height" />         // (b)

        <TabWidget                                                       // (a)
            android:id="@android:id/tabs"
            android:layout_width="match_parent"
            android:layout_height="wrap_content"
            android:visibility="gone" />

        <LinearLayout                                                    // (c)
            android:layout_width="match_parent"
            android:layout_height="@dimen/tabbar_height"                  // (b)
            android:layout_alignParentBottom="true"
            android:gravity="bottom"
            android:orientation="horizontal" >                           //标签栏整体是水平布局

            <LinearLayout                                                //"首页"标签按钮
                android:id="@+id/first_linear"
                android:layout_width="0dp"
                android:layout_height="match_parent"
                android:layout_weight="1"
                android:orientation="vertical" >                         //按钮内是垂直布局

                <TextView
                    style="@style/TabLabel"                              // (d)
                    android:drawableTop="@drawable/first_tab_selector"   // (e)
                    android:text="@string/first_tab" />                  // (f)
            </LinearLayout>
            //其他标签按钮元素的定义（略）
            ……
        </LinearLayout>
    </RelativeLayout>
</TabHost>
```

其中：

（a）**<TabHost...android:id="@android:id/tabhost"...>**、**<RelativeLayout...>**、**<FrameLayout.../>** 和**<TabWidget android:id="@android:id/tabs".../> </RelativeLayout> </TabHost>**：TabActivity 标签栏前端布局代码中，根元素必须是 TabHost（名称 tabhost）；根元素下面是一个 RelativeLayout；再下面

包含一个 FrameLayout（名称 tabcontent）；再往里还有一层 TabWidget（名称 tabs）。这些元素的类型和名称都是 TabActivity 框架固定好了的，用户**不能**修改。直到 TabWidget 之内，才开始了用户自己编写的代码。

（b）**android:layout_marginBottom="@dimen/tabbar_height"** 和 **android:layout_height="@dimen/tabbar_height"**：标签栏的高度 tabbar_height 定义在 values 目录下的 dimens.xml 文件中。内容如下：

```
<resources>
    <dimen name="tabbar_height">45dp</dimen>
</resources>
```

这里暂定为 45dp，读者也可根据需要进行调整。

（c）**<LinearLayout...android:orientation="horizontal" > <LinearLayout... android:orientation="vertical" >...</LinearLayout>......</LinearLayout>**：标签栏的设计代码全部写在框架的 TabWidget 内，本例采用双层线性布局设计，整个标签栏是水平的线性布局，而每个标签按钮内又是一个垂直的线性布局。

（d）**style="@style/TabLabel"**：标签文字直接套用定义好的风格 TabLabel。

（e）**android:drawableTop="@drawable/first_tab_selector"**：设置按钮的图标由状态列表图形控制切换。

（f）**android:text="@string/first_tab"**：标签文字的内容集中定义在 values 目录下的 strings.xml 文件中。内容如下：

```
<resources>
    <string name="app_name">标签栏</string>
    <string name="first_tab">首页</string>
    <string name="second_tab">分类</string>
    <string name="third_tab">推荐</string>
    <string name="four_tab">购物车</string>
    <string name="five_tab">我的</string>
</resources>
```

这里第一个标签文字 name 属性 "first_tab" 对应字符串为 "首页"。

2. 功能实现

实现思路：

（1）每个标签页对应界面上一个 LinearLayout 线性布局，在主程序类 TabHostActivity 中声明它们的全局变量引用，同时为各标签页定义好标记 TAG1～TAG5。

（2）在 onCreate() 方法初始化的时候就在每一个 LinearLayout 线性布局上绑定点击事件监听器，通过 getTabHost() 获取框架的标签栏并用 addTab() 向其中添加各标签按钮。

（3）当点击事件发生时，先通过 LinearLayout 线性布局引用取消所有标签按钮的选中状态，然后再根据接收点击事件的 LinearLayout 线性布局的 id 将标签栏上对应 TAG 标记的标签页设为当前页。

在 TabHostActivity.java 中实现功能，代码如下：

```
public class TabHostActivity extends TabActivity implements OnClickListener {
    private TabHost myTabHost;                                    //标签栏对象
    private LinearLayout first_linear, second_linear, third_linear, four_linear, five_linear;
    //以下是对应各个标签页的标记                    //对应 tab_host.xml 中各个标签按钮的引用
    private String TAG1 = "first";                                //第 1 页（首页）
    private String TAG2 = "second";                               //第 2 页（分类）
    private String TAG3 = "third";                                //第 3 页（推荐）
    private String TAG4 = "four";                                 //第 4 页（购物车）
```

```java
        private String TAG5 = "five";                              //第 5 页（我的）

        @Override
        protected void onCreate(Bundle savedInstanceState) {
            super.onCreate(savedInstanceState);
            setContentView(R.layout.tab_host);
            first_linear = findViewById(R.id.first_linear);
            first_linear.setOnClickListener(this);
            second_linear = findViewById(R.id.second_linear);
            second_linear.setOnClickListener(this);
            third_linear = findViewById(R.id.third_linear);
            third_linear.setOnClickListener(this);
            four_linear = findViewById(R.id.four_linear);
            four_linear.setOnClickListener(this);
            five_linear = findViewById(R.id.five_linear);
            five_linear.setOnClickListener(this);
            myTabHost = getTabHost();                              // (a)
            //初始化向标签栏添加各个标签按钮                          // (b)
            myTabHost.addTab(myTabHost.newTabSpec(TAG1).setIndicator(getString(R.string.first_tab),
getResources().getDrawable(R.drawable.first_tab_selector)).setContent(new Intent(this, FirstPageActivity.class)));
            myTabHost.addTab(myTabHost.newTabSpec(TAG2).setIndicator(getString(R.string.second_tab),
getResources().getDrawable(R.drawable.second_tab_selector)).setContent(new Intent(this, SecondPageActivity.class)));
            myTabHost.addTab(myTabHost.newTabSpec(TAG3).setIndicator(getString(R.string.third_tab),
getResources().getDrawable(R.drawable.third_tab_selector)).setContent(new Intent(this, ThirdPageActivity.class)));
            myTabHost.addTab(myTabHost.newTabSpec(TAG4).setIndicator(getString(R.string.four_tab),
getResources().getDrawable(R.drawable.four_tab_selector)).setContent(new Intent(this, FourPageActivity.class)));
            myTabHost.addTab(myTabHost.newTabSpec(TAG5).setIndicator(getString(R.string.five_tab),
getResources().getDrawable(R.drawable.five_tab_selector)).setContent(new Intent(this, FivePageActivity.class)));
            setTabView(first_linear);                              //设置初始显示"首页"页
        }

        @Override
        public void onClick(View view) {
            setTabView(view);
        }

        private void setTabView(View v) {                          // (c)
            //先将所有标签按钮置为未选中（复位）
            first_linear.setSelected(false);
            second_linear.setSelected(false);
            third_linear.setSelected(false);
            four_linear.setSelected(false);
            five_linear.setSelected(false);
            v.setSelected(true);                                   //选中当前的标签按钮
            //根据被点击的标签按钮视图设置当前要显示的标签页
            if (v == first_linear) {
                myTabHost.setCurrentTabByTag(TAG1);                //显示"首页"页
            } else if (v == second_linear) {
                myTabHost.setCurrentTabByTag(TAG2);                //显示"分类"页
            } else if (v == third_linear) {
                myTabHost.setCurrentTabByTag(TAG3);                //显示"推荐"页
```

```
        } else if (v == four_linear) {
            myTabHost.setCurrentTabByTag(TAG4);         //显示"购物车"页
        } else if (v == five_linear) {
            myTabHost.setCurrentTabByTag(TAG5);         //显示"我的"页
        }
    }
}
```

其中：

（a）**myTabHost = getTabHost();**：getTabHost()方法获取系统内置的标签栏组件，由于之前在 UI 设计文件中按照 Android 的标签栏框架套用编写代码，系统就能够根据框架固定的名称 @android:id/tabhost 找到这个工具栏。

（b）**myTabHost.addTab(myTabHost.newTabSpec(TAG1).setIndicator(getString(R.string.first_tab), getResources().getDrawable(R.drawable.first_tab_selector)).setContent(new Intent(this, FirstPageActivity.class)));** 等：addTab()方法向标签栏添加标签，在其中又调用 newTabSpec(TAG)方法生成指定规格的标签，newTabSpec(TAG)方法带一个字符串类型的参数，用于指定标签页的标记；setIndicator()设置标签按钮的文字及图标资源；最后还必须以 setContent()方法指明该标签页要显示的内容视图。由于本例侧重演示标签栏的开发，对每个标签的内容页暂且先以简单（仅含一个 TextView）的页面占位。例如，"首页"的页面文件 first_page.xml 代码如下：

```
<LinearLayout xmlns:android="http://schemas.android.com/apk/res/android"
    ...
    android:orientation="vertical"
    android:padding="10dp" >
    <TextView
        android:id="@+id/first_textview"
        ...
        android:gravity="bottom|center"        //位于底部中央（贴近标签栏）显示
        android:textColor="@android:color/holo_green_dark"
        android:textSize="24sp"
        android:textStyle="bold" />
</LinearLayout>
```

其他页面代码

其他页面的代码与之相同（仅文件名和文本视图的 id 不一样），此处略。

"首页"对应 Activity 的代码如下：

```
public class FirstPageActivity extends Activity {
    @Override
    protected void onCreate(Bundle savedInstanceState) {
        super.onCreate(savedInstanceState);
        setContentView(R.layout.first_page);
        TextView first_textview = findViewById(R.id.first_textview);
        first_textview.setText("这是 首页 界面");
    }
}
```

Activity 代码

其他页面的 Activity 代码同上（仅类名、引用页名和显示的文字信息差异），此处略。

（c）**private void setTabView(View v)**：这个是自定义的方法，它根据用户点击的标签按钮设置当前要显示的页面，用户点击的标签按钮以视图参数的形式传入该方法中。

4.2.2 轮播条

【例 4.2】 开发"易果鲜超市"App 顶部的轮播条,包含 5 幅广告图片轮番滚动显示,如图 4.5 所示。当用户点击某幅图片时,下方显示是第几幅广告。

第 3 幅广告

图 4.5 轮播条运行效果

创建 Android Studio 工程 **MyBanner**。下面先来分析轮播条的实现思路。

1. 实现思路

在 Android 中并没有现成的轮播条控件,不过通过观察图 4.5 中的轮播条,我们可以看出它大体上由两部分组成:

(1) 上部循环翻页滚动着的图片集。
(2) 底部指示当前图片所在次序的圆点指示器。

为此,我们考虑用**自定义控件**的方法自己制作一个轮播条组合控件,然后在程序中应用即可。

上部的图片翻页功能我们使用 Android 中的 ViewPager(翻页视图)控件实现,底部圆点指示器则使用 RadioGroup 和 RadioButton 控件组合实现。

在工程 layout 目录下创建自定义轮播条控件的 banner.xml 文件,设计内容如下:

```
<RelativeLayout xmlns:android="http://schemas.android.com/apk/res/android"
    ...
    android:orientation="vertical" >
    <androidx.viewpager.widget.ViewPager                    //实现上部的图片翻页
        android:id="@+id/myViewPager"
        android:layout_width="match_parent"
        android:layout_height="match_parent" />
    <RadioGroup                                              //实现底部的圆点指示器
        android:id="@+id/myRadioGroup"
        android:layout_width="wrap_content"
        android:layout_height="wrap_content"
        android:paddingBottom="2dp"
        android:orientation="horizontal"
        android:layout_alignParentBottom="true"
        android:layout_centerHorizontal="true" />
</RelativeLayout>
```

为能适应不同尺寸的广告图片,整个轮播条定义在一个相对布局中。

轮播条底部圆点指示器的每个圆点都是一个 RadioButton,平时为空心圆,当播放到某张图时,其对应圆点变为蓝色实心的外观。还是用状态列表图形来控制这种外观变化,预先在 drawable 目录下放置两种不同外观的圆点图片 icon_point_c.png（实心点）和 icon_point_n.png（空心点），然后创建 indicator_selector.xml 文件，内容如下：

```
<selector xmlns:android="http://schemas.android.com/apk/res/android">
    <item android:state_checked="false" android:drawable="@drawable/icon_point_n" />   //未轮播到
    <item android:state_checked="true" android:drawable="@drawable/icon_point_c" />    //轮播到时
</selector>
```

同时，将 5 幅水果广告图片（img1.jpg～img5.jpg）也放在 drawable 目录下。

2. 自定义轮播条

在源文件 Banner.java 中编写定义轮播条的代码，内容如下：

```
public class Banner extends RelativeLayout implements View.OnClickListener {
    private Context myContext;                                          //声明上下文对象
    private LayoutInflater myInflater;                                  //声明布局填充器
    private List<ImageView> myImageList = new ArrayList<ImageView>();   //图像视图列表存储广告图
    private ViewPager myViewPager;                                      //翻页视图
    private RadioGroup myRadioGroup;                                    //单选按钮组（实现指示器）
    private int myImageCount;                                           //广告图片张数
    private int dp15;
    private static int interval = 3000;                                 //图片轮播间隔（毫秒）

    public Banner(Context ctx) {
        this(ctx, null);
    }

    public Banner(Context ctx, AttributeSet attrs) {
        super(ctx, attrs);
        myContext = ctx;
        myInflater = ((Activity) myContext).getLayoutInflater();
        View view = myInflater.inflate(R.layout.banner, null);          // （a）
        myViewPager = view.findViewById(R.id.myViewPager);
        myRadioGroup = view.findViewById(R.id.myRadioGroup);
        addView(view);
        dp15 = dip2px(myContext, 15);
    }

    public void start() {
        myHandler.postDelayed(scrollRunnable, interval);                //延迟后启动滚动轮播
    }

    public void setBannerImage(ArrayList<Integer> imageList) {          // （b）
        for (int i = 0; i < imageList.size(); i++) {
            Integer iId = imageList.get(i).intValue();
            ImageView iView = new ImageView(myContext);
            iView.setLayoutParams(new LayoutParams(LayoutParams.MATCH_PARENT, LayoutParams.MATCH_PARENT));
            iView.setScaleType(ImageView.ScaleType.FIT_XY);
            iView.setImageResource(iId);
            iView.setOnClickListener(this);                             //绑定点击事件监听器
            myImageList.add(iView);
        }
```

```java
        myViewPager.setAdapter(new ImageAdapter());                          // (c)
        myViewPager.addOnPageChangeListener(new ViewPager.SimpleOnPageChangeListener() {
                                                                             // (d)
            @Override
            public void onPageSelected(int pos) {                            // (d)
                ((RadioButton) myRadioGroup.getChildAt(pos)).setChecked(true);
            }
        });

        myImageCount = imageList.size();
        for (int i = 0; i < myImageCount; i++) {                             // (e)
            RadioButton radioButton = new RadioButton(myContext);            // (e)
            radioButton.setLayoutParams(new RadioGroup.LayoutParams(dp15, dp15));
            radioButton.setGravity(Gravity.CENTER);
            radioButton.setButtonDrawable(R.drawable.indicator_selector);
            myRadioGroup.addView(radioButton);                               // (e)
        }
        myViewPager.setCurrentItem(0);                                       //默认显示第一幅广告
        ((RadioButton) myRadioGroup.getChildAt(0)).setChecked(true);         //默认点亮第一个指示圆点
    }

    //根据手机的分辨率从 dp 的单位转成为 px（像素）
    public int dip2px(Context ctx, float dpValue) {
        final float scale = ctx.getResources().getDisplayMetrics().density;
        return (int) (dpValue * scale + 0.5f);
    }

    private Handler myHandler = new Handler();
    private Runnable scrollRunnable = new Runnable() {
        @Override
        public void run() {
            //滚动到下一幅
            int index = myViewPager.getCurrentItem() + 1;
            if (myImageList.size() <= index) {
                index = 0;
            }
            myViewPager.setCurrentItem(index);
            myHandler.postDelayed(this, interval);                           //延迟后继续滚动
        }
    };

    private class ImageAdapter extends PagerAdapter {                        // (c) 图像翻页适配器
        @Override
        public int getCount() {                                              //获取图片张数
            return myImageList.size();
        }

        @Override
        public boolean isViewFromObject(View arg0, Object arg1) {
            return arg0 == arg1;
        }
```

```java
        @Override
        public void destroyItem(ViewGroup container, int position, Object object) {
            container.removeView(myImageList.get(position));                //销毁页面
        }

        @Override
        public Object instantiateItem(ViewGroup container, int position) {  //实例化添加页面
            container.addView(myImageList.get(position));
            return myImageList.get(position);
        }
    }

    @Override
    public void onClick(View v) {
        myBannerListener.onBannerClick(myViewPager.getCurrentItem());
    }

    public void setOnBannerListener(BannerClickListener listener) {
        myBannerListener = listener;                                        //设置图片的点击事件监听器
    }

    private BannerClickListener myBannerListener;                           //广告图片的点击事件监听器

    public interface BannerClickListener {
        void onBannerClick(int pos);
    }
}
```

其中：

（a）**View view = myInflater.inflate(R.layout.banner, null);**：自定义的控件要通过 LayoutInflater（布局填充器）获取其视图对象，先从上下文获取到 LayoutInflater，LayoutInflater 能够根据我们定义的 banner.xml 文件自动生成轮播条的视图对象，得到了轮播条视图对象，就很容易从中进一步获取组成其的子控件元素（ViewPager、RadioGroup）的引用。然后用 addView()方法将视图对象添加设定为该轮播条的视图。

（b）**public void setBannerImage(ArrayList<Integer> imageList)**：该方法的作用是把 imageList 列表中的图片都转化为图像视图。由于本例程序中的广告图片是以图像视图列表的形式存储的，故要将图片列表转换成图像视图列表。而广告图还要能响应用户点击，所以要在其中每个图像视图上绑定点击事件监听器。

（c）**myViewPager.setAdapter(new ImageAdapater());** 和 **private class ImageAdapater extends PagerAdapter**：ViewPager 是通过 PagerAdapter（翻页适配器）实现翻页滚动的效果的，这里定义一个图像翻页适配器类继承自 PagerAdapter，需要声明和重写其中一些方法。

（d）**myViewPager.addOnPageChangeListener(new ViewPager.SimpleOnPageChangeListener() { ... });** 和 **public void onPageSelected(int pos)**：ViewPager 的 SimpleOnPageChangeListener（简单页面变更监听器）实现当图片翻页切换时所要执行的动作。这里通过重写该监听器的 onPageSelected()方法，将对应图片的指示器圆点点亮为实心。

（e）**for (int i = 0; i < myImageCount; i++) { RadioButton radioButton = new RadioButton**

(myContext);......myRadioGroup.addView(radioButton);}：根据图片列表的大小动态生成指示器圆点（实为单选按钮），然后添加进单选按钮组中。

3. 使用轮播条

轮播条定义和实现好后，就可以在主程序中使用了。在设计页面的 activity_banner.xml 文件中放置我们定义的轮播条控件和一个文本视图，内容如下：

```xml
<LinearLayout xmlns:android="http://schemas.android.com/apk/res/android"
    ...
    android:orientation="vertical" >
    <com.easybooks.mybanner.Banner          //这里必须写自定义控件类在实际项目中的完整包路径名
        android:id="@+id/myBanner"
        android:layout_width="match_parent"
        android:layout_height="wrap_content" />
    <TextView
        android:id="@+id/myTextView"
        ...
        android:gravity="center"
        android:paddingTop="20dp"
        android:text="请点击推介图片"
        android:textColor="@color/black"
        android:textSize="18sp" />
</LinearLayout>
```

然后，在主程序类 BannerActivity.java 中引用轮播条控件，代码如下：

```java
public class BannerActivity extends AppCompatActivity {
    private Banner myBanner;                              //声明轮播条控件对象的引用
    private TextView myTextView;                          //文本视图（显示点了第几幅广告）
    @Override
    protected void onCreate(Bundle savedInstanceState) {
        super.onCreate(savedInstanceState);
        setContentView(R.layout.activity_banner);
        myBanner = findViewById(R.id.myBanner);
        LayoutParams lParams = (LayoutParams) myBanner.getLayoutParams();     // （a）
        lParams.height = (int) (getSreenWidth(this) * 250f / 640f);
        myBanner.setLayoutParams(lParams);
        ArrayList<Integer> imageList = new ArrayList<Integer>();
        imageList.add(Integer.valueOf(R.drawable.img1));
        imageList.add(Integer.valueOf(R.drawable.img2));
        imageList.add(Integer.valueOf(R.drawable.img3));
        imageList.add(Integer.valueOf(R.drawable.img4));
        imageList.add(Integer.valueOf(R.drawable.img5));
        myBanner.setBannerImage(imageList);                                    // （b）
        myBanner.setOnBannerListener(new Banner.BannerClickListener() {
            public void onBannerClick(int pos) {
                myTextView.setText(String.format("第 %d 幅广告", pos + 1));
            }
        });
        myBanner.start();                                 //开启轮播
        myTextView = findViewById(R.id.myTextView);
    }
```

```
        public int getSreenWidth(Context ctx) {
            WindowManager wm = (WindowManager) ctx.getSystemService(Context.WINDOW_SERVICE);
            DisplayMetrics dm = new DisplayMetrics();
            wm.getDefaultDisplay().getMetrics(dm);
            return dm.widthPixels;
        }
    }
```

其中：

（a）**LayoutParams lParams = (LayoutParams) myBanner.getLayoutParams();**：获取轮播条控件的布局参数（以 LayoutParams 类型返回），并根据手机屏幕的实际宽度调整自适应，然后再以 setLayoutParams()方法重新设置轮播条尺寸。

（b）**myBanner.setBannerImage(imageList);**：将要显示的广告图片添加进一个图片列表中，然后调用轮播条的 setBannerImage()方法将其转换成供内部处理用的图像视图列表。

4.2.3 频道栏

【例 4.3】 开发"易果鲜超市"App 首页的水果频道栏，如图 4.6 所示。

图 4.6 频道栏运行效果

1. 安装 RecyclerView

频道栏的实现需要借助 Android 的高级控件 RecyclerView（循环视图），这是一个功能极为强大的 UI 控件，能实现许多高级的界面特效，但在使用它之前需要先安装进项目工程。下面是具体的操作方法。

（1）创建空的 Android Studio 工程 MyRecyclerGrid。

（2）在默认 activity_main.xml 的设计模式下，从工具面板中找到 RecyclerView（位于 Common 下）并将其拖曳至设计界面，此时会弹出一个消息框询问是否添加依赖库，如图 4.7 所示，单击"OK"按钮，系统开始自动安装 RecyclerView。

（3）安装完毕，在工程的 app\build.gradle 文件中可看到增加了如下一行代码：

```
dependencies {
    ......
    implementation 'androidx.recyclerview:recyclerview:1.1.0'
}
```

这就表示 RecyclerView 安装成功，可以使用了。

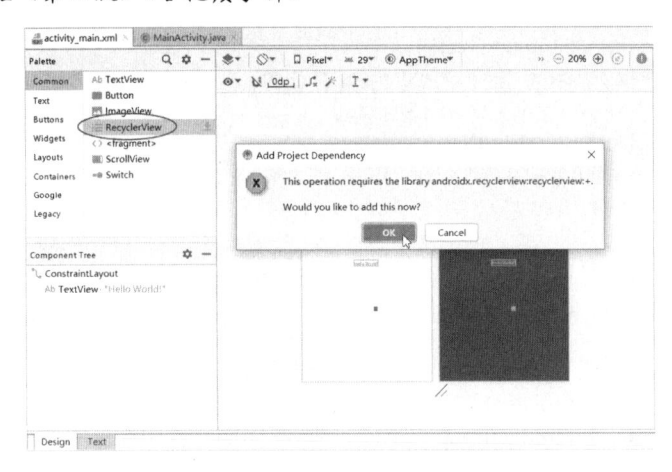

图 4.7 添加 RecyclerView 依赖库

2. 界面设计

RecyclerView 属于 Android 的视图类组件，在界面上应用这类组件进行开发遵循一定的套路模式。使用视图类控件的 App，其 UI 设计文件一般都包括两个：
- 主界面 XML 文件：当中直接放置要用的视图类控件元素。
- 具体项 XML 文件：在其中定义该视图内各个具体项的显示效果。

两者都位于工程的 layout 目录内。

下面分别设计这两个 UI 文件的内容。

（1）主界面 XML 文件。

本例主界面 XML 文件 recycler_grid.xml，其上只是简单地放置了一个 RecyclerView，代码如下：

```
<LinearLayout xmlns:android="http://schemas.android.com/apk/res/android"
    ...
    android:padding="5dp"
    android:orientation="vertical" >
    <androidx.recyclerview.widget.RecyclerView>       //注意要写出 RecyclerView 完整的包路径类名
        android:id="@+id/myRecyclerGrid"
        android:layout_width="match_parent"
        android:layout_height="wrap_content" />
</LinearLayout>
```

（2）具体项 XML 文件。

视图内的具体项定义在 item_grid.xml 中，代码如下：

```
<LinearLayout xmlns:android="http://schemas.android.com/apk/res/android"
    android:id="@+id/rcgrid_item"
    ...
    android:orientation="vertical" >
    <ImageView                                          //图像视图（显示栏目图标）
        android:id="@+id/rcgrid_picture"
        android:layout_width="match_parent"
        android:layout_height="wrap_content"
        android:paddingTop="15dp"
        android:scaleType="fitCenter" />

    <TextView                                           //文本视图（显示栏目名）
```

```xml
            android:id="@+id/rcgrid_title"
            android:layout_width="match_parent"
            android:layout_height="wrap_content"
            android:paddingTop="5dp"
            android:gravity="center"
            android:textColor="@color/gray"
            android:textSize="14sp" />
</LinearLayout>
```

由于每个栏目对应项的元素和布局都是一致的，只需定义这一个 XML 文件就可以完全确定整个频道栏的外观。

3. 数据模型与适配器

（1）定义数据模型。

当所要显示的数据项比较多而复杂时，通常建议专门创建一个数据模型类来集中存储和管理数据，这也符合 MVC 的设计理念。本例创建水果类 Fruit.java，代码如下：

```java
public class Fruit {
    public int pid;                                               //水果图标的 id
    public String title;                                          //水果栏目名称

    public Fruit(int pid, String title) {
        this.pid = pid;
        this.title = title;
    }

    private static int[] gridPictureArray = {R.drawable.f1, R.drawable.f2, R.drawable.f3
            , R.drawable.f4, R.drawable.f5, R.drawable.f6, R.drawable.f7
            , R.drawable.f8, R.drawable.f9, R.drawable.f10};         // (a)
    private static String[] gridTitleArray = {"苹果", "柑橘橙柚", "香蕉", "桃李",
            "蜜瓜/西瓜", "莓果", "杏枣", "葡萄", "热带水果", "其他"};    // (b)

    public static ArrayList<Fruit> getDefaultGrid() {                // (c)
        ArrayList<Fruit> gridList = new ArrayList<Fruit>();
        for (int i = 0; i < gridPictureArray.length; i++) {
            gridList.add(new Fruit(gridPictureArray[i], gridTitleArray[i]));
        }
        return gridList;
    }
}
```

其中：

（a）**private static int[] gridPictureArray = {R.drawable.f1, ..., R.drawable.f10};**：整型数组 gridPictureArray[]存放水果图标的 id，10 张水果图片（f1.png~f10.png）预先处理成统一的尺寸（48×48），置于工程 drawable 目录下。

（b）**private static String[] gridTitleArray = {"苹果", ..., "其他"};**：字符串数组 gridTitleArray[]存放水果栏目的名称。

（c）**public static ArrayList<Fruit> getDefaultGrid()**：该方法返回 RecyclerView 视图默认显示的内容，内容按照数据模型来组织，每一项都包含图标和栏目名，以一个 Fruit 类泛型列表（ArrayList<Fruit>）的结构返回。

(2)开发适配器。

Android 的视图类控件都必须借助适配器来显示数据,适配器既可以是默认的 Android 系统内置适配器,也可以由用户自己开发定制。本例我们自己专为 RecyclerView 视图来开发一个适配器 GridAdapter.java,代码如下:

```java
public class GridAdapter extends RecyclerView.Adapter<RecyclerView.ViewHolder> implements AdapterView.OnItemClickListener {
    private Context myContext;                              //声明上下文
    private LayoutInflater myInflater;                      //布局填充器
    private ArrayList<Fruit> myFruitsList;                  //水果类数据列表

    public GridAdapter(Context ctx, ArrayList<Fruit> lst) {
        myContext = ctx;
        myInflater = LayoutInflater.from(ctx);              //从上下文获取布局
        myFruitsList = lst;                                 //获取水果类数据
    }

    @Override
    public int getItemCount() {                             //获得栏目数
        return myFruitsList.size();
    }

    @Override
    public RecyclerView.ViewHolder onCreateViewHolder(ViewGroup vg, int viewType) {
        View view = myInflater.inflate(R.layout.item_grid, vg, false);   // (a)
        RecyclerView.ViewHolder vHolder = new ItemHolder(view);
        return vHolder;
    }

    @Override
    public void onBindViewHolder(RecyclerView.ViewHolder vh, final int pos) {
        ItemHolder iholder = (ItemHolder) vh;
        iholder.rcgrid_picture.setImageResource(myFruitsList.get(pos).pid);  //由 id 获得水果图标
        iholder.rcgrid_title.setText(myFruitsList.get(pos).title);           //设置水果栏目名

        //栏目项的点击事件需要自己实现
        iholder.rcgrid_item.setOnClickListener(new OnClickListener() {
            @Override
            public void onClick(View v) {
                if (myOnItemClickListener != null) {
                    myOnItemClickListener.onItemClick(null, v, pos, 0);
                }
            }
        });
    }

    @Override
    public int getItemViewType(int pos) {
        return 0;
    }
```

```java
        @Override
        public long getItemId(int pos) {
            return pos;
        }

        public class ItemHolder extends RecyclerView.ViewHolder {     // (b)
            public LinearLayout rcgrid_item;                           //栏目项布局
            public ImageView rcgrid_picture;                           //栏目项图标
            public TextView rcgrid_title;                              //栏目项名称

            public ItemHolder(View view) {
                super(view);
                rcgrid_item = view.findViewById(R.id.rcgrid_item);
                rcgrid_picture = view.findViewById(R.id.rcgrid_picture);
                rcgrid_title = view.findViewById(R.id.rcgrid_title);
            }
        }

        private AdapterView.OnItemClickListener myOnItemClickListener;     //声明栏目项点击事件监听器

        public void setOnItemClickListener(AdapterView.OnItemClickListener listener) {
            this.myOnItemClickListener = listener;                         //设置栏目项的点击事件监听器
        }

        @Override
        public void onItemClick(AdapterView<?> arg0, View view, int pos, long arg3) {
            String msg = String.format("点击栏目 %s", myFruitsList.get(pos).title);
            Toast.makeText(myContext, msg, Toast.LENGTH_SHORT).show();
        }
    }
```

其中：

(a) **View view = myInflater.inflate(R.layout.item_grid, vg, false);**：这里正是从之前定义的 item_grid.xml 生成栏目项的视图的。

(b) **public class ItemHolder extends RecyclerView.ViewHolder**：我们编写了一个 ItemHolder 类，继承自 RecyclerView 的 ViewHolder，专门实现对频道栏中各个栏目项的设定操作，赋予它们各自具体的水果图标和栏目名称。

4. 使用 RecyclerView

在设计好数据模型和适配器后，在主程序中就可以直接使用 RecyclerView 来显示频道栏效果。主程序 RecyclerGridActivity.java，代码如下：

```java
public class RecyclerGridActivity extends AppCompatActivity {
    private RecyclerView myRecyclerGrid;                               //循环视图对象引用
    @Override
    protected void onCreate(Bundle savedInstanceState) {
        super.onCreate(savedInstanceState);
        setContentView(R.layout.recycler_grid);
        myRecyclerGrid = findViewById(R.id.myRecyclerGrid);
        GridLayoutManager manager = new GridLayoutManager(this, 5);    // (a)
        myRecyclerGrid.setLayoutManager(manager);
```

```
        GridAdapter adapter = new GridAdapter(this, Fruit.getDefaultGrid());        // （b）
        adapter.setOnItemClickListener(adapter);                                     // （c）
        myRecyclerGrid.setAdapter(adapter);                                          //绑定适配器
        myRecyclerGrid.setItemAnimator(new DefaultItemAnimator());
    }
}
```

其中：

（a）**GridLayoutManager manager = new GridLayoutManager(this, 5);**：GridLayoutManager 是 RecyclerView 的布局管理器，实现如频道栏这样的网格布局。RecyclerView 有多种布局管理器，LinearLayoutManager 实现线性布局，StaggeredGridLayoutManager 实现瀑布流布局。用户选用不同的布局管理器，设计出来的界面会完全不一样，但是针对每一种布局管理器都必须先开发与之相适应的适配器才能够使用。

（b）**GridAdapter adapter = new GridAdapter(this, Fruit.getDefaultGrid());**：这里创建适配器对象，通过数据模型的 getDefaultGrid()获取数据，作为参数传入适配器的构造方法，然后将适配器与 RecyclerView 绑定，就可以显示内容了。

（c）**adapter.setOnItemClickListener(adapter);**：通过适配器为栏目项设置点击事件监听器。

4.3 内容的呈现

4.3.1 列表视图

【例 4.4】 开发"易果鲜超市"App 的果品展示列表，如图 4.8 所示。

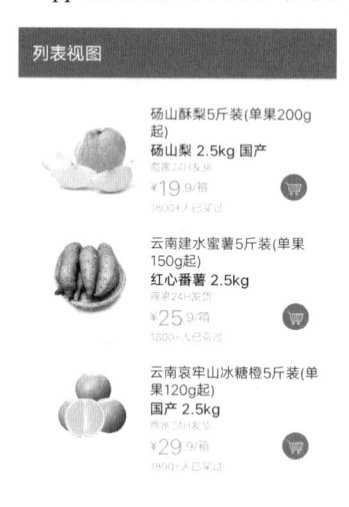

图 4.8 果品展示列表的效果

创建 Android Studio 工程 MyListView。本例用 ListView（列表视图）实现，ListView 是 Android Studio 环境内置的，无须额外安装。

1. 界面设计

既然同为视图类的控件，ListView 与 RecyclerView 一样，在使用的时候需要编写两个 UI 文件。
（1）主界面 XML 文件。
主界面 XML 文件 list_view.xml 上面也是简单地放置一个 ListView，代码如下：

```xml
<LinearLayout xmlns:android="http://schemas.android.com/apk/res/android"
    ...
    android:orientation="vertical"
    android:padding="5dp" >
    <ListView                                     //这里无须写出类的包路径全名
        android:id="@+id/myListView"
        android:layout_width="match_parent"
        android:layout_height="wrap_content" />
</LinearLayout>
```

（2）具体项 XML 文件。

视图内的具体项定义在 item_list.xml 中，代码如下：

```xml
<LinearLayout xmlns:android="http://schemas.android.com/apk/res/android"
    android:id="@+id/linear_item"                //最外层是个水平的线性布局
    android:layout_width="match_parent"
    android:layout_height="wrap_content"
    android:orientation="horizontal" >

    <ImageView                                    //图像视图（显示水果的图片）
        android:id="@+id/list_icon"
        android:layout_width="80dp"
        android:layout_height="120dp"
        android:paddingTop="20dp"
        android:layout_weight="1"
        android:scaleType="fitXY" />

    <LinearLayout                                 //其余文字部分统一置于这个垂直线性布局内
        android:layout_width="0dp"
        android:layout_height="wrap_content"
        android:paddingTop="20dp"
        android:layout_weight="3.5"
        android:orientation="vertical" >

        <TextView                                 //文本视图（显示水果名称）
            android:id="@+id/list_name"
            android:layout_width="match_parent"
            android:layout_height="wrap_content"
            android:layout_weight="1"
            android:gravity="left|center"
            android:textColor="@color/black"
            android:textSize="16sp" />

        <TextView                                 //文本视图（显示水果说明信息）
            android:id="@+id/list_note"
            android:layout_width="match_parent"
            android:layout_height="wrap_content"
            android:layout_weight="1"
            android:gravity="left|center"
            android:textColor="@color/black"
            android:textSize="16sp"
            android:textStyle="bold" />
```

```xml
        <TextView
            android:id="@+id/tv_0"
            android:layout_width="match_parent"
            android:layout_height="wrap_content"
            android:layout_weight="1"
            android:gravity="left|center"
            android:hint="商家24H 发货"
            android:textColor="@color/black"
            android:textSize="12sp" />

        <LinearLayout                                          //内嵌的水平线性布局（显示水果价格行的内容）
            android:layout_width="wrap_content"
            android:layout_height="wrap_content"
            android:layout_weight="3"
            android:orientation="horizontal">

            <TextView                                          //文本视图（显示价格前的"¥"符号）
                android:id="@+id/tv_1"
                android:layout_width="wrap_content"
                android:layout_height="match_parent"
                android:gravity="center"
                android:text="¥"
                android:textColor="@color/orange"
                android:textSize="14sp" />

            <TextView                                          //文本视图（显示水果价格）
                android:id="@+id/list_price"
                android:layout_width="wrap_content"
                android:layout_height="match_parent"
                android:gravity="center"
                android:textColor="@color/orange"
                android:textSize="24sp" />

            <TextView                                          //文本视图（显示价格.9 后缀）
                android:id="@+id/tv_2"
                android:layout_width="wrap_content"
                android:layout_height="match_parent"
                android:gravity="center"
                android:text=".9"
                android:textColor="@color/orange"
                android:textSize="14sp" />

            <TextView                                          //文本视图（显示单位）
                android:id="@+id/tv_3"
                android:layout_width="wrap_content"
                android:layout_height="match_parent"
                android:width="100dp"
                android:gravity="left|center_vertical"
                android:hint="/箱"
                android:textColor="@color/black"
```

```xml
            android:textSize="14sp" />
        <ImageView                                          //图像视图（显示购物车图标）
            android:id="@+id/cart_icon"
            android:layout_width="30dp"
            android:layout_height="30dp"
            android:scaleType="fitXY"
            android:src="@drawable/cart" />
    </LinearLayout>

    <TextView
        android:id="@+id/tv_4"
        android:layout_width="match_parent"
        android:layout_height="wrap_content"
        android:layout_weight="1"
        android:gravity="left|center"
        android:hint="1800+人已买过"
        android:textColor="@color/black"
        android:textSize="12sp" />
</LinearLayout>
</LinearLayout>
```

可以看到，这里使用了多层线性布局嵌套图像和文本视图的方式来实现界面列表项中的丰富内容。实际开发时需要实时运行 App，根据界面外观不断地修改调整设计，直至达到满意的效果。

2. 开发数据模型和适配器

（1）定义数据模型。

本例同样要创建数据模型来存储果品信息，数据模型类 Fruit.java 的代码如下：

```java
public class Fruit {
    public int icon;                                    //水果图片
    public String name;                                 //水果名称
    public String note;                                 //水果说明信息
    public String price;                                //水果价格

    public Fruit(int icon, String name, String note, String price) {
        this.icon = icon;
        this.name = name;
        this.note = note;
        this.price = price;
    }

    private static int[] iconArray = {R.drawable.l0, R.drawable.l1, R.drawable.l2};    //存放水果图片
    private static String[] nameArray = {"砀山酥梨 5 斤装(单果200g起)", "云南建水蜜薯 5 斤装(单果150g起)",
"云南哀牢山冰糖橙 5 斤装(单果 120g 起)"};                                   //存放水果名称
    private static String[] noteArray = {                                   //存放水果说明信息
            "砀山梨 2.5kg 国产",
            "红心番薯 2.5kg",
            "国产 2.5kg"
    };
    private static String[] priceArray = {"19", "25", "29"};                //存放水果价格
```

```java
    public static ArrayList<Fruit> getDefaultList() {                    //返回默认显示的内容
        ArrayList<Fruit> fruitList = new ArrayList<Fruit>();
        for (int i = 0; i < iconArray.length; i++) {
            fruitList.add(new Fruit(iconArray[i], nameArray[i], noteArray[i], priceArray[i]));
        }
        return fruitList;
    }
}
```

这里同样也是用不同类型的数组结构分门别类存放水果的各项数据信息,以泛型列表的形式返回结构化的水果项数据。

(2) 开发适配器。

下面来为 ListView 视图开发适配器,适配器类 FruitAdapter.java,代码如下:

```java
public class FruitAdapter extends BaseAdapter implements OnItemClickListener {    //(a)
    private Context myContext;                                          //声明上下文
    private LayoutInflater myInflater;                                  //布局填充器
    private int myLayoutId;                                             //布局 id
    private ArrayList<Fruit> myFruitList;                               //水果类数据列表
    private int myBackground;                                           //列表项背景

    public FruitAdapter(Context ctx, int lid, ArrayList<Fruit> flst, int bg) {
        myContext = ctx;
        myInflater = LayoutInflater.from(ctx);                          //从上下文获取布局
        myLayoutId = lid;                                               //设置布局 id
        myFruitList = flst;                                             //获取水果类数据
        myBackground = bg;                                              //设置列表项背景
    }

    @Override
    public int getCount() {                                             //获得列表项数目
        return myFruitList.size();
    }

    @Override
    public Object getItem(int arg0) {
        return myFruitList.get(arg0);
    }

    @Override
    public long getItemId(int arg0) {
        return arg0;
    }

    @Override
    public View getView(final int position, View convertView, ViewGroup parent) {    //(b)
        ViewHolder vholder = null;
        if (convertView == null) {
            vholder = new ViewHolder();
            convertView = myInflater.inflate(myLayoutId, null);                      //(b)
            vholder.linear_item = convertView.findViewById(R.id.linear_item);
            vholder.list_icon = convertView.findViewById(R.id.list_icon);
```

```java
            vholder.list_name = convertView.findViewById(R.id.list_name);
            vholder.list_note = convertView.findViewById(R.id.list_note);
            vholder.list_price = convertView.findViewById(R.id.list_price);
            convertView.setTag(vholder);
        } else {
            vholder = (ViewHolder) convertView.getTag();
        }
        Fruit fruit = myFruitList.get(position);
        vholder.linear_item.setBackgroundColor(myBackground);
        vholder.list_icon.setImageResource(fruit.icon);
        vholder.list_name.setText(fruit.name);
        vholder.list_note.setText(fruit.note);
        vholder.list_price.setText(fruit.price);
        return convertView;
    }

    public final class ViewHolder {                                         // (c)
        private LinearLayout linear_item;
        public ImageView list_icon;
        public TextView list_name;
        public TextView list_note;
        public TextView list_price;
    }

    @Override
    public void onItemClick(AdapterView<?> parent, View view, int position, long id) {
        String msg = String.format("你选了第 %d 个水果 %s", position + 1, myFruitList.get(position).name);
        Toast.makeText(myContext, msg, Toast.LENGTH_LONG).show();
    }
}
```

其中：

（a）**public class FruitAdapter extends BaseAdapter implements OnItemClickListener**：我们开发的适配器继承自 BaseAdapter，它是 Android 专为具有复杂内容项的视图提供的一种适应性很强的基本适配器，用户很容易在其基础上扩展，开发出适用于多种视图展示的适配器类。

（b）**public View getView(final int position, View convertView, ViewGroup parent) convertView = myInflater.inflate(myLayoutId, null);**：应用 BaseAdapter 关键在重写其内部的 getView()方法，该方法有一个 View 类型的 convertView 参数，通过这个参数获取布局填充器生成的内容项视图，然后再进一步得到其中的各个具体控件元素的引用。

（c）**public final class ViewHolder**：定义 ViewHolder 类用于存储从 convertView 参数获取的列表项中各控件元素的引用对象，然后实现对各控件的操作，赋予它们各自具体的水果数据项。

3. 使用 ListView

在设计好数据模型和适配器后，就可以在主程序中使用 ListView 了。主程序 ListViewActivity.java 代码如下：

```java
public class ListViewActivity extends AppCompatActivity {
    private ListView myListView;                                //列表视图对象引用
    private ArrayList<Fruit> fruitList;                         //水果数据列表
    @Override
```

```
protected void onCreate(Bundle savedInstanceState) {
    super.onCreate(savedInstanceState);
    setContentView(R.layout.list_view);
    fruitList = Fruit.getDefaultList();                                    //获取默认水果数据
    FruitAdapter adapter = new FruitAdapter(this, R.layout.item_list, fruitList, Color.WHITE);
                                                                           //创建适配器对象
    myListView = findViewById(R.id.myListView);
    myListView.setAdapter(adapter);                                        //绑定适配器
    myListView.setOnItemClickListener(adapter);                            //绑定列表项点击事件监听器
    //设定列表项的首尾分隔线及边距
    myListView.setHeaderDividersEnabled(true);
    myListView.setFooterDividersEnabled(true);
    myListView.setDividerHeight(3);
    myListView.setPadding(25, 10, 25, 0);
    }
}
```

4.3.2 网格视图

【例 4.5】 开发"易果鲜超市"App 的"推荐"页，以网格的形式展示果品，如图 4.9 所示。

创建 Android Studio 工程 MyGridView。本例用 GridView（网格视图）实现，GridView 也是 Android Studio 环境内置的，无须额外安装。

1. 自定义文本视图

在前面第 2 章的学习中，我们已经知道文本视图不仅可以显示文字信息，还能够在上下左右各个方位显示图片，甚至显示滚动字幕。但是，即便是这样，对于当前互联网应用中所要求的某些高级特效，例如带有醒目边框和图片的商品返利信息、带边框的淡化文字消息（如图 4.10 所示），制作起来

图 4.9 以网格的形式展示果品

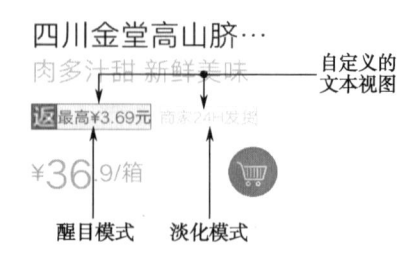

图 4.10 自定义文本视图实现的文字特效

仍然稍显吃力。这个时候就需要我们来自己定制具有特殊功能的文本视图，Android 支持包括文本视图在内的很多基本控件的自定义功能，只要继承相应的基本控件类，在此基础上重写某些方法加以扩展即可得到我们想要的功能更为强大的新控件。

在本例中，我们自定义了一个文本视图控件 BorderTextView，其源文件 BorderTextView.java 代码如下：

```java
package com.easybooks.mygridview;
...
public class BorderTextView extends AppCompatTextView {                    //（a）
    private Paint paint;                                                    //绘制外观的画笔
    private Drawable drawable;                                              //图片资源对象
    private int sroke_width = 1;                                            //边框宽度
    private boolean bgRed = false;                                          //（b）红色醒目模式

    public BorderTextView(Context context) {
        super(context);
    }

    public BorderTextView(Context context, AttributeSet attrs) {            //（c）
        super(context, attrs);
        paint = new Paint();                                                //创建画笔
        if (attrs != null) {
            TypedArray attrArray = getContext().obtainStyledAttributes(attrs, R.styleable.BorderTextView);
                                                                            //（c）获取自定义属性
            bgRed = attrArray.getBoolean(R.styleable.BorderTextView_bgRed, bgRed);
        }
        if (bgRed) {                                                        //醒目模式
            paint.setColor(Color.RED);                                      //将边框设为红色
            this.setTextColor(Color.RED);                                   //文字设为红色
            this.setBackgroundColor(context.getResources().getColor(R.color.red));  //背景设为淡红
            //文本视图左端显示返利图标
            drawable = getResources().getDrawable(R.drawable.rt);
            drawable.setBounds(0, 0, drawable.getMinimumWidth(), drawable.getMinimumHeight());
            this.setCompoundDrawables(drawable, null, null, null);          //图标在金额文本左边
        } else {                                                            //淡化模式
            paint.setColor(Color.LTGRAY);                                   //将边框设为浅灰色
            this.setTextColor(Color.LTGRAY);                                //文字设为浅灰色
            this.setText("商家 24h 发货");                                   //文字内容
        }
    }

    @Override
    protected void onDraw(Canvas canvas) {                                  //（d）
        //重绘 TextView 的 4 个边框
        canvas.drawLine(0, 0, this.getWidth() - sroke_width, 0, paint);
        canvas.drawLine(0, 0, 0, this.getHeight() - sroke_width, paint);
        canvas.drawLine(this.getWidth() - sroke_width, 0, this.getWidth() - sroke_width, this.getHeight() - sroke_width, paint);
        canvas.drawLine(0, this.getHeight() - sroke_width, this.getWidth() - sroke_width, this.getHeight() - sroke_width, paint);
        super.onDraw(canvas);
```

        }
    }
其中：
（a）**public class BorderTextView extends AppCompatTextView**：Android 要求自定义的文本视图继承自 AppCompatTextView 类而非 TextView，因为 TextView 类本身已作为 Android 内一个最终的控件元素用于实现一般的文本视图功能，不允许再对其随意进行修改。

（b）**private boolean bgRed = false;**：我们定义的这个文本视图实现了两种迥异的显示风格，一种是带图标和红色边框的醒目模式，另一种是带浅灰色框和文字的淡化模式，如图 4.10 所示。在程序中用布尔型开关变量 bgRed 进行控制。

（c）**public BorderTextView(Context context, AttributeSet attrs)** 和 **TypedArray attrArray = getContext().obtainStyledAttributes(attrs, R.styleable.BorderTextView);**：在初始化时通过参数 AttributeSet 向构造方法传入控件的自定义属性，然后用 obtainStyledAttributes()方法获取前端用户设置的属性值，属性值以一个 TypedArray 类型返回，这样程序就知道了前端所设置的属性值，再根据属性值来描绘控件的外观。要将内部变量 bgRed 设计成控件的属性，需要在工程 values 目录下配置一个 attrs.xml 文件，在其中定义属性元素，如下：

```xml
<?xml version="1.0" encoding="utf-8"?>
<resources>
    <declare-styleable name="BorderTextView">        //这里的名称是自定义控件的类名
        <attr name="bgRed" format="boolean"></attr>   //指定属性名及类型
    </declare-styleable>
</resources>
```

> 👀 **注意：**
> attrs.xml 文件只能在 Android Studio 环境下右击工程 values 目录，在弹出的快捷菜单中选择"New"→"XML"→"Values XML File"命令创建，不能直接手工创建和编写。

（d）**protected void onDraw(Canvas canvas)**：通过重写 onDraw()方法，在画布上重绘文本视图控件的边框，显示出定制的外观。

定义了新的文本视图控件之后，就可以在 UI 设计文件中像使用基本界面控件那样使用它。在网格视图项的设计文件 item_grid.xml 中引用自定义的控件（加粗处），代码如下：

```xml
<LinearLayout xmlns:android="http://schemas.android.com/apk/res/android"
    xmlns:app="http://schemas.android.com/apk/res-auto"
    android:id="@+id/linear_item"
    ...
    android:orientation="vertical">

    <ImageView                                        //图像视图（显示水果的图片）
        android:id="@+id/grid_icon"
        ... />

    <TextView                                         //文本视图（显示水果名称）
        android:id="@+id/grid_name"
        ...
        android:singleLine="true"
        ... />

    <TextView                                         //文本视图（显示水果说明信息）
```

网格项代码

```
            android:id="@+id/grid_note"
            ...
            android:singleLine="true"
            ... />

    <LinearLayout
        ...
        android:orientation="horizontal">

        <com.easybooks.mygridview.BorderTextView        //自定义文本视图（显示水果返利信息）
            android:id="@+id/btv_rebate"
            android:layout_width="wrap_content"
            android:layout_height="wrap_content"
            android:gravity="center"
            app:bgRed="true"                            //设为醒目模式
            android:textSize="10sp">
        </com.easybooks.mygridview.BorderTextView>

        <com.easybooks.mygridview.BorderTextView        //自定义文本视图（显示淡化文字消息）
            android:id="@+id/btv_note"
            android:layout_width="wrap_content"
            android:layout_height="wrap_content"
            android:layout_marginLeft="5dp"
            android:layout_gravity="center_vertical"
            app:bgRed="false"                           //设为淡化模式
            android:textSize="10sp">
        </com.easybooks.mygridview.BorderTextView>
    </LinearLayout>

    <LinearLayout
        ...
        android:orientation="horizontal">
        //其他文本视图标签元素的定义与【例4.4】列表视图类同（略）
        ......
    </LinearLayout>
</LinearLayout>
```

自定义的控件在使用时必须引用类的完整包路径名，这里是"com.easybooks.mygridview.BorderTextView"。另外，本例的主界面 XML 文件 grid_view.xml 中也只需放置一个 GridView 即可，代码略。

2. 开发数据模型和适配器

GridView 的数据模型和适配器与 ListView 的基本类同，下面仅分别罗列出它们的实现代码，不再进行详细说明。读者可与前例 ListView 的代码进行对比学习，以加深理解视图类控件的编程方式。

（1）数据模型 Fruit.java 的代码如下：

```
public class Fruit {
    public int icon;                    //水果图片
    public String name;                 //水果名称
    public String note;                 //水果说明信息
    public String rebate;               //水果返利信息
    public String price;                //水果价格
```

```java
    public Fruit(int icon, String name, String note, String rebate, String price) {
        this.icon = icon;
        this.name = name;
        this.note = note;
        this.rebate = rebate;
        this.price = price;
    }

    private static int[] iconArray = {R.drawable.g0, R.drawable.g1, R.drawable.g2, R.drawable.g3, R.drawable.g4, R.drawable.g5};      //存放水果图片
    private static String[] nameArray = {"四川金堂高山脐橙 9 斤装 (单果 150g 起)", "广西紫色百香果 3 斤装 (单果 60g 起)", "砀山酥梨 5 斤装(单果 200g 起)", "四川金艳黄金猕猴桃 20 粒装 (单果 100g 起)", "云南建水蜜薯 5 斤装 (单果 150g 起)", "陕西徐香即食猕猴桃 20 枚装  (90g 起)"};      //存放水果名称
    private static String[] noteArray = {                                    //存放水果说明信息
            "肉多汁甜  新鲜美味",
            "开箱即食，酸甜适口",
            "酥香爽脆  甜润入心  果心小肉白",
            "蒲江黄心猕猴桃  味道甜美  丰盈多汁",
            "高原蜜薯  自然成熟  软糯香甜  天然爆浆",
            "口感酸甜  鲜嫩多汁  果味浓郁  赠送猕猴桃挖勺"
    };
    private static String[] rebateArray = {"3.69", "1.99", "1.99", "3.99", "2.59", "2.99"};
                                                                            //存放水果返利信息
    private static String[] priceArray = {"36", "19", "19", "39", "25", "29"};      //存放水果价格

    public static ArrayList<Fruit> getDefaultList() {                       //返回默认显示的内容
        ArrayList<Fruit> fruitList = new ArrayList<Fruit>();
        for (int i = 0; i < iconArray.length; i++) {
            fruitList.add(new Fruit(iconArray[i], nameArray[i], noteArray[i], rebateArray[i], priceArray[i]));
        }
        return fruitList;
    }
}
```

（2）适配器 FruitAdapter.java 的代码如下：

```java
public class FruitAdapter extends BaseAdapter implements OnItemClickListener {    //继承 BaseAdapter
    private Context myContext;                  //声明上下文
    private LayoutInflater myInflater;          //布局填充器
    private int myLayoutId;                     //布局 id
    private ArrayList<Fruit> myFruitList;       //水果类数据列表
    private int myBackground;                   //网格项背景

    public FruitAdapter(Context ctx, int lid, ArrayList<Fruit> flst, int bg) {
        myContext = ctx;
        myInflater = LayoutInflater.from(ctx);  //从上下文获取布局
        myLayoutId = lid;                       //设置布局 id
        myFruitList = flst;                     //获取水果类数据
        myBackground = bg;                      //设置网格项背景
    }

    @Override
```

```java
public int getCount() {                                    //获得网格项数目
    return myFruitList.size();
}

@Override
public Object getItem(int arg0) {
    return myFruitList.get(arg0);
}

@Override
public long getItemId(int arg0) {
    return arg0;
}

@Override
public View getView(final int position, View convertView, ViewGroup parent) {    //重写 getView()方法
    ViewHolder vholder = null;
    if (convertView == null) {
        vholder = new ViewHolder();
        convertView = myInflater.inflate(myLayoutId, null);                      //获取内容项视图
        vholder.linear_item = convertView.findViewById(R.id.linear_item);
        vholder.grid_icon = convertView.findViewById(R.id.grid_icon);
        vholder.grid_name = convertView.findViewById(R.id.grid_name);
        vholder.grid_note = convertView.findViewById(R.id.grid_note);
        vholder.btv_rebate = convertView.findViewById(R.id.btv_rebate);
        vholder.grid_price = convertView.findViewById(R.id.grid_price);
        convertView.setTag(vholder);
    } else {
        vholder = (ViewHolder) convertView.getTag();
    }
    Fruit fruit = myFruitList.get(position);
    vholder.linear_item.setBackgroundColor(myBackground);
    vholder.grid_icon.setImageResource(fruit.icon);
    vholder.grid_name.setText(fruit.name);
    vholder.grid_note.setText(fruit.note);
    vholder.btv_rebate.setText("最高¥" + fruit.rebate + "元");
    vholder.grid_price.setText(fruit.price);
    return convertView;
}

public final class ViewHolder {               //存储从 convertView 获取的网格项中各控件元素的引用对象
    private LinearLayout linear_item;
    public ImageView grid_icon;
    public TextView grid_name;
    public TextView grid_note;
    public BorderTextView btv_rebate;
    public TextView grid_price;
}

@Override
public void onItemClick(AdapterView<?> parent, View view, int position, long id) {
```

```
                String msg = String.format("你选了第 %d 个水果 %s", position + 1, myFruitList.get(position).name);
                Toast.makeText(myContext, msg, Toast.LENGTH_LONG).show();
            }
        }
```

3. 使用 GridView

在主程序 GridViewActivity.java 中使用 GridView，代码如下：

```
public class GridViewActivity extends AppCompatActivity {
    private GridView myGridView;                                            //网格视图对象引用
    private ArrayList<Fruit> fruitList;                                     //水果数据列表
    @Override
    protected void onCreate(Bundle savedInstanceState) {
        super.onCreate(savedInstanceState);
        setContentView(R.layout.grid_view);
        fruitList = Fruit.getDefaultList();                                 //获取默认水果数据
        FruitAdapter adapter = new FruitAdapter(this, R.layout.item_grid, fruitList, Color.WHITE);
                                                                            //创建适配器对象
        myGridView = findViewById(R.id.myGridView);
        myGridView.setAdapter(adapter);                                     //绑定适配器
        myGridView.setOnItemClickListener(adapter);                         //绑定网格项点击事件监听器
        //设定网格项的显示属性
        myGridView.setBackgroundColor(Color.WHITE);
        myGridView.setHorizontalSpacing(0);
        myGridView.setVerticalSpacing(0);
        myGridView.setStretchMode(GridView.STRETCH_COLUMN_WIDTH);           //图片拉伸充满网格
        myGridView.setColumnWidth(250);
        myGridView.setPadding(0, 0, 0, 0);
    }
}
```

4.3.3 类别标签列表

【例 4.6】 开发"易果鲜超市"App"分类"页的果品分类列表，以类别标签的形式切换水果类别，如图 4.11 所示。

图 4.11 果品分类列表的效果

本例我们采用 RecyclerView（循环视图）与 TabHost（Tab 导航栏）相结合的方式来实现类别标签列表。创建 Android Studio 工程 MyRecyclerList。由于要使用 RecyclerView，当然先要安装，具体操作见 4.2.3 小节；TabHost 是 Android Studio 的内置控件，无须安装。

1. 界面开发

本例界面一共有 3 个 UI 文件，一个包含 TabHost 的文件 tabhost_main.xml，用于实现类别标签页；另外两个是 RecyclerView 的 UI 设计文件（recycler_list.xml 和 item_list.xml），用于实现标签下的列表效果。

（1）tabhost_main.xml 文件。

类别标签页设计文件，其中放置一个 TabHost，这个控件在 Android Studio 工具面板里就有，在设计模式下直接拖曳到界面上即可。系统自动生成代码框架如下：

```xml
<?xml version="1.0" encoding="utf-8"?>
<LinearLayout xmlns:android="http://schemas.android.com/apk/res/android"
    android:layout_width="match_parent"
    android:layout_height="match_parent">

    <TabHost
        android:id="@+id/myTabHost"                    //这是我们自己命名的
        android:layout_width="match_parent"
        android:layout_height="match_parent">

        <LinearLayout
            android:layout_width="match_parent"
            android:layout_height="match_parent"
            android:orientation="vertical">

            <TabWidget
                android:id="@android:id/tabs"
                android:layout_width="match_parent"
                android:layout_height="wrap_content" />

            <FrameLayout
                android:id="@android:id/tabcontent"
                android:layout_width="match_parent"
                android:layout_height="match_parent">

                <FrameLayout                           //标签页的具体内容置于框架布局内
                    android:layout_width="match_parent"
                    android:layout_height="match_parent"
                    android:id="@+id/frameLayout">
                </FrameLayout>
            </FrameLayout>
        </LinearLayout>
    </TabHost>
</LinearLayout>
```

将 TabHost 控件命名为 myTabHost，以便后面程序中引用。

> **注意：**
> 细心的读者会发现，上面这个代码框架与 4.2.1 小节标签栏的几乎完全一样！都包含名为 tabs 的 TabWidget 以及名为 tabcontent 的 FrameLayout，但 TabHost 与基于 TabActivity 的标签栏却是两种不同类型的控件，它们最大的区别就是 TabHost 支持用户自定义名称，而 TabActivity 却只能使用内部名称 tabhost，且编程时必须继承 TabActivity 父类。

(2) recycler_list.xml 文件。

循环视图的主界面 XML 文件,其上简单放置一个 RecyclerView 即可,过程略。

(3) item_list.xml 文件。

定义循环视图内列表项的具体布局和外观,用线性布局嵌套图像和文本视图的方式设计,过程略。

2. 开发数据模型与适配器

本例与之前的频道栏同样使用的循环视图实现,两者的数据模型和适配器程序结构也是大体一样的,只不过布局方式由网格改为了线性布局。

(1) 数据模型。

因为要在多个标签页上显示多组水果的列表,我们采用多组数组来分别存放每个标签页的数据。以下是数据模型 Fruit.java 的代码,为节省篇幅,这里只列出了其中一组数组的数据:

```java
public class Fruit {
    public int pid;                                            //水果图标的 id
    public String title;                                       //水果条目名称
    public String note;                                        //水果价格标注

    public Fruit(int pid, String title, String note) {
        this.pid = pid;
        this.title = title;
        this.note = note;
    }
    // "苹果/梨类" 标签页的数据
    ...
    // "莓果类" 标签页的数据
    private static int[] listPictureArray2 = {R.drawable.fruit_21, R.drawable.fruit_22, R.drawable.fruit_23};
    private static String[] listTitleArray2 = {
            "江苏红颜草莓 2 斤装", "智利红车厘子 JJ 级  900g", "秘鲁蓝莓 6 盒装"};
    private static String[] listNoteArray2 = {
            "¥89/箱",
            "¥159/箱",
            "¥89.9/箱"};
    public static ArrayList<Fruit> getDefaultList2() {
        ArrayList<Fruit> arrayList = new ArrayList<Fruit>();
        for (int i = 0; i < listPictureArray2.length; i++) {
            arrayList.add(new Fruit(listPictureArray2[i], listTitleArray2[i], listNoteArray2[i]));
        }
        return arrayList;
    }
    // "柑橘橙柚" 标签页的数据
    ...
}
```

当然,读者也可以尝试使用多维数组或者其他数据结构来存储数据。

(2) 适配器。

适配器 ListAdapter.java,代码如下:

```java
public class ListAdapter extends RecyclerView.Adapter<RecyclerView.ViewHolder> implements AdapterView.OnItemClickListener {
    private Context myContext;                                 //声明上下文
```

```java
    private LayoutInflater myInflater;                                      //布局填充器
    private ArrayList<Fruit> myFruitsList;                                  //水果类数据列表

    public ListAdapter(Context ctx, ArrayList<Fruit> lst) {
        myContext = ctx;
        myInflater = LayoutInflater.from(ctx);                              //从上下文获取布局
        myFruitsList = lst;                                                 //获取水果类数据
    }

    @Override
    public int getItemCount() {                                             //获得列表条目数
        return myFruitsList.size();
    }

    @Override
    public RecyclerView.ViewHolder onCreateViewHolder(ViewGroup vg, int viewType) {
        View view = myInflater.inflate(R.layout.item_list, vg, false);      //生成列表条目项的视图
        RecyclerView.ViewHolder vholder = new ItemHolder(view);
        return vholder;
    }

    @Override
    public void onBindViewHolder(RecyclerView.ViewHolder vh, final int pos) {
        ItemHolder iholder = (ItemHolder) vh;
        iholder.rclist_picture.setImageResource(myFruitsList.get(pos).pid); //由 id 获得条目前的图标
        iholder.rclist_title.setText(myFruitsList.get(pos).title);          //设置条目名称
        iholder.rclist_note.setText(myFruitsList.get(pos).note);            //设置价格标注

        //列表条目项的点击事件需要自己实现
        iholder.rclist_item.setOnClickListener(new OnClickListener() {
            @Override
            public void onClick(View v) {
                if (myOnItemClickListener != null) {
                    myOnItemClickListener.onItemClick(null, v, pos, 0);
                }
            }
        });
    }

    @Override
    public int getItemViewType(int pos) {
        return 0;
    }

    @Override
    public long getItemId(int pos) {
        return pos;
    }

    //该类负责对列表条目项中控件元素的设定操作
    public class ItemHolder extends RecyclerView.ViewHolder {
```

```java
            public LinearLayout rclist_item;                          //条目项中的线性布局
            public ImageView rclist_picture;                          //条目项前的图标
            public TextView rclist_title;                             //条目名称
            public TextView rclist_note;                              //条目价格标注

            public ItemHolder(View view) {
                super(view);
                rclist_item = view.findViewById(R.id.rclist_item);
                rclist_picture = view.findViewById(R.id.rclist_picture);
                rclist_title = view.findViewById(R.id.rclist_title);
                rclist_note = view.findViewById(R.id.rclist_note);
            }
        }

        private AdapterView.OnItemClickListener myOnItemClickListener;  //声明条目项点击事件监听器

        public void setOnItemClickListener(AdapterView.OnItemClickListener listener) {
            this.myOnItemClickListener = listener;                    //设置条目项的点击事件监听器
        }

        @Override
        public void onItemClick(AdapterView<?> arg0, View view, int pos, long arg3) {
            String msg = String.format("%s 加入购物车", myFruitsList.get(pos).title);
            Toast.makeText(myContext, msg, Toast.LENGTH_SHORT).show();
        }
    }
```

3. 开发标签页内容框架

从类别标签页设计文件 tabhost_main.xml 的代码框架中看到，标签页的具体内容是置于一个框架布局内的，故要开发对应该布局的一个框架类，在其中使用 RecyclerView 完成列表展示。本例框架布局类 FruitFragment.java，代码如下：

```java
public class FruitFragment extends Fragment {                         // (a)
    protected View myView;                                            // (b)
    private RecyclerView myRecyclerList;                              //循环视图对象引用
    protected Context myContext;                                      //声明上下文
    private int curgroup;                                             //类别标签编号（分类号）
    private ListAdapter adapter;                                      //适配器对象引用

    @Override
    public View onCreateView(LayoutInflater inflater, ViewGroup container, Bundle savedInstanceState) {
        myView = inflater.inflate(R.layout.recycler_list, container, false);   // (b)
        myRecyclerList = myView.findViewById(R.id.myRecyclerList);

        myContext = getActivity();
        LinearLayoutManager manager = new LinearLayoutManager(myContext);  // (c)
        manager.setOrientation(RecyclerView.VERTICAL);
        myRecyclerList.setLayoutManager(manager);

        curgroup = getArguments().getInt("group");                    //获得当前水果分类号
        if (curgroup == 0) adapter = new ListAdapter(myContext, Fruit.getDefaultList1());
        if (curgroup == 1) adapter = new ListAdapter(myContext, Fruit.getDefaultList2());
```

```
            if (curgroup == 2) adapter = new ListAdapter(myContext, Fruit.getDefaultList3());
            adapter.setOnItemClickListener(adapter);                //绑定条目项点击事件监听器
            myRecyclerList.setAdapter(adapter);                     //绑定适配器
            myRecyclerList.setItemAnimator(new DefaultItemAnimator());
            return myView;
        }
    }
```

其中:

(a) **public class FruitFragment extends Fragment**: 我们编写的框架布局类继承自 Android 的 Fragment, Fragment 是板块类, 它可以挂载到页面上, 也可以从页面剥离销毁, 有着类似于 Activity 的生命周期, 但比之 Activity 占用的系统资源更少、更灵活。

(b) **protected View myView;** 和 **myView = inflater.inflate(R.layout.recycler_list, container, false);**: 这个视图来自布局填充器, 根据 recycler_list.xml 生成, 显示特定类别标签页下的水果条目列表。

(c) **LinearLayoutManager manager = new LinearLayoutManager(myContext);**: 这里改用 RecyclerView 的 LinearLayoutManager (线性布局管理器) 生成垂直的列表布局。

4. 标签页切换控制

最后, 在主程序文件 RecyclerListActivity.java 中实现对不同类别标签页的切换控制功能, 代码如下:

```java
public class RecyclerListActivity extends AppCompatActivity {
    private TabHost myTabHost;                                      //Tab 导航栏控件引用
    @Override
    protected void onCreate(Bundle savedInstanceState) {
        super.onCreate(savedInstanceState);
        setContentView(R.layout.tabhost_main);
        myTabHost = findViewById(R.id.myTabHost);
        myTabHost.setup();                                          //装载 TabHost
        //创建各个类别标签                                           // (a)
        myTabHost.addTab(myTabHost.newTabSpec("tab_1").setIndicator("苹果/梨类", getResources().getDrawable(R.drawable.f1)).setContent(R.id.frameLayout));
        myTabHost.addTab(myTabHost.newTabSpec("tab_2").setIndicator("莓果类", getResources().getDrawable(R.drawable.f2)).setContent(R.id.frameLayout));
        myTabHost.addTab(myTabHost.newTabSpec("tab_3").setIndicator("柑橘橙柚", getResources().getDrawable(R.drawable.f3)).setContent(R.id.frameLayout));
        myTabHost.setCurrentTab(1);                                 // (b)
        previewFruit(1);                                            // (c)
        myTabHost.setOnTabChangedListener(new TabHost.OnTabChangeListener() {
            @Override
            public void onTabChanged(String tabId) {                //标签改变时切换页面
                switch (myTabHost.getCurrentTabTag()) {
                    case "tab_1":
                        previewFruit(0);                            //显示"苹果/梨类"页
                        break;
                    case "tab_2":
                        previewFruit(1);                            //显示"莓果类"页
                        break;
                    case "tab_3":
                        previewFruit(2);                            //显示"柑橘橙柚"页
                        break;
```

```
                    default:
                        break;
                }
            }
        });
    }

    public void previewFruit(int group) {                          // （c）
        FragmentManager manager = getSupportFragmentManager();     //创建板块管理类
        FragmentTransaction transaction = manager.beginTransaction(); //开启事务
        FruitFragment fruitFragment = new FruitFragment();         //新建"水果板块"
        Bundle bundle = new Bundle();
        bundle.putInt("group", group);
        fruitFragment.setArguments(bundle);                        //向板块传递水果分类号
        Fragment fragment = manager.findFragmentById(R.id.frameLayout);
        if (fragment == null) transaction.add(R.id.frameLayout, fruitFragment);   //挂载板块
        else transaction.replace(R.id.frameLayout, fruitFragment); //替换板块
        transaction.commit();                                      //提交事务
    }
}
```

其中：

（a）**myTabHost.addTab(myTabHost.newTabSpec("tab_1").setIndicator("苹果/梨类", getResources().getDrawable(R.drawable.f1)).setContent(R.id.frameLayout));** 等：创建类别标签使用的方法，与标签栏的一样，addTab()方法添加标签；newTabSpec(TAG)方法生成指定规格的标签，参数 TAG 是字符串形式，为类别标签的唯一标识；setIndicator()设置标签按钮的文字及图标资源，前一个参数是标签文字，后一个是其图标，如果前一个参数设为空字符串，则系统以图片作为标签页头，如果设置了标签文字内容，则默认只显示标签文字而隐去图标；最后 setContent()方法指明该标签页要显示的板块内容。

（b）**myTabHost.setCurrentTab(1);**：setCurrentTab(int)方法设置当前要显示的标签页，参数指定标签页的索引序号（从 0 开始）。

（c）**previewFruit(1);** 和 **public void previewFruit(int group)**：这是自定义的方法，根据用户点击 Tab 选择的类别，向"水果板块"传递 group 参数以显示不同类别水果的预览列表，在该方法中实现了对页面板块的挂载、替换等管理操作。

4.4 整合为完整 App

至此，"易果鲜超市" App 所需的全部界面 UI 元素我们都已经分别单独开发完成了。接下来要做的工作就是将它们整合在一起，成为一个完整的 App。

整合思路：

（1）将原来各个分立项目中的.java 源文件、界面设计 XML 文件、res 资源文件复制进单独的一个工程中，便于整合引用。

（2）以【例 4.1】的标签栏作为基础，在其之上进行整合，将它的每个标签内容页替换成实现相应功能例子工程的主 Activity 类。

（3）对于"首页"来说，由于其上需要同时集成轮播条、频道栏和列表视图，我们以轮播条实例工程作为基准（当然读者也可以选其他两个），修改其 Activity 主程序（BannerActivity），将频道栏和列表视图的功能代码添加到其中，同时修改相应的界面设计 XML 文件，添置集成的控件。

下面详细介绍整合的步骤和方法。

4.4.1 界面元素集成

创建一个空的 Android Studio 工程 EasyFruit，在其中安装好 RecyclerView。

1. 整理资源

对之前开发的各个工程的资源进行统一归档整理，这里笔者建议大家建立 3 个文件夹（java、layout、res）：java 文件夹下集中存放所有的 .java 源文件；layout 文件夹下存放所有界面设计的 XML 文件；res 文件夹下则存放所有资源（即各工程 drawable 目录中的内容）。在归档之前要仔细检查文件有没有重名，若有，要改成不一样的名称以避免整合在一起时发生冲突。下面将各个工程中需要重命名的文件以及新的名称列于表 4.1。

表 4.1 需要重命名的文件

所 在 工 程	原 文 件	重命名后的文件
MyRecyclerGrid	Fruit.java	Fruit_RG.java
MyRecyclerGrid	item_grid.xml	item_rcgrid.xml
MyListView	Fruit.java	Fruit_LV.java
MyListView	FruitAdapter.java	FruitAdapter_LV.java
MyGridView	Fruit.java	Fruit_GV.java
MyGridView	FruitAdapter.java	FruitAdapter_GV.java
MyRecyclerList	Fruit.java	Fruit_RL.java
MyRecyclerList	item_list.xml	item_rclist.xml

在对文件改名的时候，最好是打开工程在 Android Studio 环境下进行。右击要改名的文件，在弹出的快捷菜单中选择 "Refactor" → "Rename" 菜单项，在弹出的对话框中输入新的文件名，如图 4.12 所示。系统会出现 "Rename Variables" 窗口，其中罗列出此次更名所涉及工程中所有的引用变量，选中后单击 "OK" 按钮，将更名同步至整个工程，这样就无须人工逐一去搜索需要修改的地方，既节省时间也减少了出错的可能。

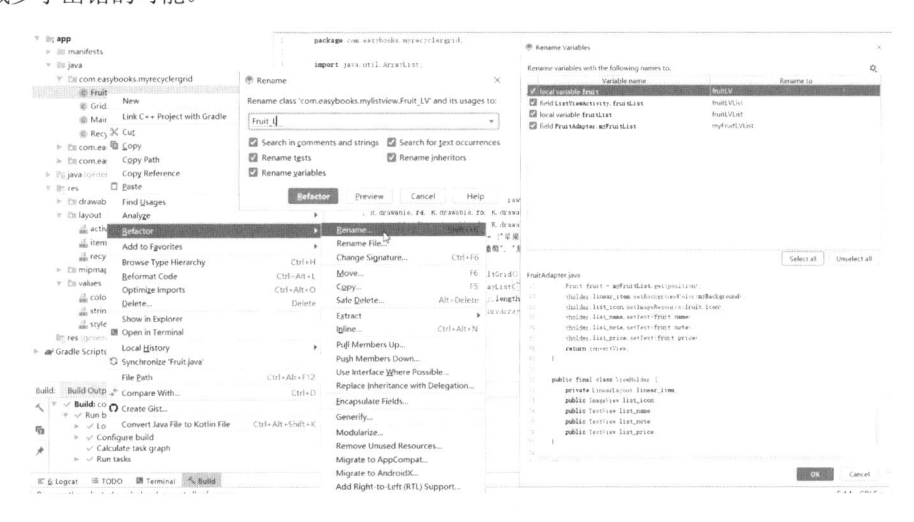

图 4.12 使用 Android Studio 同步更名引用变量

2. 复制资源

将归档整理好的文件复制到工程 EasyFruit 的各对应目录下,其中 java 文件夹中的源程序文件复制到 java\com\easybooks\easyfruit 目录;layout 文件夹下的 XML 文件复制到 layout 目录;res 文件夹下的资源文件复制到 drawable 目录。复制完之后还要做下面两件事:

(1)对所有的.java 源文件修改其 package 包路径为 com.easybooks.easyfruit。

(2)检查 layout 下所有 XML 设计文件,对于那些需要写出完整包路径全名定义的控件元素,要将其包路径名改为当前项目对应的包路径名,如图 4.13 所示是对轮播条控件的修改,Android Studio 会自动给出输入提示。

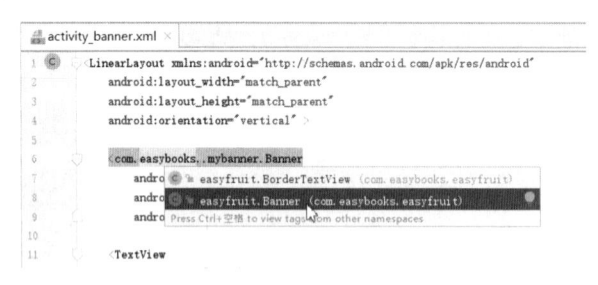

图 4.13 修改控件的包路径名

3. 整合代码

本例集成 App 时,以标签栏为基准,需要修改添加以下几处代码。

(1)TabHostActivity.java 文件。

在其中添加标签页内容,替换为对应的 UI 组件(加粗处),内容如下:

```
public class TabHostActivity extends TabActivity implements OnClickListener {
    ......
    @Override
    protected void onCreate(Bundle savedInstanceState) {
        ...
        myTabHost = getTabHost();
        myTabHost.addTab(myTabHost.newTabSpec(TAG1).setIndicator(getString(R.string.first_tab),
getResources().getDrawable(R.drawable.first_tab_selector)).setContent(new Intent(this, BannerActivity.class)));
                                                            //集成轮播条(MyBanner)
        myTabHost.addTab(myTabHost.newTabSpec(TAG2).setIndicator(getString(R.string.second_tab),
getResources().getDrawable(R.drawable.second_tab_selector)).setContent(new Intent(this, RecyclerListActivity.class)));
                                                            //集成类别标签列表(MyRecyclerList)
        myTabHost.addTab(myTabHost.newTabSpec(TAG3).setIndicator(getString(R.string.third_tab),
getResources().getDrawable(R.drawable.third_tab_selector)).setContent(new Intent(this, GridViewActivity.class)));
                                                            //集成网格视图(MyGridView)
        myTabHost.addTab(myTabHost.newTabSpec(TAG4).setIndicator(getString(R.string.four_tab),
getResources().getDrawable(R.drawable.four_tab_selector)).setContent(new Intent(this, FourPageActivity.class)));
        myTabHost.addTab(myTabHost.newTabSpec(TAG5).setIndicator(getString(R.string.five_tab),
getResources().getDrawable(R.drawable.five_tab_selector)).setContent(new Intent(this, FivePageActivity.class)));
        setTabView(first_linear);
    }
    ......
}
```

可见，只要将原来用于占位的内容页 Activity 替换为对应控件元素的主 Activity 类即可。

（2）BannerActivity.java 文件。

因为"首页"上要集成轮播条、频道栏和列表视图这 3 个控件，其中轮播条 Activity 作为"首页"的内容已经集成在 TabHostActivity.java 中了，频道栏和列表视图则必须放到轮播条本身的 Activity 中才能实现"首页"布局效果，为此修改轮播条的 BannerActivity.java 代码如下：

```
public class BannerActivity extends AppCompatActivity {
    private Banner myBanner;
    private RecyclerView myRecyclerGrid;                //集成频道栏（MyRecyclerGrid）
    private ListView myListView;                        //集成列表视图（MyListView）
    private ArrayList<Fruit_LV> fruitLVList;
    @Override
    protected void onCreate(Bundle savedInstanceState) {
        super.onCreate(savedInstanceState);
        setContentView(R.layout.activity_banner);
        ......
        myBanner.start();
        //集成频道栏（MyRecyclerGrid）
        myRecyclerGrid = findViewById(R.id.myRecyclerGrid);
        GridLayoutManager manager = new GridLayoutManager(this, 5);
        myRecyclerGrid.setLayoutManager(manager);
        GridAdapter gadapter = new GridAdapter(this, Fruit_RG.getDefaultGrid());
        gadapter.setOnItemClickListener(gadapter);
        myRecyclerGrid.setAdapter(gadapter);
        myRecyclerGrid.setItemAnimator(new DefaultItemAnimator());
        //集成列表视图（MyListView）
        fruitLVList = Fruit_LV.getDefaultList();
        FruitAdapter_LV fadapter = new FruitAdapter_LV(this, R.layout.item_list, fruitLVList, Color.WHITE);
        myListView = findViewById(R.id.myListView);
        myListView.setAdapter(fadapter);
        myListView.setOnItemClickListener(fadapter);
        myListView.setHeaderDividersEnabled(true);
        myListView.setFooterDividersEnabled(true);
        myListView.setDividerHeight(3);
        myListView.setPadding(25, 10, 25, 0);
    }
    ......
}
```

可以看到，这里是将原频道栏 RecyclerGridActivity 与列表视图 ListViewActivity 中的功能放到轮播条 BannerActivity 中，这样原来的 RecyclerGridActivity.java 与 ListViewActivity.java 这两个源文件就没用了，可以将它们从项目工程中删除。

（3）activity_banner.xml 文件。

由于现在将频道栏和列表视图都显示在轮播条页上，故要对原轮播条界面设计文件 activity_banner.xml 进行修改，添加这两个控件：

```
<LinearLayout xmlns:android="http://schemas.android.com/apk/res/android"
    android:layout_width="match_parent"
    android:layout_height="match_parent"
    android:orientation="vertical" >
```

```xml
    <com.easybooks.easyfruit.Banner
        android:id="@+id/myBanner"
        ... />

    <androidx.recyclerview.widget.RecyclerView
        android:id="@+id/myRecyclerGrid"
        android:layout_width="match_parent"
        android:layout_height="wrap_content" />              <!--集成频道栏（MyRecyclerGrid）-->

    <ListView
        android:id="@+id/myListView"
        android:layout_width="match_parent"
        android:layout_height="wrap_content" />              <!--集成列表视图（MyListView）-->
</LinearLayout>
```

4. 配置 Activity

最后，不要忘了在工程 AndroidManifest.xml 中配置所有的 Activity 以及设置启动页，内容如下：

```xml
<?xml version="1.0" encoding="utf-8"?>
<manifest xmlns:android="http://schemas.android.com/apk/res/android"
    package="com.easybooks.easyfruit">
    <application
        ......
        <activity android:name=".TabHostActivity">                //启动页（标签栏）
            <intent-filter>
                <action android:name="android.intent.action.MAIN" />
                <category android:name="android.intent.category.LAUNCHER" />
            </intent-filter>
        </activity>
        <activity android:name=".BannerActivity" />              //轮播条
        <activity android:name=".RecyclerListActivity" />        //类别标签列表
        <activity android:name=".GridViewActivity" />            //网格视图
        <activity android:name=".FourPageActivity" />            //占位页
        <activity android:name=".FivePageActivity" />            //占位页
    </application>
</manifest>
```

由于频道栏和列表视图的功能已经集成于首页（轮播条）上，并没有独立的 Activity，故这里无须专门配置。

至此，整合工作全部完成。运行工程 EasyFruit，我们就得到了一个完整的"易果鲜超市"App，效果如图 4.14 所示。

4.4.2 通知消息计数

现在很多 App 都有通知消息计数功能，这种功能最初应用在微信消息提示，现已被许多购物网站广泛采用。如图 4.15 所示，左边是京东 App 界面，当顾客选中某件商品加入购物车后，标签栏购物车图标右上角会显示一个带数字的通知圆点，这样用户无须切换到购物车页也可以一目了然地知道自己选了几件商品。我们在"易果鲜超市"App 中也来开发这样一个通知消息计数功能，如图 4.15 右边所示。

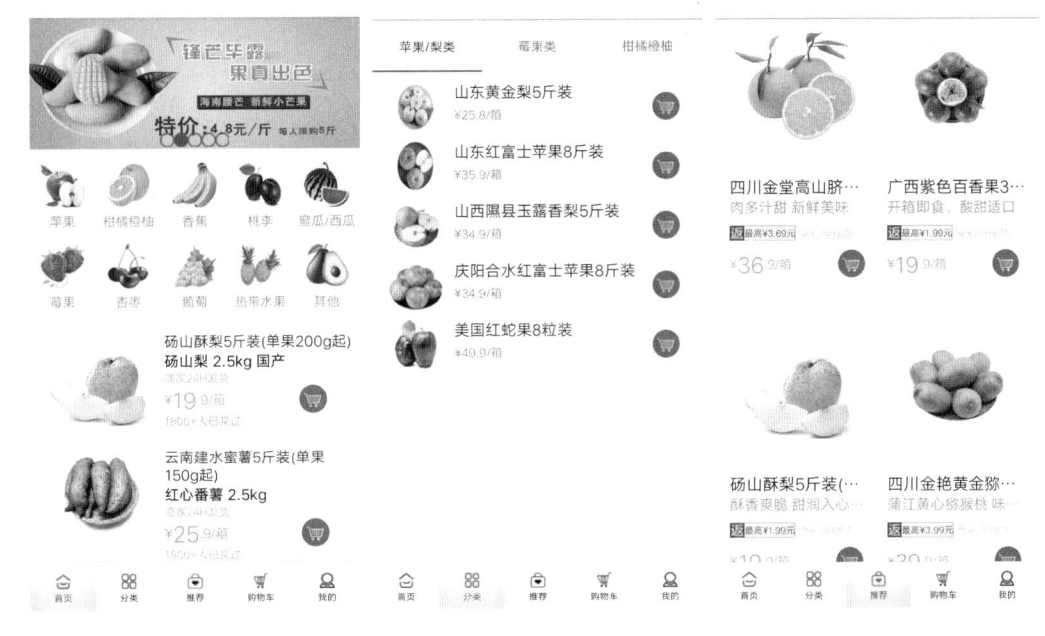

图 4.14 完整的"易果鲜超市"App 运行效果

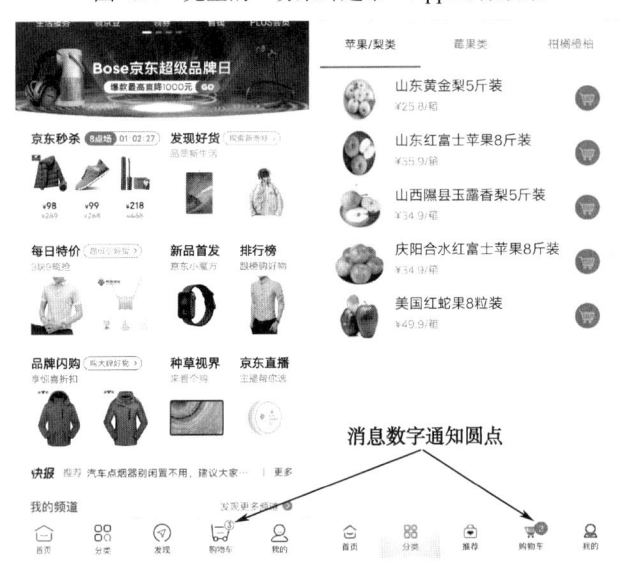

图 4.15 数字消息通知圆点

实现思路：

（1）该功能采用自定义视图的方式实现。在工程 EasyFruit 中定义视图类 NotePointView，在其中实现数字圆点绘制及计数功能。

（2）外部程序使用这个视图并通过 Android 系统的线程消息通知句柄机制，实时地刷新圆点内的数字。

1. 自定义视图

消息数字通知圆点的视图 NotePointView.java，代码如下：

```
public class NotePointView extends AppCompatTextView {
    private boolean initBgFlag;
```

```java
        private int backgroundColor = Color.parseColor("#f43530");        //填充颜色
        public NotePointView(Context context, AttributeSet attrs) {
            super(context, attrs);
            setGravity(Gravity.CENTER);                                    //数字居中显示
            getViewTreeObserver().addOnPreDrawListener(new ViewTreeObserver.OnPreDrawListener() {
                @Override
                public boolean onPreDraw() {
                    if (!initBgFlag) {
                        StateListDrawable bg = new StateListDrawable();
                        GradientDrawable gradientStateNormal = new GradientDrawable();
                        gradientStateNormal.setColor(backgroundColor);     //设置背景色
                        gradientStateNormal.setShape(GradientDrawable.RECTANGLE);
                        gradientStateNormal.setCornerRadius(50);           //设置圆角弧度
                        gradientStateNormal.setStroke(0, 0);
                        bg.addState(View.EMPTY_STATE_SET, gradientStateNormal);
                        NotePointView.this.setBackgroundDrawable(bg);      //bg 为一个 StateListDrawable 对象
                        initBgFlag = true;
                        return false;
                    }
                    return true;
                }
            });
        }
    }
```

2. 主程序引用视图

在标签栏主程序 TabHostActivity.java 中使用这个视图，代码如下：

```java
public class TabHostActivity extends TabActivity implements OnClickListener {
    ......
    //集成消息数字通知圆点
    public static int count = 0;                                           //消息计数值
    private NotePointView myNotePointView;                                 //通知圆点视图的引用

    @Override
    protected void onCreate(Bundle savedInstanceState) {
        super.onCreate(savedInstanceState);
        setContentView(R.layout.tab_host);
        ...
        setTabView(first_linear);
        //集成消息数字通知圆点
        myNotePointView = findViewById(R.id.note_num);
        if (count == 0) {
            myNotePointView.setVisibility(View.GONE);                      //用户选购数为 0 时圆点不可见
        } else {
            myNotePointView.setVisibility(View.VISIBLE);
            myNotePointView.setText(String.valueOf(count));                //显示圆点计数值
        }
        new Thread(new Runnable() {                                        //UI 刷新要通过线程消息句柄
            @Override
            public void run() {
                while (true) {
```

```java
                    try {
                        Thread.sleep(500);
                        Message msg = Message.obtain();
                        msg.what = 1000;
                        msg.obj = "note";
                        myHandler.sendMessage(msg);
                    } catch (InterruptedException e) {
                    }
                }
            }
        }).start();
    }
    ......
    private Handler myHandler = new Handler() {
        public void handleMessage(Message message) {      //收到句柄消息后刷新圆点数字
            if (count == 0) {
                myNotePointView.setVisibility(View.GONE);
            } else {
                myNotePointView.setVisibility(View.VISIBLE);
                myNotePointView.setText(String.valueOf(count));
            }
        }
    };
}
```

当用户在"分类"页上点选水果条目时，为使圆点数值增加，需要在条目的点击事件处理中改变 TabHostActivity 中的 count 值，语句写在 ListAdapter.java 中，内容如下：

```java
public class ListAdapter extends RecyclerView.Adapter<RecyclerView.ViewHolder> implements AdapterView.OnItemClickListener {
    private Context myContext;
    private LayoutInflater myInflater;
    private ArrayList<Fruit_RL> myFruitsList;
    ......
    @Override
    public void onItemClick(AdapterView<?> arg0, View view, int pos, long arg3) {
        String msg = String.format("%s 加入购物车", myFruitsList.get(pos).title);
        Toast.makeText(myContext, msg, Toast.LENGTH_SHORT).show();
        TabHostActivity.count++;                          //集成消息数字通知圆点
    }
}
```

3. 在界面上放置视图

最后，将圆点视图放置到 UI 界面上显示，需要修改 tab_host.xml，在其中对应"购物车"标签的线性布局元素中嵌套一个相对布局（RelativeLayout），里面放文本视图和圆点视图即可，代码如下：

```xml
<TabHost xmlns:android="http://schemas.android.com/apk/res/android"
    android:id="@android:id/tabhost"
    ... >
    <RelativeLayout
        ... >
        <FrameLayout
            android:id="@android:id/tabcontent"
```

计数界面代码

```xml
        ... />
        <TabWidget
            android:id="@android:id/tabs"
            ... />
    <LinearLayout
            ...
            android:gravity="bottom"
            android:orientation="horizontal" >
        ......
        <LinearLayout                                              // "购物车"标签页
            android:id="@+id/four_linear"
            ...
            android:orientation="vertical" >
            <RelativeLayout
                android:layout_width="match_parent"
                android:layout_height="match_parent" >
                <TextView
                    style="@style/TabLabel"
                    android:drawableTop="@drawable/four_tab_selector"
                    android:text="@string/four_tab" />
                <com.easybooks.easyfruit.NotePointView            //放置我们自定义的圆点视图
                    android:id="@+id/note_num"
                    android:layout_width="16dp"
                    android:layout_height="16dp"
                    android:layout_gravity="center"
                    android:layout_marginTop="2dp"
                    android:layout_marginLeft="40dp"
                    android:textColor="@android:color/white"
                    android:textSize="10sp"
                    android:visibility="invisible" />             //初始默认为不可见
            </RelativeLayout>
        </LinearLayout>
        ......
    </LinearLayout>
    </RelativeLayout>
</TabHost>
```

如此，就实现了购物车通知消息计数功能。运行程序，在"分类"页上点选水果条目项，购物车右上角出现红色带数字的圆点且数值动态增加计数，效果如图 4.15 右边所示。

本章介绍了当前移动 App 开发中常用界面元素效果的实现技术。在实际开发中若能综合运用，就能制作出复杂、美观、酷炫多彩的 App 界面出来。希望读者在今后的开发中能灵活应用。

第 5 章 Android 服务与广播程序设计

5.1 Service（服务）程序设计

5.1.1 Service 概述

因为手机屏幕尺寸的限制，通常 Android 系统仅允许一个应用程序处于激活状态并显示在前端手机屏上。但是，系统同样需要有程序能处理其他一些任务（如提供某些服务、界面更新或处理意外事件），因此要有一种机制允许在没有用户界面的情况下，使程序能够长时间地运行于后台。例如，音乐播放器软件希望在关闭界面后仍能保持音乐持续播放。为此，Android 提供了 Service（服务）组件。

1. Service 基本概念

Service 是一个没有用户界面，也不直接与用户交互，却能够长期在后台运行的应用组件。因为没有界面，占用系统资源少，而且 Service 比 Activity 具有更高的优先级，即使在系统资源紧张的情况下也不会被轻易终止，故可以认为它是一种能在系统中永久运行的组件。Service 除了可以实现后台服务功能，还能用于进程间通信（IPC 机制），解决不同应用程序进程之间的调用和通信问题。

2. Service 基类方法

要创建一个 Service，先要创建一个 Java 类，扩展 Service 基类或其子类。Service 基类定义了各种方法，如表 5.1 所示。

表 5.1 Service 基类的方法

方　　法	说　　明
onStartCommand()	当另一个组件（如 Activity）调用 startService()方法请求服务启动时调用。若开发人员实现该方法，则需要在任务完成时调用 stopSelf()或 stopService()方法停止服务；若仅想提供绑定，则不必实现该方法
onBind()	当其他组件通过 bindService()绑定服务时调用。在该方法的实现中，必须提供客户端与服务通信的接口。该方法必须实现，但如果不允许绑定，则返回 null
onCreate()	当服务第一次创建时调用。如果服务已在运行，该方法不被调用
onDestroy()	当服务不再使用并将被销毁时调用。在该方法内清理线程注册的监听器、接收器等资源

3. Service 创建与注册

创建一个 Android Studio 工程，在其中创建一个继承自 Service 类的子类 LocalService，该类有一个抽象方法 onBind()，必须在子类中实现。

Service 需要在 AndroidManifest.xml 中进行注册，注册类 LocalService 的语句如下：

`<service android: name=".LocalService"/>`

该语句位于 application 节点内，与 Activity 位于同一层次。

4. Service 启动流程

Service 有启动和绑定两种使用方式。

（1）启动方式。

该方式通过 startService()方法启动服务，一旦启动，Service 就在后台运行，即使启动它的组件已被销毁也不受影响。这种方式下启动 Service 的组件**不能**获取 Service 的对象实例，因此无法调用 Service 的方法获取 Service 中的任何状态和数据信息。因此，以启动方式使用的 Service 必须具备自管理能力，例如下载或上传一个文件，当操作完成时 Service 能够停止它本身。该方式下的 Service 工作流程如图 5.1 所示。

说明：

① 如果 Service 还没有运行，则先调用 onCreate()，然后调用 onStart()；如果 Service 已经运行，则只调用 onStart()。所以，一个 Service 的 onStart()方法可能会重复调用多次。

② 如果直接退出而没有调用 stopService()方法，Service 会一直在后台运行，它的调用者在启动后可以通过 stopService()来关闭 Service。

（2）绑定方式。

该方式通过 bindService()方法启动服务，绑定 Service 的组件通过服务链接（Connection）使用 Service，服务链接能够获取 Service 的对象实例，因此可以调用 Service 的方法直接获取 Service 中的状态和数据信息，可以向 Service 发送请求、获取返回结果，还可以通过跨进程通信（IPC）来交互。但是，这种方式下的 Service 只有在与应用组件绑定后才能运行。同一个 Service 可以绑定多个服务链接，这样就可以同时为多个不同的组件提供服务。使用 Service 的组件通过 bindService()建立服务链接，如果在绑定过程中 Service 没有启动，bindService()方法会自动启动 Service；通过 unbindService()方法停止服务链接，unbind()方法销毁 Service。该方式下的 Service 工作流程如图 5.2 所示。

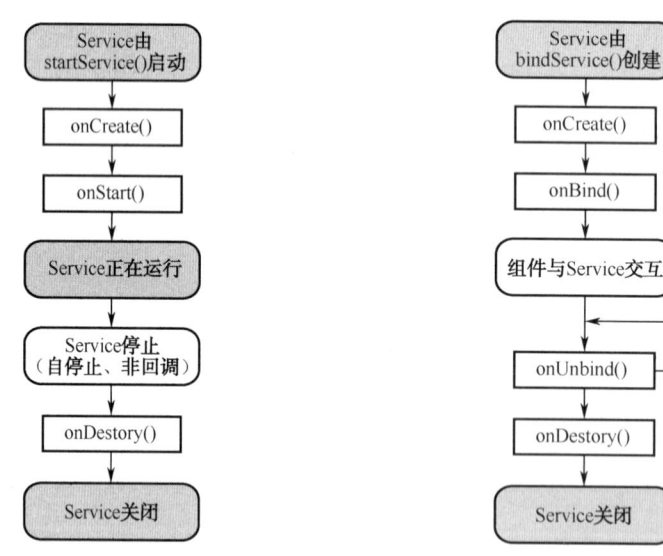

图 5.1　启动方式 Service 工作流程

图 5.2　绑定方式 Service 工作流程

说明：

① onBind()方法返回给调用者组件一个 IBind 接口实例，允许调用者组件回调 Service 中的方法，从而得到 Service 的实例对象、运行状态、内部数据或进行其他与 Service 的互动操作。这时候调用者组件会与 Service 绑定在一起，调用者组件一旦退出，Service 也会调用自身的 onUnbind()→onDestroy() 随之退出。

② 在 Service 每一次的开启、关闭过程中，只有 onStart 可被多次调用（通过多次 startService 调用），其他 onCreate、onBind、onUnbind、onDestory 在一个生命周期中只能被调用一次。

以上两种方式并不是完全独立的，在某些情况下需要结合使用。还是以音乐播放器为例，开启播放某个音乐文件是通过 startService()以启动方式执行的，但在播放的过程中如果用户想暂停，则要以绑定方式通过 bindService()方法获取服务链接和 Service 的对象实例，才能进而调用对象实例中的方法来暂停播放过程。但如果只是想要启动一个程序在后台长期执行某项任务，那么使用启动方式就可以了。

5.1.2　启动方式使用 Service

【例 5.1】　本例演示用启动方式使用服务的过程。在服务运行过程中，主程序无法直接（只能通过句柄提交线程 Runnable 接口）与服务进行交互，两者是相对独立的。

1．设计界面

创建 Android Studio 工程 CircleStartService，设计界面如图 5.3 所示。

图 5.3　设计界面

界面上放置"开始"和"停止"两个按钮，用户通过它们启停服务，中央区的 FrameLayout 用于显示后台服务所绘出的圆。

content_main.xml 文件代码如下：

```
<?xml version="1.0" encoding="utf-8"?>
<RelativeLayout xmlns:android="http://schemas.android.com/apk/res/android"…>

    <LinearLayout
        android:orientation="vertical"
        android:layout_width="match_parent"
        android:layout_height="match_parent">

        <LinearLayout
            android:orientation="horizontal"
            android:layout_width="match_parent"
            android:layout_height="wrap_content">
```

```xml
<TextView
    android:layout_width="wrap_content"
    android:layout_height="wrap_content"
    android:text="生成随机圆"
    android:id="@+id/textView"
    android:textSize="@dimen/abc_text_size_large_material"
    android:textColor="@android:color/black" />

<Button
    android:layout_width="wrap_content"
    android:layout_height="wrap_content"
    android:text="开始"
    android:id="@+id/buttonStart"
    android:textSize="@dimen/abc_text_size_medium_material"
    android:onClick="start" />

<Button
    android:layout_width="wrap_content"
    android:layout_height="wrap_content"
    android:text="停止"
    android:id="@+id/buttonStop"
    android:textSize="@dimen/abc_text_size_medium_material"
    android:onClick="stop" />
</LinearLayout>

<FrameLayout                                           //用于显示后台服务所绘出的圆
    android:layout_width="match_parent"
    android:layout_height="330dp"
    android:id="@+id/circleFrame"></FrameLayout>

<LinearLayout
    android:orientation="horizontal"
    android:layout_width="match_parent"
    android:layout_height="match_parent">

    <TextView
        android:layout_width="wrap_content"
        android:layout_height="wrap_content"
        android:text="半径："
        android:id="@+id/textRadius"
        android:textSize="@dimen/abc_text_size_medium_material"
        android:textColor="@android:color/black" />
    </LinearLayout>
  </LinearLayout>
</RelativeLayout>
```

2. 创建服务

实现思路：

创建服务类 DrawCircle，该服务以线程的形式在后台运行。先执行 onCreate()方法创建线程对象，接着创建 MyCircle 对象（该对象的 onDraw()方法完成画圆功能）；接着用 onStartCommand()方法启动线程。在 Runnable 的 run()方法中，调用主程序 MainActivity 的 updateGUI()方法刷新界面。
DrawCircle.java 代码如下：

```java
public class DrawCircle extends Service {
    private Thread drawThread;                                  //声明线程对象
    private View circle;                                        //声明 MyCircle 对象
    private int radius;                                         //圆半径
    private static final String TAG = "绘圆服务（DrawCircle）";  //标识该服务

    @Override
    public IBinder onBind(Intent intent) {                      //服务必须实现的方法
        return null;
    }

    private Runnable drawWork = new Runnable() {                // (a)
        @Override
        public void run() {
            try {
                while (!drawThread.isInterrupted()) {
                    radius = (int) (Math.random() * 400);
                    MainActivity.updateGUI(circle, radius);
                    drawThread.sleep(1000);
                }
            } catch (InterruptedException e) {
                e.printStackTrace();
            }
        }
    };

    @Override
    public void onCreate() {                                    // (b)
        super.onCreate();
        drawThread = new Thread(null, drawWork, "DrawCircle");
        circle = new MyCircle(this);
        Log.i(TAG, "onCreate");
    }

    @Override
    public int onStartCommand(Intent intent, int flags, int startId) {  //启动服务
        super.onStartCommand(intent, flags, startId);
        if (!drawThread.isAlive()) drawThread.start();
        Log.i(TAG, "onStartCommand");
        return START_STICKY;
    }

    @Override
    public void onDestroy() {                                   //销毁服务
        super.onDestroy();
        drawThread.interrupt();
        Log.i(TAG, "onDestroy");
    }
}
```

```
public class MyCircle extends View {                    // (c)
    public MyCircle(Context context) {
        super(context);
    }

    @Override
    protected void onDraw(Canvas canvas) {
        super.onDraw(canvas);
        Paint paint = new Paint();
        canvas.drawCircle(400, 400, radius, paint);
    }
}
```

其中：

（a）**private Runnable drawWork = new Runnable(){ ... }**：创建线程运行时要执行的方法，通过重写该接口的 run()方法，在 while 循环中线程不断地随机生成 0～400 之间数值的圆半径，调用主程序 updateGUI()刷新界面，每刷新一次，线程会休眠 1000 毫秒后继续。

（b）**public void onCreate()**：这是服务生命周期开始执行的第一个方法，在其中创建服务线程及绘圆类对象。

（c）**public class MyCircle extends View**：绘圆类 MyCircle 继承自 View 类，通过重写其 onDraw()方法来实现自定义画圆的功能，画圆通过调用 Canvas（画布）对象的 drawCircle()方法实现。

3. 主程序启动

实现思路：

在主程序的 updateGUI()方法中通过 Handler（句柄）对象的 post()方法提交界面刷新，而界面刷新的具体操作在 Runnable 接口 refreshWork 的 run()方法中执行。

MainActivity.java 代码如下：

```
public class MainActivity extends AppCompatActivity {
    private static FrameLayout circleFrame;
    private static TextView textRadius;
    private static Handler handler = new Handler();        //界面刷新句柄
    private static View mycircle;                          //圆视图对象
    private static int myradius;                           //圆半径

    @Override
    protected void onCreate(Bundle savedInstanceState) {
        super.onCreate(savedInstanceState);
        setContentView(R.layout.activity_main);
        findViews();
        …
    }

    private void findViews() {
        circleFrame = findViewById(R.id.circleFrame);
        textRadius = findViewById(R.id.textRadius);
    }
```

```java
    public static void updateGUI(View c, int r) {                    // (a)
        mycircle = c;
        myradius = r;
        handler.post(refreshWork);
    }

    private static Runnable refreshWork = new Runnable() {           // (b)
        @Override
        public void run() {
            //刷新界面
            circleFrame.removeAllViews();                            // (b)
            circleFrame.addView(mycircle);                           // (b)
            textRadius.setText("半径：" + String.valueOf(myradius));  // (b)
        }
    };

    public void start(View view) {                                   //开启服务
        Intent intent = new Intent(this, DrawCircle.class);
        startService(intent);
    }

    public void stop(View view) {                                    //停止服务
        Intent intent = new Intent(this, DrawCircle.class);
        stopService(intent);
    }
    …
}
```

其中：

（a）**public static void updateGUI(View c, int r)**：静态方法 updateGUI()负责刷新界面。它主要完成两个工作：一是获取服务中的圆视图对象及半径值；二是通过 Handler 提交到线程 Runnable 接口中。

（b）**private static Runnable refreshWork = new Runnable(){ ... };**、**circleFrame.removeAllViews();**、**circleFrame.addView(mycircle);**和 **textRadius.setText("半径：" + String.valueOf(myradius));**：在线程 Runnable 接口中完成刷新的任务。要做以下三件事：一是用 FrameLayout 的 removeAllViews()方法清除界面上旧有的圆；二是用 addView()方法添加线程新绘制的圆；三是用 setText()方法在界面上显示圆的半径。

4．服务配置

在 AndroidManifest.xml 中，将开发的服务（DrawCircle）配置其中（加粗处），代码如下：

```xml
<?xml version="1.0" encoding="utf-8"?>
<manifest xmlns:android="http://schemas.android.com/apk/res/android"
    package="com.easybooks.android.circlestartservice">
    <application
        …
        <activity
            android:name=".MainActivity"
            …
        </activity>
```

```
<service android:name=".DrawCircle"/>
    </application>
</manifest>
```

5. 运行效果

运行程序，点击"开始"按钮启动服务线程，在屏幕中央每隔 1000 毫秒会出现一个不同大小的圆，左下角对应显示该圆的半径值。点击"停止"按钮关闭服务，绘圆的过程就此终止，界面停住不再有变化，效果如图 5.4 所示。

与此同时，在开发环境底部的日志窗口还可观察到服务启停整个生命周期中所调用的方法，如图 5.5 所示。

图 5.4　启动绘圆服务运行效果

图 5.5　服务启停整个生命周期中所调用的方法

5.1.3　绑定方式使用 Service

【例 5.2】　本例演示用绑定方式使用服务的过程。与启动方式最大的不同在于，主程序可随时直接主动去获取服务中的数据，无须依赖于 Runnable 接口。此方式适用于主程序与服务需要频繁交互和共享数据的场合。

1. 设计界面

创建 Android Studio 工程 CircleBindService，设计界面如图 5.6 所示。与上例的界面布局基本相同，只是在底部加了一个"计算"按钮以及显示圆周长和面积的 TextView，设计代码略。

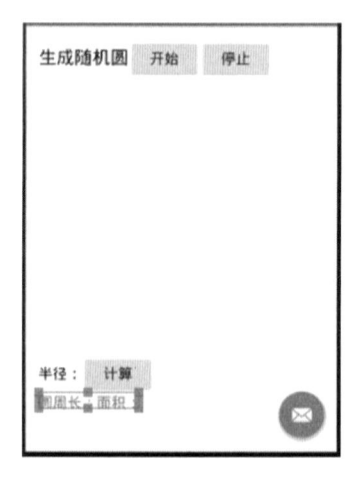

图 5.6　设计界面

界面完整代码

2. 创建服务

实现思路：

创建服务类 DrawCircle，该服务以线程的形式在后台运行。先执行 onCreate()方法创建线程对象，接着创建 MyCircle 对象（该对象的 onDraw()方法完成画圆功能）；接着 onBind()方法绑定线程。在 Runnable 的 run()方法中，调用主程序 MainActivity 的 updateGUI()方法刷新界面。

DrawCircle.java 代码如下：

```java
public class DrawCircle extends Service {
    private Thread drawThread;                                    //声明线程对象
    private View circle;                                          //声明 MyCircle 对象
    private int radius;                                           //圆半径
    private static final String TAG = "绘圆服务（DrawCircle）";    //标识该服务

    private Runnable drawWork = new Runnable() {
        @Override
        public void run() {
            try {
                while (!drawThread.isInterrupted()) {
                    radius = (int) (Math.random() * 400);
                    MainActivity.updateGUI(circle, radius);
                    drawThread.sleep(1000);
                }
            } catch (InterruptedException e) {
                e.printStackTrace();
            }
        }
    };

    @Override
    public void onCreate() {
        super.onCreate();
        drawThread = new Thread(null, drawWork, "DrawCircle");
        circle = new MyCircle(this);
        Log.i(TAG, "onCreate");
    }

    public class CalcuCircletor extends Binder {                  // (a)
        DrawCircle getService() {
            return DrawCircle.this;
        }
    }

    @Override
    public IBinder onBind(Intent intent) {                        // (b)
        if (!drawThread.isAlive()) drawThread.start();
        Log.i(TAG, "onBind");
        return new CalcuCircletor();
    }

    @Override
    public boolean onUnbind(Intent intent) {                      // (c)
        drawThread.interrupt();
        Log.i(TAG, "onUnbind");
        return super.onUnbind(intent);
    }

    public class MyCircle extends View {
```

```
            public MyCircle(Context context) {
                super(context);
            }

            @Override
            protected void onDraw(Canvas canvas) {
                super.onDraw(canvas);
                Paint paint = new Paint();
                canvas.drawCircle(400, 400, radius, paint);
            }
        }
        /**
         * 公开提供给服务调用者使用的方法                          // (d)
         */
        public float getCircu() {                              // (d)
            return (float) (2 * Math.PI * radius);             //计算圆周长
        }

        public float getArea() {                               // (d)
            return (float) (Math.PI * radius * radius);        //计算圆面积
        }
    }
```

其中：

（a）**public class CalcuCircletor extends Binder**：CalcuCircletor 继承自 Binder，在绑定方式使用的服务中必须实现这个类，它的作用在于以对象形式返回服务本身，以供主程序调用服务中对外公开的方法。

（b）**public IBinder onBind(Intent intent)**：onBind()方法不仅以绑定方式开启服务，还返回给主程序一个 CalcuCircletor 对象，主程序就是通过它获取服务对象的。

（c）**public boolean onUnbind(Intent intent)**：onUnbind()方法解除绑定，即终止服务线程。

（d）**public float getCircu(){ ... }** 和 **public float getArea(){ ... }**：主程序就是通过这些方法与绑定服务交互的。这里对外提供 getCircu()和 getArea()两个方法，分别实现计算圆周长和面积的功能。

3. 主程序绑定

实现思路：

在主程序的 updateGUI()方法中通过 Handler（句柄）对象的 post()方法提交界面刷新，而界面刷新的具体操作在 Runnable 接口 refreshWork 的 run()方法中执行。与上例不同的是，在 onCreate()中要创建 ServiceConnection 对象，并重写其 onServiceConnected()方法来获取服务的实例，这样就可以在点击"计算"按钮执行事件方法 getCircleParam()中调用服务所公开的计算圆周长和面积的 getCircu()和 getArea()方法。

MainActivity.java 代码如下：

```
public class MainActivity extends AppCompatActivity {
    private static FrameLayout circleFrame;
    private static TextView textRadius;
    private static TextView textCircleParam;
    private static Handler handler = new Handler();              //界面刷新句柄
    private static View mycircle;                                //圆视图对象
```

```java
    private static int myradius;                                    //圆半径
    private DrawCircle drawCircle;                                  //声明服务对象
    private ServiceConnection myconn;

    @Override
    protected void onCreate(Bundle savedInstanceState) {
        super.onCreate(savedInstanceState);
        setContentView(R.layout.activity_main);
        findViews();
        …
        myconn = new ServiceConnection() {                          // (a)
            @Override
            public void onServiceConnected(ComponentName name, IBinder service) {
                drawCircle = ((DrawCircle.CalcuCircletor) service).getService();
                                                                    // (a) 一旦绑定成功就获取服务实例
                Log.i("ServiceConnection", "已获取所绑定的服务实例,可进行数据交互");
            }

            @Override
            public void onServiceDisconnected(ComponentName name) {  // (a)
                drawCircle = null;
            }
        };
    }

    private void findViews() {
        circleFrame = findViewById(R.id.circleFrame);
        textRadius = findViewById(R.id.textRadius);
        textCircleParam = findViewById(R.id.textCircleParam);
    }

    public static void updateGUI(View c, int r) {
        mycircle = c;
        myradius = r;
        handler.post(refreshWork);
    }

    private static Runnable refreshWork = new Runnable() {
        @Override
        public void run() {
            //刷新界面
            circleFrame.removeAllViews();
            circleFrame.addView(mycircle);
            textRadius.setText("半径: " + String.valueOf(myradius));
        }
    };

    public void bind(View view) {                                   //绑定服务
        Intent intent = new Intent(this, DrawCircle.class);
        bindService(intent, myconn, BIND_AUTO_CREATE);
    }
```

```java
    public void unbind(View view) {                                          //解绑服务
        unbindService(myconn);
        drawCircle = null;
        Log.i("MainActivity", "已销毁所解绑的服务实例，并断开数据连接");
    }

    //获取当前圆周长和面积数值并显示在界面下方
    public void getCircleParam(View view) {
        if (drawCircle != null) {                                            // (b)
            float circu = drawCircle.getCircu();
            float area = drawCircle.getArea();
            textCircleParam.setText("圆周长：" + String.format("%.2f", circu) + "\n" + "面积：" + String.format("%.4f", area));
        }
        else
            textCircleParam.setText("圆周长：  面积：");
    }
    …
}
```

其中：

（a）**myconn = new ServiceConnection(){ ... };**、**drawCircle = ((DrawCircle.CalcuCircletor) service).getService();** 和 **public void onServiceDisconnected(ComponentName name)**：创建 ServiceConnection 对象的主要作用就是获取所绑定的服务实例，服务实例通过 DrawCircle.CalcuCircletor 的 getService()方法获取。ServiceConnection 对象还有一个 onServiceDisconnected()方法,在其中销毁已获取的服务实例。

（b）**if (drawCircle != null)**：因为之前已经获取了服务实例，只要服务实例不为空，主程序就可以直接调用服务中公开的方法。

4. 服务配置

在 AndroidManifest.xml 中配置服务 DrawCircle，方法同前，此处略。

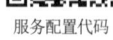

5. 运行效果

点击"开始"按钮，服务线程开始以随机半径绘圆；点击"计算"按钮，主程序从绑定的服务实例中获得当前圆的周长和面积值，并显示在屏幕下方，如图 5.7 所示。

图 5.7 绑定绘圆服务的运行效果

读者同样也可由开发环境底部日志窗口的输出，观察到绑定服务的整个生命周期过程，如图 5.8 所示。

图 5.8 绑定服务的整个生命周期过程

5.1.4 多 Service 交互及生命周期

以不同方式使用的服务，其生命周期也不一样，归纳起来有如下几种情形：

(1) 被启动的服务。

如果一个 Service 被某个 Activity 调用 startService()方法所启动，那么不管是否有 Activity 使用 bindService()绑定或 unbindService()解绑该服务，它都会始终在后台运行。如果一个 Service 被 startService()方法多次启动，那么 onCreate()方法只会调用一次，onStart()则每次都会调用，并且系统只会创建该 Service 的一个实例。该 Service 将会一直在后台运行，而不管对应程序的 Activity 是否还在运行，直到被调用 stopService()或自身的 stopSelf()方法停止。无论调用了多少次启动服务的方法，只要调用一次 stopService()方法便可以停止服务。

(2) 被绑定的服务。

如果一个 Service 被某个 Activity 调用 bindService()方法绑定启动，不管调用 bindService()几次，onCreate()方法都只会调用一次，同时 onStart()方法始终不会被调用。当连接建立之后，Service 将会一直运行，除非调用 unbindService()断开连接或者之前调用 bindService()的 Activity 退出了，系统将会自动停止该 Service，调用其 onDestroy()方法。

(3) 被启动又被绑定的服务。

如果一个 Service 既被启动又被绑定，则它将会一直在后台运行。并且不管如何调用，onCreate()始终只会调用一次，而对应 startService()调用了多少次，onStart()便会调用多少次。调用 unbindService()将不会停止服务，而必须调用 stopService()或 Service 自身的 stopSelf()来停止服务。

(4) 当服务被停止时清除服务。

当一个 Service 被停止时，onDestroy()方法将会被调用，在其中可以做一些清理工作，例如停止在 Service 中创建的线程等。

> **注意：**
> ① 调用 bindService()绑定到 Service 时，应保证在某处调用 unbindService()解除绑定。
> ② 使用 startService()启动服务之后，不管是否使用 bindService()，都要用 stopService()停止服务。
> ③ 同时使用 startService()与 bindService()时，不管调用顺序如何，需要同时调用 unbindService()与 stopService()才能停止服务。
> ④ 新 SDK 版本 onStart()方法已经变为 onStartCommand()，但原方法仍然有效。

【例 5.3】 创建包含两个服务 PowerMonitor（电源监控）和 App（应用），主程序 MainActivity 初始默认以启动方式运行 PowerMonitor 服务，该服务在整个程序运行期间一直在后台，用户可点击屏幕上的"开启应用"按钮以启动方式运行 App 服务。App 启动后自动与 PowerMonitor 服务绑定，通过绑定来使用 PowerMonitor 所管理的电源。App 服务在运行期间也会定时查看 PowerMonitor 剩余的电量，若发现电量耗尽，就自行与 PowerMonitor 解除绑定并退出。用户可随时点击"关闭应用"按钮来停止 App 服务。

1. 设计界面

创建 Android Studio 工程 PowerMonitorNormal，其上的"开启应用"和"关闭应用"按钮是用来控制 App 服务的（如图 5.9 所示）。content_main.xml 文件代码如下：

图 5.9 设计界面

```xml
<?xml version="1.0" encoding="utf-8"?>
<RelativeLayout xmlns:android="http://schemas.android.com/apk/res/android"...>

    <TextView
        android:layout_width="wrap_content"
        android:layout_height="wrap_content"
        android:text="电量："
        android:id="@+id/textView"
        android:textColor="@android:color/black"
        android:textSize="@dimen/abc_text_size_large_material" />

    <ProgressBar                                            //模拟显示电量的进度条
        style="?android:attr/progressBarStyleHorizontal"
        android:layout_width="match_parent"
        android:layout_height="100dp"
        android:id="@+id/progressBarPower"
        android:layout_alignBottom="@+id/textView"
        android:layout_toRightOf="@+id/textView"
        android:layout_toEndOf="@+id/textView"
        android:max="100"
        android:background="@android:color/holo_green_light" />

    <Button
        android:layout_width="match_parent"
        android:layout_height="wrap_content"
        android:text="开    启    应    用"
        android:id="@+id/buttonStartApp"
        android:layout_below="@+id/textView"
        android:layout_alignParentLeft="true"
        android:layout_alignParentStart="true"
        android:textSize="@dimen/abc_text_size_medium_material"
        android:layout_marginTop="20dp"
        android:onClick="start" />

    <Button
        android:layout_width="match_parent"
        android:layout_height="wrap_content"
        android:text="关    闭    应    用"
        android:id="@+id/buttonStopApp"
        android:layout_below="@+id/buttonStartApp"
        android:layout_alignParentLeft="true"
        android:layout_alignParentStart="true"
        android:textSize="@dimen/abc_text_size_medium_material"
        android:onClick="stop" />

</RelativeLayout>
```

2. 创建服务

（1）创建服务类 PowerMonitor。

实现思路：

在它的 onCreate()中创建该服务线程并赋值初始电量为 100，线程启动后通过 Runnable 接口的 run()

方法每 100 毫秒更新一次界面进度条的电量值。由于 App 服务需要与 PowerMonitor 绑定，故该服务中同时也实现了 onBind()、onUnbind()，并对外提供了一些公开使用的方法：checkPower()和 usePower()。

PowerMonitor.java 的代码如下：

```java
public class PowerMonitor extends Service {
    private Thread monitorThread;                                          //声明监控服务线程
    private int power;                                                     //剩余电量值
    private static final String TAG = "电源监控服务（PowerMonitor）";        //服务标识

    private Runnable monitorWork = new Runnable() {                        // （a）
        @Override
        public void run() {
            try {
                while (!monitorThread.isInterrupted()) {
                    MainActivity.updateGUI(power);                         // （a）
                    monitorThread.sleep(100);                              // （a）
                }
            } catch (InterruptedException e) {
                e.printStackTrace();
            }
        }
    };

    @Override
    public void onCreate() {
        super.onCreate();
        monitorThread = new Thread(null, monitorWork, "PowerMonitor");
        power = 100;
        Log.i(TAG, "onCreate");
    }

    @Override
    public int onStartCommand(Intent intent, int flags, int startId) {
        super.onStartCommand(intent, flags, startId);
        if (!monitorThread.isAlive()) monitorThread.start();
        Log.i(TAG, "onStartCommand");
        return START_STICKY;
    }

    @Override
    public void onDestroy() {
        super.onDestroy();
        monitorThread.interrupt();
        Log.i(TAG, "onDestroy");
    }

    public class Plug extends Binder {                                     // （b）
        PowerMonitor getService() {
            return PowerMonitor.this;                                      //模拟一个插头插到插座上获取电能
        }
    }
```

```java
    @Override
    public IBinder onBind(Intent intent) {
        if (!monitorThread.isAlive()) monitorThread.start();
        Log.i(TAG, "onBind");
        return new Plug();
    }

    @Override
    public boolean onUnbind(Intent intent) {
        Log.i(TAG, "onUnbind");
        return super.onUnbind(intent);
    }

    /**
     * 公开提供给服务调用者（App 服务）使用的方法              // (c)
     */
    public int checkPower() {                                // (c)
        return power;                                        //获取并查看电量值
    }

    public void usePower(int quantity) {                     // (c)
        power -= quantity;                                   //修改电量值（模拟 App 服务耗电）
        if (power < 0) power = 0;
    }
}
```

其中：

（a）**private Runnable monitorWork = new Runnable(){ ... };**、**MainActivity.updateGUI(power);** 和 **monitorThread.sleep(100);**：电源监控服务与之前的服务一样，也是通过 Runnable 接口机制更新界面的，每隔 100 毫秒调用一次主程序界面的 updateGUI() 方法。

（b）**public class Plug extends Binder**：定义类 Plug 继承自 Binder，其作用与上例的 CalcuCircletor 一样，也是为了让其他程序或服务能通过它来获取本服务的实例。

（c）**public int checkPower(){ ... }** 和 **public void usePower(int quantity){ ... }**：这里定义了两个方法，旨在公开让 App 服务调用，模拟查看剩余电量和耗电。

（2）创建服务类 App。

实现思路：

该服务模拟其他耗电的手机应用，在 onCreate() 中创建 ServiceConnection 对象获取电源监控服务实例，在 onStartCommand() 中与电源监控服务绑定。App 服务运行后，在 Runnable 接口中每秒调用一次监控服务的 usePower 要耗电量 5，同时调用 checkPower() 检查剩余电量值，一旦发现电量值小于 0，App 会自己调用 unbindService() 与监控服务解除绑定，这样就不再消耗监控服务中的电能了。

App.java 代码如下：

```java
public class App extends Service {
    private Thread appThread;                                        //声明应用服务线程
    private static final String TAG = "应用服务（App）";              //服务标识
    private PowerMonitor powerMonitor;                               //电源监控服务对象
    private ServiceConnection myconn;
```

```java
@Override
public IBinder onBind(Intent intent) {
    return null;
}

private Runnable appWork = new Runnable() {                              // (a)
    @Override
    public void run() {
        try {
            while (!appThread.isInterrupted()) {
                if (powerMonitor != null) {
                    if (powerMonitor.checkPower() > 0)
                        powerMonitor.usePower(5);            //每秒耗电量为5
                    else {
                        unbindService(myconn);
                        powerMonitor = null;
                        Log.i(TAG, "应用监测到手机电已耗尽,自行退出");
                    }
                }
                appThread.sleep(1000);
            }
        } catch (InterruptedException e) {
            if (powerMonitor != null) {
                unbindService(myconn);
                powerMonitor = null;
            }
            Log.i(TAG, "应用被关闭");
            e.printStackTrace();
        }
    }
};

@Override
public void onCreate() {
    super.onCreate();
    appThread = new Thread(null, appWork, "App");
    myconn = new ServiceConnection() {
        @Override
        public void onServiceConnected(ComponentName name, IBinder service) {
            powerMonitor = ((PowerMonitor.Plug) service).getService();
                                            //一旦绑定成功就获取服务实例
            Log.i("ServiceConnection", "应用已获取电源监控服务实例,可使用手机电源");
        }

        @Override
        public void onServiceDisconnected(ComponentName name) {
            powerMonitor = null;
        }
    };
    Log.i(TAG, "onCreate");
}
```

```java
@Override
public int onStartCommand(Intent intent, int flags, int startId) {
    super.onStartCommand(intent, flags, startId);
    Intent i = new Intent(this, PowerMonitor.class);
    bindService(i, myconn, BIND_AUTO_CREATE);          // (b)
    if (!appThread.isAlive()) appThread.start();
    Log.i(TAG, "onStartCommand");
    return START_STICKY;
}

@Override
public void onDestroy() {
    super.onDestroy();
    appThread.interrupt();
    Log.i(TAG, "onDestroy");
}
}
```

其中：

（a）**private Runnable appWork = new Runnable(){ ... };**：App 服务也是通过 Runnable 接口机制运行，不断耗电和检查剩余电量，直到发现电量耗尽后与电源监控服务解绑。

（b）**bindService(i, myconn, BIND_AUTO_CREATE);**：App 服务就是在这里绑定电源监控服务 PowerMonitor 的。

3. 主程序启动

实现思路：

主程序在 findViews() 中启动 PowerMonitor 服务，界面进度条动态更新的机制与之前一样也是采用句柄提交 Runnable 接口的方式。通过界面上的"开启应用"按钮执行 start() 方法启动 App 服务，通过"关闭应用"按钮执行 stop() 方法停止 App 服务。

MainActivity.java 代码如下：

```java
public class MainActivity extends AppCompatActivity {
    private static ProgressBar progressBarPower;            //界面电量进度条
    private static Button buttonStartApp;
    private static Button buttonStopApp;
    private static Handler handler = new Handler();          //界面刷新句柄
    private static int mypower;                              //电量值

    @Override
    protected void onCreate(Bundle savedInstanceState) {
        super.onCreate(savedInstanceState);
        setContentView(R.layout.activity_main);
        findViews();
        ...
    }

    private void findViews() {
        progressBarPower = findViewById(R.id.progressBarPower);
        buttonStartApp = findViewById(R.id.buttonStartApp);
        buttonStopApp = findViewById(R.id.buttonStopApp);
```

```
            Intent intent = new Intent(this, PowerMonitor.class);
            startService(intent);                              //初始启动 PowerMonitor 服务
        }

        public static void updateGUI(int p) {
            mypower = p;
            handler.post(refreshWork);
        }

        private static Runnable refreshWork = new Runnable() {
            @Override
            public void run() {
                //刷新界面
                progressBarPower.setProgress(mypower);
            }
        };

        public void start(View view) {                         //开启应用服务
            Intent intent = new Intent(this, App.class);
            startService(intent);
        }

        public void stop(View view) {                          //停止应用服务
            Intent intent = new Intent(this, App.class);
            stopService(intent);
        }
        …
}
```

4. 服务配置

在 AndroidManifest.xml 中，将开发好的两个服务（PowerMonitor 和 App）配置其中（加粗处），代码如下：

```
<application
    …
    <activity
        android:name=".MainActivity"
        …
    </activity>
    <service android:name=".PowerMonitor"/>
    <service android:name=".App"/>
</application>
```

5. 运行效果

运行程序后，点击"开启应用"按钮，上方显示电量（红色进度条）开始逐步减少。点击"关闭应用"按钮，电量停止减少。再次点击"开启应用"按钮，电量值又继续减少，直到红色条减退消失（表示电量耗尽），应用自行退出。此时若再点击"开启应用"按钮，应用刚一启动就会马上退出，反复启动皆是如此（因为已经没电了）。如图 5.10 所示。

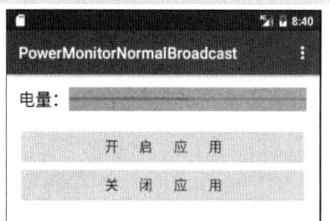

图 5.10 多服务交互运行效果

整个过程可从下面两个服务的生命周期日志中很清楚地看出来，如图 5.11 所示。

```
12-08 08:41:24.339 17810-17810/com.easybooks.android.powermonitornormalbroadcast I/电源监控服务(PowerMonitor): onCreate
12-08 08:41:24.339 17810-17810/com.easybooks.android.powermonitornormalbroadcast I/电源监控服务(PowerMonitor): onStartCommand
                                                [ 12-08 08:41:24.369 17810:17854 D/ ]
                                                HostConnection::get() New Host Connection established 0xaa3fafa0, tid 17854
12-08 08:41:24.372 17810-17854/com.easybooks.android.powermonitornormalbroadcast I/OpenGLRenderer: Initialized EGL, version 1.4
12-08 08:41:24.395 17810-17854/com.easybooks.android.powermonitornormalbroadcast W/EGL_emulation: eglSurfaceAttrib not implemented
12-08 08:41:24.395 17810-17854/com.easybooks.android.powermonitornormalbroadcast W/OpenGLRenderer: Failed to set EGL_SWAP_BEHAVIOR on surface 0xaa00fa80, error=EGL_SUCCESS
12-08 08:41:28.455 17810-17810/com.easybooks.android.powermonitornormalbroadcast I/应用服务(App): onCreate
12-08 08:41:28.456 17810-17810/com.easybooks.android.powermonitornormalbroadcast I/应用服务(App): onStartCommand
12-08 08:41:28.456 17810-17810/com.easybooks.android.powermonitornormalbroadcast I/电源监控服务(PowerMonitor): onBind
12-08 08:41:28.457 17810-17810/com.easybooks.android.powermonitornormalbroadcast I/ServiceConnection: 应用已获取电源监控服务实例,可使用手机电源
12-08 08:41:33.137 17810-17810/com.easybooks.android.powermonitornormalbroadcast I/应用服务(App): onDestroy
12-08 08:41:33.140 17810-17810/com.easybooks.android.powermonitornormalbroadcast I/电源监控服务(PowerMonitor): onUnbind
12-08 08:41:33.140 17810-17918/com.easybooks.android.powermonitornormalbroadcast I/应用服务(App): 应用被关闭
12-08 08:41:38.419 17810-17810/com.easybooks.android.powermonitornormalbroadcast I/应用服务(App): onCreate
12-08 08:41:38.420 17810-17810/com.easybooks.android.powermonitornormalbroadcast I/应用服务(App): onStartCommand
12-08 08:41:38.420 17810-17810/com.easybooks.android.powermonitornormalbroadcast I/ServiceConnection: 应用已获取电源监控服务实例,可使用手机电源
12-08 08:41:53.439 17810-18066/com.easybooks.android.powermonitornormalbroadcast I/应用服务(App): 应用监测到手机电已耗尽,自行退出
12-08 08:41:57.302 17810-17810/com.easybooks.android.powermonitornormalbroadcast I/应用服务(App): onStartCommand
12-08 08:41:57.318 17810-17810/com.easybooks.android.powermonitornormalbroadcast I/ServiceConnection: 应用已获取电源监控服务实例,可使用手机电源
12-08 08:41:57.444 17810-18066/com.easybooks.android.powermonitornormalbroadcast I/应用服务(App): 应用监测到手机电已耗尽,自行退出
12-08 08:42:03.452 17810-17810/com.easybooks.android.powermonitornormalbroadcast I/应用服务(App): onStartCommand
12-08 08:42:03.452 17810-17810/com.easybooks.android.powermonitornormalbroadcast I/ServiceConnection: 应用已获取电源监控服务实例,可使用手机电源
12-08 08:42:04.453 17810-18066/com.easybooks.android.powermonitornormalbroadcast I/应用服务(App): 应用监测到手机电已耗尽,自行退出
```

图 5.11 两个服务的生命周期日志

5.2 广播（BroadcastReceiver）

5.2.1 BroadcastReceiver 概述

在 Android 系统中，广播体现在方方面面，例如，当开机完成后系统会产生一条广播，接收到这条广播就能实现开机启动服务的功能；当网络状态改变时系统会产生一条广播，接收到这条广播就能及时地做出提示和保存数据等操作；当电池电量改变时，系统会产生一条广播，接收到这条广播就能在电量低时告知用户及时保存数据。Android 中的广播机制非常出色，很多事情原本需要开发者亲自操作，现在只需等待广播告知就可以了，大大减少了开发的工作量。

1. 基本概念

BroadcastReceiver（广播接收者）用于异步接收广播，广播的发送是通过调用 sendBroadcast()、sendOrderedBroadcast()、sendStickyBroadcast()等方法来实现的。通常一个广播可以被订阅了此 Intent 的多个广播接收者所接收。广播接收器通过调用 BroadcastReceiver()方法接收广播，对广播的通知做出反应。广播接收器没有用户界面，但是它可以为它们接收到信息启动一个 Activity 或者使用 NotificationManager 来通知用户。

BroadcastReceiver 是一个系统全局的监听器，用于监听系统全局的广播消息，所以它可以很方便地进行组件之间的通信。BroadcastReceiver 虽然是一个监听器，但是它和之前用到的 OnXxxListener 监听器不同，那些只是程序级别的监听器，运行在指定程序的所在进程中，当程序退出时，OnXxxListener 监听器也就随之关闭了。但是 BroadcastReceiver 属于系统级的监听器，它拥有自己的进程，只要存在与之匹配的广播消息被以 Intent 的形式发送出来，它就会被激活。

2. 生命周期

虽然 BroadcastReceiver 属于 Android 的四大组件之一，也有自己独立的生命周期，但是和 Activity 与 Service 又不同。当在系统注册一个 BroadcastReceiver 之后，每次系统以一个 Intent 的形式发布广播时，系统都会创建与之对应的 BroadcastReceiver 广播接收者实例，并自动触发它的 onReceive()方法。当 onReceive()方法被执行完成之后，BroadcastReceiver 实例就会被销毁。虽然它独自享用一个单独的进程，但也不是没有限制的，如果 BroadcastReceiver.onReceive()方法不能在几秒内执行完成，Android

系统就会认为该 BroadcastReceiver 对象无响应，然后弹出"Application No Response"对话框。所以，不要在 BroadcastReceiver.onReceive()方法内执行一些耗时的操作。

BroadcastReceiver 使用的基本流程如下：注册 receiver（静态或动态）→调用 sendBroadcast()方法发送广播→调用 onReceive(Context context，Intent intent)方法处理广播→调用 startService()方法启动服务→调用 stopService()方法关闭服务。

一个 BroadcastReceiver 对象只有在被调用 onReceive(Context，Intent)方法时才有效，当从该方法返回后，该对象就无效了，其生命周期结束。从这个特征可见，在所调用的 onReceive(Context，Intent)方法里，不能有过于耗时的操作。如果确实需要根据广播的内容完成一些耗时操作，一般考虑通过 Intent 启动一个 Service 而不是在 BroadcastReceiver 内开启一个新线程来完成该操作。因为 BroadcastReceiver 本身的生命周期很短，可能出现的情形是子线程还没有结束，BroadcastReceiver 就已经退出了，而如果 BroadcastReceiver 所在的进程结束了，该线程就会被标记为一个空线程。根据 Android 的内存管理策略，在系统内存紧张的时候，会按照优先级结束线程，而空线程是优先级最低的，这样就可能导致 BroadcastReceiver 所启动的子线程永远无法执行完毕。

3. 注册方式

BroadcastReceiver 本质上是一个监听器，所以使用它的方法也非常简单，只需要继承 BroadcastReceiver 类，在其中重写 onReceive(Context context,Intent intent)方法，就可以在系统的任何位置，通过 sendBroadcast()、sendOrderedBroadcast()方法发送广播给这个 BroadcastReceiver。但是仅仅这么做是不够的，BroadcastReceiver 作为系统组件，还必须在 Android 系统中注册才能使用。有以下两种注册的方式。

（1）静态注册。

在 AndroidManifest.xml 文件的<application/>节点中用标签<receiver>进行注册，并在标签内用<intent- filter>设置过滤器。例如：

```
<receiver android:name="myReceiver">
    <intent-filter>
        <action android:name="android.intent.action.myBroadcast"/>      //（a）
        <category android:name="android.intent.category.DEFAULT"/>
    </intent-filter>
</receiver>
```

其中：

（a）**<action android:name="android.intent.action.myBroadcast"/>**：android:name 属性指定注册的 BroadcastReceiver 对象，通过<Intent-filter>指定<action>和<category>，并可通过 android:priority 属性设定 BroadcastReceiver 的优先级，数值越大优先级越高。

此后，只要是发往 android.intent.action.myBroadcast 地址的广播，myReceiver 都能接收到。

> 👀 注意：
> 用这种方式注册的 BroadcastReceiver 是常驻型的，也就是说当应用关闭后，如果有广播信息传来，BroadcastReceiver 也照样会被系统调用而自动运行。

静态注册方式由系统来管理 receiver，而且程序里的所有 receiver 都配置在 XML 文件中，一目了然。

（2）动态注册。

动态注册需要在代码中动态地指定广播地址并注册，先定义并设置好一个 IntentFilter 对象，然后在需要注册的地方调用 registerReceiver()方法，需要取消时调用 unregisterReceiver()方法。当用动态方式注册的 BroadcastReceiver 的 Context（上下文）对象被销毁时，BroadcastReceiver 也就自动取消注册了。

另外，如果在使用 sendBroadcast()方法时指定了接收权限，则只有在 AndroidManifest.xml 文件中用<uses-permission>标签声明了拥有此权限的 BroadcastReceiver 才能接收到广播。同样，如果在注册 BroadcastReceiver 时指定了可接收的广播的权限，则只有在 AndroidManifest.xml 文件中用<uses-permission>标签声明了拥有此权限的 Context 对象所发送的广播才能被这个 BroadcastReceiver 收到。

例如：
```
myReceiver receiver = new myReceiver();
IntentFilter filter = new IntentFilter();
Filter.addAction("android.intent.action.myBroadcast");
registerReceiver(receiver, filter);                                          // (a)
```

其中：

（a）**registerReceiver(receiver, filter);**：registerReceiver 是 android.content.ContextWrapper 类中的方法，Activity 和 Service 都继承了 ContextWrapper 类，所以可以直接调用。

在实际应用中，我们在 Activity 或 Service 中注册了一个 BroadcastReceiver，当这个 Activity 或 Service 被销毁时如果没有解除注册，系统会报一个异常提示。所以，编程时要记得在特定的地方执行解除注册操作，代码如下：

```
@override
Protected void onDestroy(){
    super.onDestroy();
    unregisterReceiver(receiver);                                            //解除注册
}
```

动态注册方式不是常驻型的，BroadcastReceiver 会随着程序生命周期的结束而关闭。

编写程序时一般在 Activity 的 onStart()方法里进行注册，在 onStop()里注销。在退出程序前要记得调用 unregisterReceiver()方法。如果在 Activity 的 onResume()里注册了，就必须在 onPause()里注销。

虽然 Android 提供了两种 BroadcastReceiver 注册方式，但在实际的开发中大多还是会在 AndroidManifest.xml 中进行注册。

4. 广播类型

根据传播方式的不同，在系统中有两种类型的广播。

（1）普通广播。

普通广播对于多个接收者来说是完全异步的，通常每个接收者都无须等待即可接收到此类广播，接收者之间互不影响。对于这种广播，接收者无法终止，即无法阻止其他接收者的接收动作。当一个普通广播被发出后，所有与之匹配的 BroadcastReceiver 都能同时收到它。这种方式的优点是传递效率比较高，缺点是一个 BroadcastReceiver 不能影响其他响应这个广播的 BroadcastReceiver。发送普通广播使用 sendBroadcast()方法，它也是 android.content.ContextWrapper 类中的方法，可以将一个指定地址和参数信息的 Intent 对象以广播形式发送出去。

（2）有序广播。

有序广播是同步执行的，也就是说它的接收器会按照预先声明的优先级依次接收这类广播，接收器以链式结构组织起来，优先级越高越先被执行。因为是顺序执行，所有优先级高的接收者可以把执行结果传递到优先级低的接收者，优先级高的接收者有能力终止这个广播，通过 abortBroadcast()方法终止广播的传播。一旦广播被终止，优先级低于它的接收器就不会再接收到这条广播了。发送有序广播使用 sendOrderedBroadcast()方法。

5.2.2 普通广播举例

【例 5.4】 在前面多 Service 交互实例的基础上，应用普通广播在手机电量快耗尽（20%以下）时给予提示，告知用户电量不足。

1. 创建 PowerMonitor 服务和 App 服务

创建 Android Studio 工程 PowerMonitorNormalBroadcast，其他页面和类代码与 PowerMonitorNormal 完全相同，此处略。

2. 创建广播接收器

在工程中添加一个广播接收器类 PowerBroadcast，PowerBroadcast.java 代码如下：

```java
public class PowerBroadcast extends BroadcastReceiver {
    @Override
    public void onReceive(Context context, Intent intent) {
        Toast.makeText(context, intent.getStringExtra("msg"), Toast.LENGTH_SHORT).show();
    }
}
```

该类继承自 BroadcastReceiver，实现了其 onReceive()方法，其中仅有一条语句用于显示广播消息，消息经由 Intent 类型的参数传递。

3. 注册广播接收器

在 AndroidManifest.xml 中，注册该广播接收器（加粗处），代码如下：

```xml
<application
    ...
    <service android:name=".PowerMonitor"/>
    <service android:name=".App"/>
    <receiver android:name=".PowerBroadcast">
        <intent-filter>
            <action android:name="com.easybooks.android.powermonitornormalbroadcast"/>
        </intent-filter>
    </receiver>
</application>
```

4. 发送广播

在服务类 PowerMonitor 中添加发送广播的代码，位于 PowerMonitor.java 中（加粗处），代码如下：

```java
public class PowerMonitor extends Service {
    private Thread monitorThread;
    private int power;
    private boolean isSend = false;                              //是否已发送过广播
    private static final String TAG = "电源监控服务（PowerMonitor）";

    private Runnable monitorWork = new Runnable() {
        @Override
        public void run() {
            try {
                while (!monitorThread.isInterrupted()) {
                    MainActivity.updateGUI(power);
```

```
                    monitorThread.sleep(100);
                //电量不足 20%时以广播方式告知主程序
                if (power < 20 && !isSend) {
                    Intent intent = new Intent();
                    intent.setAction("com.easybooks.android.powermonitornormalbroadcast");
                                                                            //（a）
                    intent.putExtra("msg","电池电量不足 20%！");
                    sendBroadcast(intent);                        //（b）
                    isSend = true;
                }
            }
        } catch (InterruptedException e) {
            e.printStackTrace();
        }
    }
};
…
}
```

其中：

（a）**intent.setAction("com.easybooks.android.powermonitornormalbroadcast");**：参数字符串 "com.easybooks.android.powermonitornormalbroadcast" 为广播消息的接收地址，该地址必须与之前 AndroidManifest.xml 中配置的 "<action android:name="com.easybooks.android.powermonitornormalbroadcast"/>" 中的 name 值相一致。

（b）**sendBroadcast(intent);**：该方法发送广播消息。

5. 运行效果

运行程序，点击"开启应用"按钮，当电量条减少到 20%时，系统弹出消息提示电量不足。如图 5.12 所示。

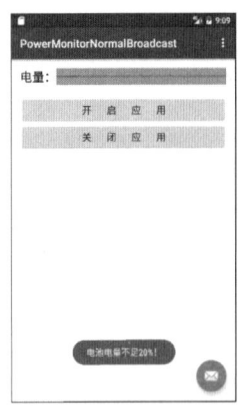

图 5.12 普通广播的运行效果

5.2.3 有序广播举例

【例 5.5】 虽然系统存在两种类型的广播，但是一般系统发送出来的广播均是有序广播，所以可以通过优先级的控制，在系统内置的程序响应前，对有序广播提前进行响应。这就是骚扰短信和电话拦截器的工作原理。

1. 设计界面

创建 Android Studio 工程 SmsFilterOrderedBroadcast，在 content_main.xml 设计页面上仅放一个 id 为 textSmsMsg 的 TextView 用于显示短信消息，代码略。

2. 创建服务

实现思路：

创建服务类 SmsSender 用于发送短信，该类模拟的是一个基站，通过 Runnable 接口不断地产生和发送短信。在 onCreate()方法中创建服务线程，在 onStartCommand()中启动线程。

SmsSender.java 代码如下：

```
public class SmsSender extends Service {
    private Thread sendThread;                                    //声明短信发送线程
```

```java
    private String[] telHead = {"138", "170"};                          //短信号码头
    private String[] smsContent = {"周何骏先生，祝生日快乐！", "恭喜您中了 500 万大奖！"};
                                                                        //短信内容
    private String[] smsMessages = new String[2];                       //存储短信结构（号码+信息内容）
    private static final String TAG = "短信发送服务（SmsSender）";      //服务标识

    @Override
    public IBinder onBind(Intent intent) {
        return null;
    }

    private Runnable sendWork = new Runnable() {
        @Override
        public void run() {
            try {
                while (!sendThread.isInterrupted()) {
                    for (int i = 0; i < 2; i++) {
                        int number = new Random().nextInt(89999999) + 10000000;       // （a）
                        smsMessages[0] = telHead[i] + String.valueOf(number);
                        smsMessages[1] = smsContent[i];                               // （b）
                        Intent intent = new Intent();
                        intent.setAction("com.easybooks.android.smsfilterorderedbroadcast");
                        intent.putExtra("tel", smsMessages[0]);                       // （b）
                        intent.putExtra("sms", smsMessages[1]);                       // （b）
                        sendOrderedBroadcast(intent, null);                           //发送有序广播
                        sendThread.sleep(5000);
                    }
                }
            } catch (InterruptedException e) {
                e.printStackTrace();
            }
        }
    };

    @Override
    public void onCreate() {
        super.onCreate();
        sendThread = new Thread(null, sendWork, "SmsSender");
        Log.i(TAG, "onCreate");
    }

    @Override
    public int onStartCommand(Intent intent, int flags, int startId) {
        super.onStartCommand(intent, flags, startId);
        if (!sendThread.isAlive()) sendThread.start();
        Log.i(TAG, "onStartCommand");
        return START_STICKY;
    }

    @Override
    public void onDestroy() {
        super.onDestroy();
```

```
            sendThread.interrupt();
            Log.i(TAG, "onDestroy");
        }
    }
```

其中：

（a）**int number = new Random().nextInt(89999999) + 10000000;**：短信号码是随机产生的，加上 138 或 170 前缀，形成完整的模拟手机号。

（b）**smsMessages[1] = smsContent[i];**、**intent.putExtra("tel", smsMessages[0]);** 和 **intent.putExtra("sms", smsMessages[1]);**：短信号码和信息内容分别作为一个字段，存储在一个包含两个元素的字符串数组 smsMessages 中，并进一步封装到 Intent 中待发送。

3. 创建短信接收器

创建广播接收器类 SmsReceiver 用于接收短信，SmsReceiver.java 代码如下：

```
public class SmsReceiver extends BroadcastReceiver {
    @Override
    public void onReceive(Context context, Intent intent) {
        String smsMsg = "来自" + intent.getStringExtra("tel") + "：\n" + intent.getStringExtra("sms");
        MainActivity.updateGUI(smsMsg);
    }
}
```

在 onReceive()方法中调用主程序的 updateGUI()刷新界面，在主界面的 TextView 中显示短信。

4. 开发短信拦截器

实现思路：

创建广播接收器类 SmsFilter 用于拦截 170 打头（现实生活中可能为诈骗）的短信，通过 abortBroadcast()方法拦截短信。

SmsFilter.java 代码如下：

```
public class SmsFilter extends BroadcastReceiver {
    @Override
    public void onReceive(Context context, Intent intent) {
        String telHead = intent.getStringExtra("tel").substring(0, 3);
        if (telHead.equals("170")) {
            abortBroadcast();                          //中断广播，不会再向比它优先级低的接收器传播下去了
            Toast.makeText(context, "拦截到一个诈骗短信！\n 来自" + intent.getStringExtra("tel"), Toast.LENGTH_SHORT).show();
        }
    }
}
```

5. 主程序

实现思路：

主程序在 findViews()方法中启动短信发送服务 SmsSender，通过 updateGUI()刷新界面上显示的短信，界面刷新同样采用的是 Runnable 机制。

MainActivity.java 代码如下：

```
public class MainActivity extends AppCompatActivity {
    private static TextView textSmsMsg;
    private static Handler handler = new Handler();                          //界面刷新句柄
```

```java
    private static String mySmsMsg;                                    //短信内容

    @Override
    protected void onCreate(Bundle savedInstanceState) {
        super.onCreate(savedInstanceState);
        setContentView(R.layout.activity_main);
        findViews();
            …
    }

    private void findViews() {
        textSmsMsg = (TextView) findViewById(R.id.textSmsMsg);
        Intent intent = new Intent(this, SmsSender.class);
        startService(intent);                                          //启动短信发送服务
    }

    public static void updateGUI(String msg) {                         //刷新短信
        mySmsMsg = msg;
        handler.post(refreshWork);
    }

    private static Runnable refreshWork = new Runnable() {
        @Override
        public void run() {
            textSmsMsg.setText(mySmsMsg);                              //刷新界面
        }
    };
        …
}
```

6. 接收与拦截配置

在 AndroidManifest.xml 中，短信拦截器（SmsFilter）的优先级必须配置得高于接收器（SmsReceiver）才能发挥拦截作用，代码如下：

```xml
<application
    …
    <service android:name=".SmsSender"/>
    <!--priority 优先级：数字越大优先级越高-->
    <receiver android:name=".SmsFilter">
        <intent-filter android:priority="10000">
            <action android:name="com.easybooks.android.smsfilterorderedbroadcast"/>
        </intent-filter>
    </receiver>
    <receiver android:name=".SmsReceiver">
        <intent-filter android:priority="10">
            <action android:name="com.easybooks.android.smsfilterorderedbroadcast"/>
        </intent-filter>
    </receiver>
</application>
```

7. 运行效果

运行程序，当不配置拦截器时，主程序将同时收到 138 和 170 打头的短信，如图 5.13 所示。而当配置了拦截器后，主程序只收到 138 短信，同时弹出拦截 170 短信的通知，如图 5.14 所示。

图 5.13　未配置拦截器　　　　　　　　　　图 5.14　弹出拦截 170 短信的通知

第6章 Android 数据存储与共享

Android 系统提供多种数据存储方式，包括 SharedPreferences、文件存储、SQLite 数据库以及跨应用的数据共享方法。不同的数据存储方式适用于不同应用领域，本章我们通过举例对比分别加以介绍。

6.1 SharedPreferences（共享优先）存储

6.1.1 SharedPreferences 概述

SharedPreferences 是一种轻量级的数据存储方式，通过调用方法就可以实现对名称/值对（NVP）型数据的保存和读取，它不仅能够保存数据，还能够实现不同应用程序间的数据共享。SharedPreferences 支持 3 种访问模式：私有（MODE_PRIVATE）、全局读（MODE_WORLD_READABLE）和全局写（MODE_WORLD_WRITEABLE）。如果定义为私有模式，仅创建的程序有权限对其进行读取或写入；如果定义为全局读模式，所有程序都可以读取，但仅有创建程序具有写入权限；如果定义为全局写模式，则所有程序都可以对其进行写入操作，但没有读权限。

使用 SharedPreferences 通行的编程步骤如下：

（1）定义访问模式。例如：

```
public static int MODE = MODE_PRIVATE;                        //定义为私有模式
public static int MODE = Context.MODE_WORLD_READABLE + Context.MODE_WORLDWRITEABLE;
                                                              //访问模式设定为既可以全局读，也可以全局写
```

（2）定义 SharedPreferences 的名称。这个名称也是它在 Android 文件系统中保存的文件名称。一般声明为字符串常量，这样可以在代码中多次使用。例如：

```
public static final String PREFERENCE NAME = "SaveSetting";
```

（3）获取 SharedPreferences 实例。将访问模式和名称作为参数传递给 getSharedPreferences()方法就可以获取到 SharedPreferences 的实例。例如：

```
SharedPreferences sharedPreferences = getSharedPreferences (PREFERENCENAME, MODE);
```

（4）修改和保存数据。在获取实例后，通过 SharedPreferences.Editor 类进行修改，最后执行 commit()保存修改内容。例如：

```
SharedPreferences.Editor editor = sharedPreferences.edit();
editor.putString("Name", "Tom");
editor.putInt("Age", 20);
editor.putFloat("Height", 1.81f);
editor.commit();
```

SharedPreferences 广泛支持各种基本数据类型，包括整型、布尔型、浮点型和长整型等。

如果需要从已保存的 SharedPreferences 中读取数据，同样是先调用 getSharedPreferences()方法，在第 1 个参数中指明要访问的 SharedPreferences 名称，然后通过 get<Type>()获取 SharedPreferences 中保存的 NVP。get<Type>()的第 1 个参数是 NVP 的名称，第 2 个参数是默认值，在无法获取数值时使用。例如：

```
SharedPreferences sharedPreferences = getSharedPreferences (PREFERENCE NAME, MODE);
String name = sharedPreferences .getString ("Name ", "Default Name ");
```

```
int age = sharedPreferences.getInt("Age ", 20);
float height = sharedPreferences.getFloat("Height", 1. 8lf);
```

6.1.2 SharedPreferences 举例

【例 6.1】 用 SharedPreferences 存取用户表单提交的注册信息。

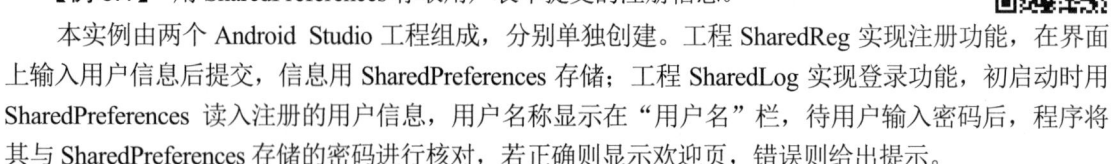

本实例由两个 Android Studio 工程组成，分别单独创建。工程 SharedReg 实现注册功能，在界面上输入用户信息后提交，信息用 SharedPreferences 存储；工程 SharedLog 实现登录功能，初启动时用 SharedPreferences 读入注册的用户信息，用户名称显示在"用户名"栏，待用户输入密码后，程序将其与 SharedPreferences 存储的密码进行核对，若正确则显示欢迎页，错误则给出提示。

为便于多种存储方式的比较，下面先来创建注册和登录应用的程序框架，后面演示其他存储方式时，直接在这个框架的基础上进行修改，就可以很清晰地看出 Android 各种数据存储方式的异同。

1. 创建注册框架

创建 Android Studio 工程 SharedReg。注册页面输入注册信息，点击"提交"按钮，将注册的用户信息以特定存储方式保存起来，效果如图 6.1 和图 6.2 所示。

图 6.1 注册页面 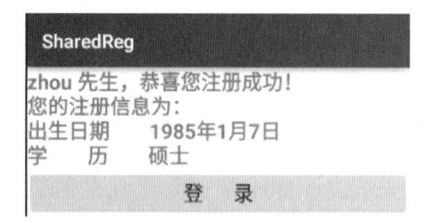   图 6.2 注册成功页面

（1）设计页面。

用第 3 章学过的表格布局设计注册页面 register.xml，代码如下：

```
<TableLayout xmlns:android="http://schemas.android.com/apk/res/android"
    ...>
    ...
    <TableRow
        android:layout_gravity="center"
        android:gravity="center">
        <TextView
            android:text="请填写注册表单"
            android:textColor="@android:color/holo_blue_dark"
            android:textSize="@dimen/abc_text_size_headline_material"
            android:textStyle="bold" />
    </TableRow>
    <TableRow>
        <TextView
            android:text="用  户  名："
            android:textSize="@dimen/abc_text_size_large_material"
```

```xml
            android:gravity="right" />
        <EditText
            android:inputType="textPersonName"
            android:id="@+id/myEditName"
            android:gravity="left"  />
    </TableRow>
    <TableRow>
        <TextView
            android:text="密      码："
            android:textSize="@dimen/abc_text_size_large_material"
            android:gravity="right" />
        <EditText
            android:inputType="numberPassword"                            //（a）
            android:id="@+id/myEditPwd"
            android:hint="必须全部为数字"                                   //（a）
            android:gravity="left" />
    </TableRow>
    <TableRow>
        <TextView
            android:text="重输确认："
            android:textSize="@dimen/abc_text_size_large_material"
            android:gravity="right" />
        <EditText
            android:inputType="numberPassword"                            //（a）
            android:id="@+id/myEditRePwd"
            android:gravity="left"  />
    </TableRow>
    <TableRow>
        <TextView
            android:text="性      别："
            android:textSize="@dimen/abc_text_size_large_material"
            android:gravity="right" />
        <ToggleButton                                                     //（b）
            android:id="@+id/myToggleButtonSex"
            android:textOn="男"
            android:textOff="女"
            android:checked="true" />
    </TableRow>
    <TableRow>
        <TextView
            android:text="出生日期："
            android:textSize="@dimen/abc_text_size_large_material"
            android:gravity="right" />
        <EditText
            android:inputType="date"                                      //（c）
            android:id="@+id/myEditBirth"
            android:gravity="left" />
    </TableRow>
    <TableRow>
        <TextView
            android:text="学      历："
```

```
                    android:textSize="@dimen/abc_text_size_large_material"
                    android:gravity="right" />
                <Spinner
                    android:id="@+id/mySpinnerDegree"/>                              //（d）
            </TableRow>
            <TableRow>
                <CheckBox
                    android:text="已阅读并接受\n 网站服务条款"                         //（e）
                    android:id="@+id/myCheckBoxAccept"
                    android:layout_gravity="center|right"                            //（e）
                    android:textStyle="bold"
                    android:onClick="onCheckBoxClick" />                             //（e）
                <Button
                    style="?android:attr/buttonStyleSmall"                           //（f）
                    android:text="提      交"
                    android:id="@+id/myButtonSubmit"
                    android:textSize="@dimen/abc_text_size_large_material"
                    android:enabled="false"                                          //（f）
                    android:gravity="left"
                    android:textAlignment="center" />
            </TableRow>
        </TableLayout>
```

其中：

（a）**android:inputType="numberPassword"** 和 **android:hint="必须全部为数字"**：文本编辑框输入内容均显示".", 初始提示信息为"必须全部为数字"。

（b）**<ToggleButton...android:checked="true" />**：开关按钮控件属性 checked="true"（默认），显示"男"；属性 checked="false"，显示"女"。

（c）**android:inputType="date"**：文本显示内容为日期。

（d）**<SpinnerSpinner android:id="@+id/mySpinnerDegree"/>**：控件选择输入内容，选择项目"博士""硕士""学士"在 findViews()中添加。

（e）**android:text="已阅读并接受\n 网站服务条款"**、**android:layout_gravity="center|right"** 和 **android:onClick="onCheckBoxClick"**：提示信息包含"\n"表示换行；上下居中水平右对齐；点击按钮，执行 onCheckBoxClick()方法。

（f）**style="?android:attr/buttonStyleSmall"** 和 **android:enabled="false"**：命令按钮采用 buttonStyleSmall 风格，初始状态不可用。在 myCheckBoxAccept 控件的点击事件中判断，若两次密码相同，使该命令按钮可用。

注册成功页面 success.xml 内容很简单，仅放置一个文本视图（用于显示成功信息）和一个"登录"按钮，代码略。

（2）开发注册功能。

实现思路：

findViews()方法中获取界面表单上各控件对象的引用并初始化，onCreate()中在需要的控件上绑定事件监听器，在监听器接口方法中实现相应的功能即可。

创建实现注册功能的 RegisterActivity.java，代码如下：

```
public class RegisterActivity extends AppCompatActivity {
    private EditText myName;                                  //编辑框（填写用户名）
    private EditText myPwd;                                   //编辑框（输入密码）
```

```java
        private EditText myRePwd;                                      //编辑框（重输密码）
        private ToggleButton mySex;                                    //开关按钮（选择性别）
        private EditText myBirth;                                      //编辑框（填写出生日期）
        private Spinner myDegree;                                      //下拉框（选择学历）
        private CheckBox myAccept;                                     //复选框（接受服务条款）
        private Button mySubmit;                                       //按钮（提交注册表单）
        private int myYear, myMonth, myDay;                            //保存出生年、月、日
        private String mydegreeTemp;                                   //保存学历
        @Override
        protected void onCreate(Bundle savedInstanceState) {
            super.onCreate(savedInstanceState);
            setContentView(R.layout.register);
            findViews();
            //设置日期编辑框点击事件监听器
            myBirth.setOnClickListener(new View.OnClickListener() {    // (a)
                @Override
                public void onClick(View view) {                       //点击日期编辑框事件
                    DatePickerDialog datePickerDialog = new DatePickerDialog(RegisterActivity.this, DateListener,
myYear, myMonth, myDay);                                               //创建 DatePickerDialog 对象，设置初始日期
                    datePickerDialog.show();                           //显示日期选择对话框
                }
            });
            //设置学历下拉框选择事件监听器
            Spinner.OnItemSelectedListener listener = new Spinner.OnItemSelectedListener() {
                @Override
                public void onItemSelected(AdapterView<?> parent, View view, int pos, long id) {
                    mydegreeTemp = parent.getItemAtPosition(pos).toString();
                }                                                      //将选择项字符串作为值

                @Override
                public void onNothingSelected(AdapterView<?> parent) {
                    mydegreeTemp = "未知";                              //未选择时的值为未知
                }
            };
            myDegree.setOnItemSelectedListener(listener);
        }
        private void findViews() {
            myName = findViewById(R.id.myEditName);
            myPwd = findViewById(R.id.myEditPwd);
            myRePwd = findViewById(R.id.myEditRePwd);
            mySex = findViewById(R.id.myToggleButtonSex);
            myBirth = findViewById(R.id.myEditBirth);
            myDegree = findViewById(R.id.mySpinnerDegree);
            myAccept = findViewById(R.id.myCheckBoxAccept);
            mySubmit = findViewById(R.id.myButtonSubmit);
            //初始化日历                                                  // (b)
            Calendar calendar = Calendar.getInstance(Locale.CHINA);    //创建 Calendar 对象
            Date date = new Date();                                    //获取当前日期 Date 对象
            calendar.setTime(date);                                    //为 Calendar 对象设置当前日期
            myYear = calendar.get(Calendar.YEAR) - 20;                 //获取 Calendar 对象中的前 20 年
            myMonth = calendar.get(Calendar.MONTH);                    //获取 Calendar 对象中的月
```

```java
            myDay = calendar.get(Calendar.DAY_OF_MONTH);        //获取 Calendar 对象中的日
            myBirth.setText("点击这里选择...");
            //初始化学历                                            // (c)
            List<String> list = new ArrayList<String>();         //创建字符串数组
            list.add("博士");
            list.add("硕士");
            list.add("学士");
            ArrayAdapter<String> adapter = new ArrayAdapter<String>(this, android.R.layout.simple_spinner_item, list);
                                                                 //创建字符串数组适配器
            adapter.setDropDownViewResource(android.R.layout.simple_spinner_dropdown_item);
            myDegree.setAdapter(adapter);                        //将适配器与学历下拉框关联起来
        }
        //监听日期选择对话框的点击事件
        private DatePickerDialog.OnDateSetListener DateListener = new DatePickerDialog.OnDateSetListener() {
            @Override
            public void onDateSet(DatePicker view, int year, int month, int day) {
                myYear = year;                                   //当前选择的日期年赋值给全局年变量
                myMonth = month;                                 //当前选择的日期月赋值给全局月变量
                myDay = day;                                     //当前选择的日期日赋值给全局日变量
                myBirth.setText(myYear + "年" + (myMonth + 1) + "月" + myDay + "日");
                                                                 //更新日期
            }
        };
        //myCheckBoxAccept 控件点击事件
        public void onCheckBoxClick(View view) {
            if (myAccept.isChecked()) {
                if (isValid(myRePwd)) mySubmit.setEnabled(true); // (d) 密码一致时"提交"按钮才可用
                else myAccept.setChecked(false);
            } else mySubmit.setEnabled(false);
        }
        //判断两个密码是否一致,一致返回 true,否则 false
        private boolean isValid(EditText editText) {             // (d)
            String pwd = myPwd.getText().toString();
            String repwd = editText.getText().toString();
            if (!repwd.equals(pwd)) {
                editText.setError("两次输入不一致!");              //密码 myRePwd 控件中显示错误提示
                return false;
            } else
                return true;
        }

        public void onSubmitClick(View view) { }                 // (e)
    }
```

其中：

（a）**myBirth.setOnClickListener(new View.OnClickListener() { ... });**：通过 View.OnClickListener 监听日期编辑框的点击事件，在其中创建 DatePickerDialog 显示日期选择对话框，通过 DatePickerDialog.OnDateSetListener 监听日期选择对话框的点击事件，把所选日期对应的年、月、日保存到相应的变量中，同时连接成日期字符串放入日期编辑框中。

（b）**Calendar calendar = Calendar.getInstance(Locale.CHINA);、Date date = new Date();、calendar.setTime(date);和 myYear = calendar.get(Calendar.YEAR) - 20;**等：通过临时创建的 Calendar 组件得到

当前日期，赋给年、月、日变量，以使在输入初始日期时以当前日期前 20 年作为进入后的初始状态。

（c）**List<String> list = new ArrayList<String>();**、**list.add("博士");**……**ArrayAdapter<String> adapter = new ArrayAdapter<String>(this, android.R.layout.simple_spinner_item, list);**、**adapter.setDropDown ViewResource(android.R.layout.simple_spinner_dropdown_item);** 和 **myDegree.setAdapter(adapter);**：监听学历下拉框，先对字符串数组元素赋值"博士""硕士""学士"，然后将字符串数组加入 ArrayAdapter 中，再将 ArrayAdapter 加入学历选择控件中。

（d）**if (isValid(myRePwd)) mySubmit.setEnabled(true);** 和 **private boolean isValid(EditText editText):** 通过 private boolean isValid() {…}方法判断两个密码控件输入密码是否一致，只有一致才能具有选择"已阅读并接受网站服务条款"控件（CheckBox）的条件，同时"提交"按钮才能操作。

（e）**public void onSubmitClick(View view) { }**："提交"按钮的事件方法，内容暂时为空，后面在测试 Android 的每一种数据存储方式时，只要在这个方法中编写特定存储方式的代码即可，框架的其他程序不用改动。现在，先在 register.xml 的"提交"按钮上设定 android:onClick="onSubmitClick" 属性，将其与该事件方法关联起来。

注册成功页面对应的 Activity 代码 SuccessActivity.java 如下：

```
public class SuccessActivity extends AppCompatActivity {
    private TextView mySuccess;               //文本视图（显示成功信息）
    private Button myEnter;                   //"登录"按钮
    @Override
    protected void onCreate(Bundle savedInstanceState) {
        super.onCreate(savedInstanceState);
        setContentView(R.layout.success);
        findViews();
        showSuccess();
    }
    private void findViews() {
        mySuccess = findViewById(R.id.myLabelSuccess);
        myEnter = findViewById(R.id.myButtonEnter);
    }
    private void showSuccess() { }            //显示成功信息的代码，稍后给出
}
```

2. 创建登录框架

创建 Android Studio 工程 SharedLog。在登录页面输入用户名和密码，点击"登录"按钮，将用户信息与之前以特定存储方式保存的注册信息（包括用户名和密码）进行比对，不一致则显示"用户名或密码错！请重输。"提示，否则显示登录成功的欢迎页，效果如图 6.3 和图 6.4 所示。

图 6.3　登录页面　　　　　　图 6.4　登录成功的欢迎页

（1）设计页面。

界面设计复用【例 3.2】LoginPage 工程的登录及欢迎页面，将对应 XML 文件复制到本例工程 layout 目录下，同时将图片资源 androidwelcomer.gif 复制到 drawable 目录下。

（2）开发登录功能。

编写 LoginActivity.java，代码如下：

```java
public class LoginActivity extends AppCompatActivity {
    private EditText myName;                                //编辑框（输入用户名）
    private EditText myPwd;                                 //编辑框（输入密码）
    private String myusername;                              //全局变量（用户名）
    private String mypassword;                              //全局变量（密码）

    @Override
    protected void onCreate(Bundle savedInstanceState) {
        super.onCreate(savedInstanceState);
        setContentView(R.layout.login);
        findViews();
        loadUser();
    }

    private void findViews() {
        myName = findViewById(R.id.myTextName);
        myPwd = findViewById(R.id.myTextPwd);
    }

    private void loadUser() {    }                          // （a）

    //点击"登录"按钮执行事件方法
    public void onLoginClick(View view) {
        String name = myName.getText().toString();
        String pwd = myPwd.getText().toString();
        //判断当前界面输入的登录名和密码与注册信息是否相同
        if (name.equals(myusername) && pwd.equals(mypassword)) {    // （b）
            Intent intent = new Intent(this, WelcomeActivity.class);
            Bundle bundle = new Bundle();
            bundle.putString("name", name);
            intent.putExtras(bundle);
            startActivity(intent);
        } else {
            Toast.makeText(this, "用户名或密码错！请重输。", Toast.LENGTH_SHORT).show();
            myName.setText(myusername);
            myPwd.setText("");
        }
    }
}
```

其中：

（a）**private void loadUser() {   }**：该方法的作用是获取以特定数据存储方式保存的用户信息（用户名和密码）并赋给全局变量 myusername 和 mypassword。本章所有实例对数据的存取操作都集中于这一个方法中，复用同一个登录框架来演示 Android 中不同存储方式下数据的共享，对不同的存储方式只须重写 loadUser() 的代码并修改对应全局变量的声明语句即可，框架的其他程序不用再改动，直接将文件复制过去即可。

（b）**if (name.equals(myusername) && pwd.equals(mypassword))**：点击"登录"按钮，如果当前界面输入的登录名和密码与注册信息相同，则显示欢迎页面，否则提示错误信息。

编写欢迎页的 WelcomeActivity.java，代码如下：

```java
public class WelcomeActivity extends AppCompatActivity {
    private TextView myWelcome;                            //文本视图（显示欢迎信息）
    @Override
    protected void onCreate(Bundle savedInstanceState) {
        super.onCreate(savedInstanceState);
        setContentView(R.layout.welcome);
        findViews();
        showWelcome();
    }
    private void findViews() {
        myWelcome = findViewById(R.id.myLabelWelcome);
    }
    private void showWelcome() {
        Bundle bundle = getIntent().getExtras();
        String name = bundle.getString("name");
        myWelcome.setText("\n" + name + " 您好！\n    欢迎光临");
    }
    public void onBackClick(View view) {
        finish();
    }
}
```

对后面所有实例，欢迎页 WelcomeActivity 的代码也都完全一样，无须改动。

至此，注册和登录两大程序框架都搭建好了，接下来在其上实现本例的功能，用 SharedPreference 保存用户信息。

3. 用 SharedPreference 保存用户信息

实现思路：

采用 SharedPreference 实现用户信息的持久化存取功能，主要包括以下几方面：
① 声明一个 SharedPreferences 对象用于保存数据。
② 设计 User 类方法 saveUser()，以名/值方式将用户注册的 User 属性数据保存起来。
③ 设计 User 类方法 getUserData()，以名/值方式读取注册用户的一组属性数据。

为使读者清楚程序的结构，我们将以上 3 步在下面的程序代码中标示出来。

（1）在工程中创建一个 User 类，User.java 代码如下：

```java
public class User implements Serializable {
    private String name;
    private String pwd;
    private int sex;
    private String birth;
    private String degree;
    //①
    private SharedPreferences sharedPreferences = null;
    private static final String SHARED_NAME = "SharedUserInfo";    //记录用户信息的文件名
    private static int MODE = Context.MODE_WORLD_READABLE + Context.MODE_WORLD_WRITEABLE;
                                                                    //权限一定要设为允许读写
```

```java
        public User() {
        }

        public User(String name, String pwd, int sex, String birth, String degree) {
            this.name = name;
            this.pwd = pwd;
            this.sex = sex;
            this.birth = birth;
            this.degree = degree;
        }

        @Override
        public String toString() {
            String userInfo = name + " " + ((sex == 1) ? "先生" : "女士") + ", 恭喜您注册成功!\n 您的注册信息为:
\n 出生日期          " + birth + "\n 学       历        " + degree;
            return userInfo;
        }

        public void saveUser(Context context) {                                    //②
            sharedPreferences = context.getSharedPreferences(SHARED_NAME, MODE);
                                                                          //创建 SharedPreferences 对象
            SharedPreferences.Editor editor = sharedPreferences.edit();   //得到 Editor 对象
            //保存数据
            editor.putString("name", name);
            editor.putString("pwd", pwd);
            editor.putInt("sex", sex);
            editor.putString("birth", birth);
            editor.putString("degree", degree);
            editor.commit();                                              //提交
        }

        public void getUserData(Context context) {                                 //③
            sharedPreferences = context.getSharedPreferences(SHARED_NAME, MODE);
                                                                          //得到 SharedPreferences 对象
            //获取注册的用户数据
            name = sharedPreferences.getString("name", "");
            pwd = sharedPreferences.getString("pwd", "");
            sex = sharedPreferences.getInt("sex", 1);
            birth = sharedPreferences.getString("birth", "");
            degree = sharedPreferences.getString("degree", "");
        }
    }
```

（2）在 RegisterActivity.java 中实现 onSubmitClick()事件方法，将前端界面输入的注册信息通过 SharedPreferences 形式保存起来，代码如下：

```java
    public class RegisterActivity extends AppCompatActivity {
        ...
        public void onSubmitClick(View view) {
            String name = myName.getText().toString();               //用户名
            String pwd = myPwd.getText().toString();                 //密码
```

```
            int sex = Integer.parseInt(mySex.getText().toString().equals("男") ? "1" : "0");    //性别
            String birth = myBirth.getText().toString();                                          //出生日期
            String degree = mydegreeTemp;                                                         //学历
            User user = new User(name, pwd, sex, birth, degree);                                  // (a)
            user.saveUser(this);                                                                  // (b)
            //跳转页面
            Intent intent = new Intent(this, SuccessActivity.class);
            startActivity(intent);
        }
    }
```

其中：

（a）**User user = new User(name, pwd, sex, birth, degree);**：创建自定义的 User 对象存储用户信息。

（b）**user.saveUser(this);**：调用其 saveUser()方法来保存用户注册的个人信息。

（3）在 LoginActivity.java 中实现 loadUser()方法，读取存储在 SharedPreferences 中的用户数据，代码如下（加粗处为添加的内容）：

```
public class LoginActivity extends AppCompatActivity {
    private EditText myName;
    private EditText myPwd;

    // (a)
    private final static String SHARED_PACKAGE = "com.easybooks.android.sharedreg";
                                                                        //外部应用所在的包
    private static final String SHARED_NAME = "SharedUserInfo";
                                                                        //记录用户信息的文件名
    private Context context = null;                                     //Context 对象用于保存外部应用的上下文
    private SharedPreferences sharedPreferences = null;
    private static int MODE = Context.MODE_WORLD_READABLE + Context.MODE_WORLD_WRITEABLE;
                                                                        //权限一定要设为允许读写
    ...
    private String myusername;                                          //全局变量（用户名）
    private String mypassword;                                          //全局变量（密码）
    @Override
    protected void onCreate(Bundle savedInstanceState) {
        ...
        loadUser();                                                     //调用 loadUser()方法
    }

    ......
    private void loadUser() {                                           // (b)
        try {
            context = createPackageContext(SHARED_PACKAGE, CONTEXT_IGNORE_SECURITY);
            //得到注册应用的 SharedPreferences 对象
            sharedPreferences = context.getSharedPreferences(SHARED_NAME, MODE);
            //获取已注册的用户名和密码并加载显示在界面上
            myusername = sharedPreferences.getString("name", "");
            myName.setText(myusername);
            mypassword = sharedPreferences.getString("pwd", "");
        } catch (PackageManager.NameNotFoundException e) {
            e.printStackTrace();
        }
```

```
        }
        ......
}
```

其中：

（a）**private final static String SHARED_PACKAGE = "com.easybooks.android.sharedreg";**、**private static final String SHARED_NAME = "SharedUserInfo";……private static int MODE = Context.MODE_WORLD_READABLE + Context.MODE_WORLD_ WRITEABLE;**：声明相关的全局变量，指定 SharedPreferences 的工作方式，包括名称、读和写的方式。

（b）**private void loadUser() {...}**：实现 loadUser()方法，在其中获得 SharedPreferences 数据并将用户名和密码赋给全局变量。

（4）在 SuccessActivity.java 中获取 User 对象存储的用户信息，在 showSuccess()中添加代码如下：

```
public class SuccessActivity extends AppCompatActivity {
    private TextView mySuccess;
    private Button myEnter;
    ......
    private void showSuccess() {
        User user = new User();
        user.getUserData(this);                    // (a)
        mySuccess.setText(user.toString());        // (b)
    }
}
```

其中：

（a）**user.getUserData(this);**：把调用 getUserData()方法获取 SharedPreferences 形式保存的用户注册信息保存到 User 类属性中。

（b）**mySuccess.setText(user.toString());**：将 User 类属性信息显示出来。

可见，这里将数据存取操作的实现细节**完全封装**于自定义的 User 类中，本章接下来的实例只需修改 User 类中的代码即可，而外部 RegisterActivity 和 SuccessActivity 的代码完全不用改动，直接将文件复制过去，修改命名空间即可。

4. 运行效果

先打开工程 SharedReg 运行，输入注册信息后提交。结束运行，关闭此工程。然后打开本例的另一个工程 SharedLog 运行。可以看到，程序在初始启动的界面上就已经自动加载显示出了刚刚注册的用户名，这个用户名是由前一个应用 SharedReg 通过 SharedPreferences 存储到共享目录的，故应用 SharedLog 也能访问到。

当用户输入密码登录时，程序会与之前注册时填写的密码进行核对，若不一致，自动清空密码框并提示用户重新输入；若正确，则跳转到欢迎页。

5. 观察存储的数据文件

用共享优先方式存储的数据文件保存在仿真器/data/data/<工程包名>/shared_prefs 目录下，打开工程 SharedReg，启动仿真器。

在 Android Studio 内，选择主菜单"Tools"→"Android"→"Android Device Monitor"命令，打开仿真器的文件管理器，在相应的路径下可找到这个文件，如图 6.5 所示。

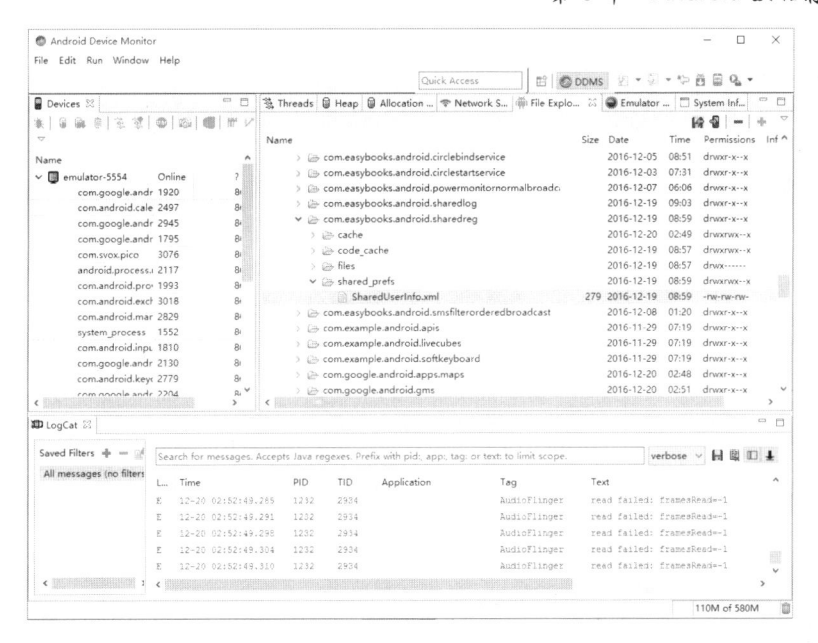

图 6.5 用 SharedPreferences 存储到共享目录中的数据文件

单击图 6.5 窗口右上角的"Pull a file from the device"（图标为 ）按钮，可将该文件导出并保存至本地计算机硬盘，以记事本打开可查看其中的内容（正是用户刚刚注册时输入的信息），如图 6.6 所示。

图 6.6 SharedPreferences 存储文件的内容

6.2 内部文件存储

虽然 SharedPreferences 能够为开发人员简化数据存储和访问过程，但直接使用文件系统保存数据仍然是 Android 数据存储中不可或缺的组成部分。Android 使用 Linux 的文件系统，开发人员可以建立和访问程序自身的私有文件。

6.2.1 Android 系统文件访问

Android 系统允许应用程序创建仅能够被自身访问的私有文件，文件保存在设备的内部存储器上 /data/data/<package name>/files 目录中。Android 系统不仅支持标准 Java 的 IO 类和方法，还提供了能够简化读写流式文件过程的方法。这里主要介绍两个方法：openFileOutput() 和 openFileInput()。

1. openFileOutput()方法

openFileOutput()方法为写入数据做准备而打开文件。如果指定的文件存在，直接打开文件准备写入数据；如果不存在，则创建一个新的文件。该方法声明为：

public FileOutputStream openFileOutput (String name, int mode)

其中，第 1 个参数是文件名称（不可包含描述路径的斜杠）；第 2 个参数是操作模式，Android 支持 4 种文件操作模式，如表 6.1 所示。

表 6.1 Android 的 4 种文件操作模式

模　　式	说　　明
MODE_PRIVATE	私有模式（缺省模式），文件仅能够被创建它（或具有相同 UID）的程序访问
MODE_APPEND	追加模式，如果文件已经存在，则在文件的结尾处添加新数据
MODE_WORLD_READABLE	全局读模式，允许任何程序读取私有文件
MODE_WORLD_WRITEABLE	全局写模式，允许任何程序写入私有文件

该方法的返回值是一个 FileOutputStream 类型，例如：

```
String FILE_NAME = "fileDemo.txt";                            // (a)
FileOutputStream fos = openFileOutput(FILE_NAME,Context.MODE_PRIVATE);
String text = "Some data";
fos.write(text.getBytes());
fos.flush();                                                  // (b)
fos.close();
```

其中：

（a）**String FILE_NAME = "fileDemo.txt";**：定义文件的名称为 fileDemo.txt，以私有模式建立，调用 write()方法将数据写入文件，调用 flush()方法将缓冲数据写入文件，最后调用 close()方法关闭 FileOutputStream。

（b）**fos.flush();**：为了提高文件系统的性能，一般调用 write()方法时，如果写入的数据量较小，系统会把数据保存在缓冲区中，等数据量积累到一定程度时再将其一次性写入文件。因此，在调用 close()方法关闭文件前，务必先调用 flush()方法，将缓冲区内所有的数据写入文件。如果开发人员在调用 close()方法前未调用 flush()，则可能导致部分数据丢失。

2. openFileInput()方法

openFileInput()方法为读取数据做准备而打开文件。该方法声明为：

public FileInputStream openFileInput (String name)

其中，第 1 个参数也是文件名称（不可包含描述路径的斜杠）。例如，以二进制方式读取数据：

```
String FILE_NAME = "fileDemo.txt";
FileInputStream fis = openFileInput(FILE_NAME);
byte[] readBytes = new byte[fis.available()];
while(fis.read(readBytes) != -1){ ... }
```

6.2.2 文件存储举例

【例 6.2】 用内部文件来保存用户注册的信息。

由于文件方式存储的数据只能由创建它的程序自身访问，不支持跨应用访问，故本例只包含一个 Android Studio 工程，实现注册及信息回显的功能。

1. 创建注册应用

创建 Android Studio 工程 InterReg，直接套用【例 6.1】的注册框架，将相关的源文件复制到本工程对应的目录即可。

2. 实现文件存取

需要修改重新实现的只有 User 类，本例改用使用内部文件的方式来实现存取，代码如下：

```java
public class User implements Serializable {
    ...
    //声明 FileOutputStream/FileInputStream 流类用于访问手机内存
    private FileOutputStream fos = null;
    private FileInputStream fis = null;
    private static final String INTER_NAME = "InterUserInfo";        //记录用户信息的文件名
    private static int MODE = Context.MODE_WORLD_READABLE + Context.MODE_WORLD_WRITEABLE;     //权限一定要设为允许读写
    ......
    public void saveUser(Context context) {
        try {
            fos = context.openFileOutput(INTER_NAME, MODE);          //获得文件输出流
            //保存数据
            String userdata = name;
            userdata += ";" + pwd;
            userdata += ";" + String.valueOf(sex);
            userdata += ";" + birth;
            userdata += ";" + degree;
            fos.write(userdata.getBytes());
            fos.flush();                                              //清除缓存
        } catch (FileNotFoundException e) {
            e.printStackTrace();
        } catch (IOException e) {
            e.printStackTrace();
        } finally {
            if (fos != null) {
                try {
                    fos.close();                                      //及时关闭文件输出流
                } catch (IOException e) {
                    e.printStackTrace();
                }
            }
        }
    }

    public void getUserData(Context context) {
        try {
            fis = context.openFileInput(INTER_NAME);                  //获得文件输入流
            byte[] buffer = new byte[fis.available()];                //定义暂存数据的数组
            fis.read(buffer);                                         //从输入流中读取数据
            //获取注册的用户数据
            String data = new String(buffer);
            name = data.split(";")[0];
            pwd = data.split(";")[1];
            sex = Integer.parseInt(data.split(";")[2]);
```

```
                birth = data.split(";")[3];
                degree = data.split(";")[4];
        } catch (FileNotFoundException e) {
                e.printStackTrace();
        } catch (IOException e) {
                e.printStackTrace();
        } finally {
                if (fis != null) {
                        try {
                                fis.close();                    //及时关闭文件输入流
                        } catch (IOException e) {
                                e.printStackTrace();
                        }
                }
        }
}
```

3. 运行效果

运行效果同【例 6.1】的注册框架，如图 6.1 和图 6.2 所示。

4. 观察存储的数据文件

用内部文件方式存储的数据文件保存在仿真器/data/data/<工程包名>/files 目录下，打开工程 InterReg，启动仿真器，打开其文件管理器，在相应的路径下可找到这个文件，如图 6.7 所示。

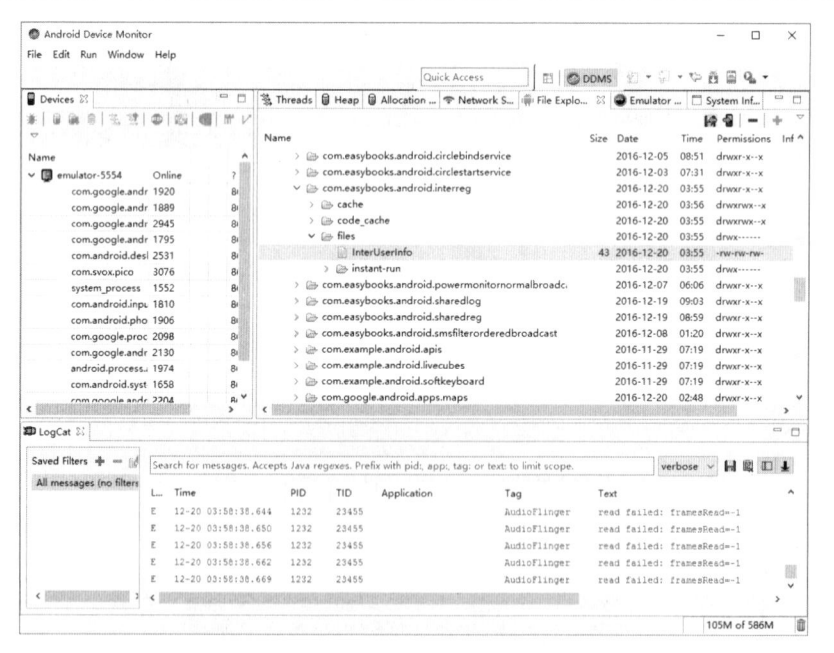

图 6.7 用内部文件方式存储的数据文件

单击窗口右上角图标 按钮，可将该文件导出并保存至本地计算机硬盘，以记事本打开可查看其中的内容（也正是用户刚刚注册时输入的信息），如图 6.8 所示。

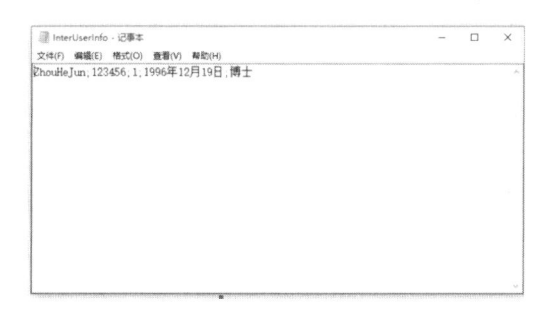

图 6.8　内部文件的内容

6.3　SQLite 数据库存储与共享

6.3.1　SQLite 概述

传统应用程序所使用的数据库管理系统比较庞大复杂，会占用较多的系统资源。随着嵌入式应用程序的大量出现，2000 年，D. Richard Hipp 开发了开源嵌入式关系数据库 SQLite，它占用资源少，运行高效可靠，可移植性强，并且提供了零配置运行模式。它屏蔽了数据库使用和管理的复杂性，应用程序仅做最基本的数据操作，其他操作则交给进程内部的数据库引擎完成。同时，因为客户端和服务器在同一进程空间运行，所以完全不需要进行网络配置和管理，减少了网络调用所造成的额外开销。以这种方式简化的数据库管理过程，使应用程序更加易于部署和使用，程序开发人员仅需要把 SQLite 正确编译到应用程序中即可。

SQLite 采用了模块化设计，将复杂的查询过程分解为细小的工作进行处理。它由 8 个独立的模块构成，这些模块又构成了 3 个主要的子系统：编译器、核心模块和后端。SQLite 数据库体系结构如图 6.9 所示。

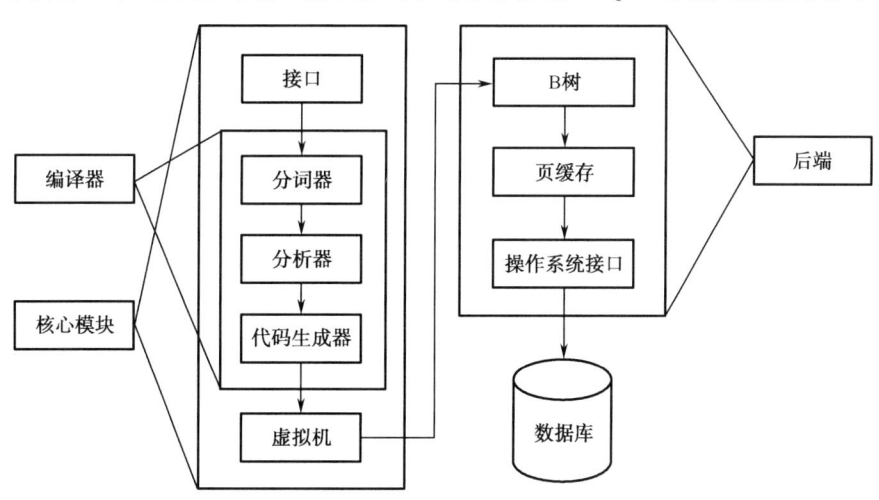

图 6.9　SQLite 数据库体系结构

（1）编译器。

在编译器中，分词器和分析器对 SQL 语句进行语法检查，然后把 SQL 语句转化为便于底层处理的分层数据结构，这种分层的数据结构称为"语法树"。然后把语法树传给代码生成器进行处理，生成一种用于 SQLite 的汇编代码，最后由虚拟机执行。

（2）核心模块。

核心模块囊括了编译器，对外提供接口，经编译器再由虚拟机接往后端。

接口由 SQLite C API 组成，无论是应用程序、脚本，还是库文件，最终都是通过接口与 SQLite 交互。

SQLite 数据库体系结构中最核心的部分是虚拟机，也称为虚拟数据库引擎（Virtual Database Engine，VDBE）。与 Java 虚拟机相似，VDBE 用来解释并执行字节代码。字节代码由 128 个操作码构成，用来对数据库进行操作，每一条指令都可以完成特定的操作，或以特定的方式处理栈的内容。

（3）后端。

后端由 B 树、页缓存和操作系统接口构成，B 树和页缓存共同对数据进行管理。B 树的主要功能就是索引，它维护着各个页面之间复杂的关系，以便快速找到所需数据。页缓存的主要作用是通过操作系统接口在 B 树和磁盘之间传递页面。

SQLite 数据库具有很强的可移植性，既可以运行在 Windows、Linux、BSD、Mac OS 以及其他一些商用 UNIX 系统（比如 Sun 的 Solaris 和 IBM 的 AIX）中，也可以工作在许多嵌入式操作系统下，比如 QNX、VxWorks、Palm OS、Symbian 和 Windows CE。SQLite 的核心大约有 3 万行标准 C 代码，因为模块化的设计而使这些代码非常易于理解。

6.3.2 SQLite 应用举例

【例 6.3】 用 SQLite 数据库存取注册信息。

对于 SQLite 方式存储的数据，Android 官方建议使用 ContentProvider 来实现与其他应用的共享，本例先开发一个 Android Studio 工程实现注册回显功能。稍后在介绍了 ContentProvider 后，再运用 ContentProvider 将数据库中的表数据对外公开（通过 URI 地址）。开发一个登录程序去访问数据库的数据。

1. 创建注册应用

创建 Android Studio 工程 SqliteReg，直接套用【例 6.1】开发好的注册框架，将相关的源文件复制到本工程对应的目录即可。

2. 创建 SQLite 适配器类

在工程中创建 SQLiteAdapter 类，用于封装 SQLite 的创建、打开、关闭及获取数据库实例的操作。
SQLiteAdapter.java 代码如下：

```java
public class SQLiteAdapter {
    //声明数据库的基本信息
    private static final int DB_VERSION = 1;                        //数据库版本
    private static final String TABLE_NAME = "users";               //记录用户信息的表名
    //表中各个字段定义
    private static final String _ID = "_id";                        //保存 id 值
    private static final String NAME = "name";                      //用户名
    private static final String PWD = "pwd";                        //密码
    private static final String SEX = "sex";                        //性别
    private static final String BIRTH = "birth";                    //出生日期
    private static final String DEGREE = "degree";                  //学历
    //声明操作 Sqlite 数据库的实例
    private SQLiteDatabase sqliteDb;
    private DBOpenHelper sqliteHelper;
```

```java
//构造方法
public SQLiteAdapter(Context context, String dbname) {
    sqliteHelper = new DBOpenHelper(context, dbname, null, DB_VERSION);
    sqliteDb = sqliteHelper.getWritableDatabase();                    //获得可写的数据库
}

//自定义的帮助类
private static class DBOpenHelper extends SQLiteOpenHelper {
    public DBOpenHelper(Context context, String dbname, SQLiteDatabase.CursorFactory factory, int version)
    {
        super(context, dbname, factory, version);
    }

        private static final String CREATE_TABLE = "create table " + TABLE_NAME
            + "(" + _ID + " integer primary key autoincrement,"
            + NAME + " text not null,"
            + PWD + " text not null,"
            + SEX + " integer not null,"
            + BIRTH + " text,"
            + DEGREE + " text);";                                     //预定义创建表的 SQL 语句

    @Override
    public void onCreate(SQLiteDatabase db) {
        db.execSQL(CREATE_TABLE);                                     //创建表
    }

    @Override
    public void onUpgrade(SQLiteDatabase db, int oldVersion, int newVersion) {
        db.execSQL("drop table if exists " + TABLE_NAME);             //删除老表
        onCreate(db);                                                 //创建新表
    }
}

//获取 SQLite 数据库实例
public SQLiteDatabase getSqliteDb() {
    return sqliteDb;
}
}
```

3. 用 SQLite 保存用户信息

修改重新实现 User 类，本例改用 SQLite 来实现存取，代码如下：

```java
public class User implements Serializable {
    private int _id;                                                  // (a)
    private String name;
    private String pwd;
    private int sex;
    private String birth;
    private String degree;
    private SQLiteAdapter sqLiteAdapter = null;                       // (b)
    private SQLiteDatabase sqLiteDb = null;                           //数据库实例
    private static final String SQLITE_NAME = "SqliteUserInfo";       //数据库名称
    private static final String TABLE_NAME = "users";                 //记录用户信息的表名
    ......
```

```java
        public void saveUser(Context context) {
            sqLiteAdapter = new SQLiteAdapter(context, SQLITE_NAME);    //创建 SQLiteAdapter 对象
            sqLiteDb = sqLiteAdapter.getSqliteDb();                     //得到 SQLite 实例
            ContentValues values = new ContentValues();                 //构造 ContentValues 实例
            //保存数据
            values.put("name", name);
            values.put("pwd", pwd);
            values.put("sex", sex);
            values.put("birth", birth);
            values.put("degree", degree);
            sqLiteDb.insert(TABLE_NAME, null, values);                  //添加数据
        }

        public void getUserData(Context context) {
            sqLiteAdapter = new SQLiteAdapter(context, SQLITE_NAME);    //创建 SQLiteAdapter 对象
            sqLiteDb = sqLiteAdapter.getSqliteDb();                     //得到 SQLite 实例
            Cursor cursor = sqLiteDb.query(TABLE_NAME, new String[]{"name", "pwd", "sex", "birth", "degree"},
null, null, null, null, null);                                          //获取注册的用户数据
            if (cursor.getCount() > 0) {
                cursor.moveToFirst();
                name = cursor.getString(0);
                pwd = cursor.getString(1);
                sex = cursor.getInt(2);
                birth = cursor.getString(3);
                degree = cursor.getString(4);
            }
            cursor.close();
        }
    }
```

其中：

（a）**private int _id;**：保存用户的 id。如果计划使用 ContentProvider 来共享表，就必须具有唯一的 id 字段。

（b）**private SQLiteAdapter sqLiteAdapter = null;**：声明一个 SQLiteAdapter 对象，作为访问 SQLite 数据库的中介，也就是刚刚创建的 SQLite 适配器。

4．运行效果

运行效果同【例 6.1】的注册框架，如图 6.1 和图 6.2 所示。

5．观察存储的数据

用 SQLite 数据库方式存储的数据文件保存在仿真器/data/data/<项目包名>/databases 目录下，打开工程 SqliteReg，启动仿真器，打开其文件管理器，在相应的路径下可找到这个文件，如图 6.10 所示。

查看 SQLite 数据库表的内容一般通过 Android 自带的 sqlite3 进行，在 Android SDK 的 tools 目录中有 sqlite3 工具，可通过 Windows 命令行启动。操作前要确保已经配置了正确的环境变量，以便 Windows 能够找到 sqlite3，如图 6.11 所示。

打开 Windows 命令行，输入"adb shell"命令，进入本工程目录/data/data/com.easybooks.android.sqlitereg/databases。

输入"ls"命令并回车，可看到工程运行所生成的 SQLite 数据库文件 SqliteUserInfo。

输入"sqlite3 SqliteUserInfo"命令并回车，启动 sqlite3 工具。

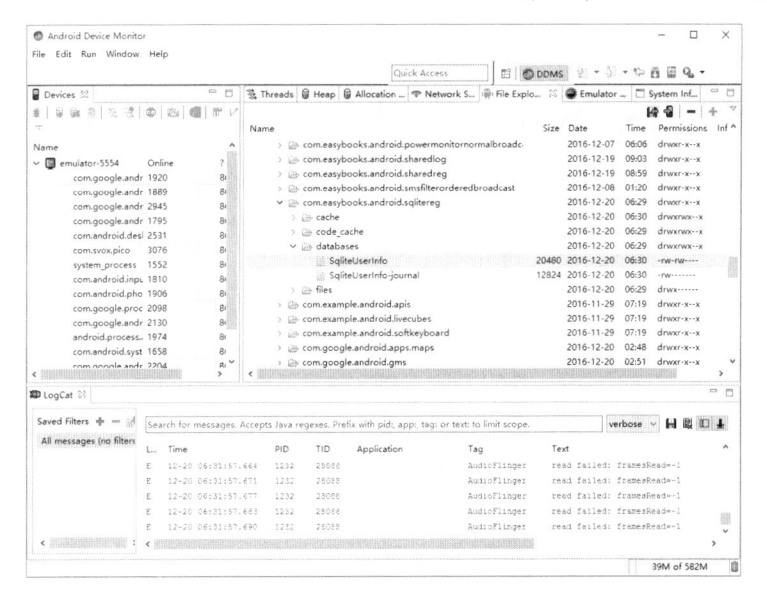

图 6.10　用 SQLite 方式存储的数据文件

图 6.11　配置环境变量

输入".tables"命令并回车，查看数据库中的表，这时可看到本例程序所创建的 users 表，该表中存有注册用户的信息，用"select * from users"语句直接查询即可看到内容。

6.4　ContentProvider 数据共享组件

6.4.1　ContentProvider 组件

ContentProvider 组件实现了应用程序之间共享数据的一种接口机制。应用程序运行在不同的进程中，因此数据和文件在不同的应用程序之间是不能直接访问的。通过 ContentProvider，应用程序可以指定需要共享的数据，而其他应用程序则可以在不知道数据来源、路径的情况下，对共享数据进行查询、添加、删除和更新等操作。

在创建 ContentProvider 前，首先要实现底层的数据源。数据源可以是数据库、文件系统或网络等，然后继承 ContentProvider 类中实现基本数据操作的接口方法，包括添加、删除、查找和更新等功能。调用者不能直接调用 ContentProvider 的接口方法，而需要使用 ContentResolver 对象，通过 URI 间接调用，调用关系如图 6.12 所示。

在 ContentResolver 对象与 ContentProvider 进行交互时，通过 URI 确定要访问的 ContentProvider 数据集。在发起一个请求的过程中，系统根据 URI 确定处理这个查询的 ContentProvider，然后初始化 ContentProvider 所需的资源。一般情况下只有一个 ContentProvider 对象，但却可以同时与多个 ContentResolver 进行交互。

```
                    URI
ContentResolver ──────→ ContentProvider
                            │
          ┌─────────────────┼─────────────────┐
          ↓                 ↓                 ↓
      文件系统            数据库              网络
```

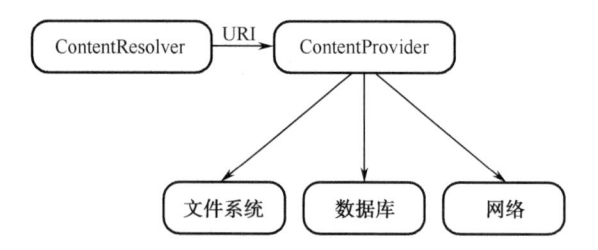

图 6.12 ContentProvider 调用关系

ContentProvider 完全屏蔽了数据提供者的具体数据存储方式。数据提供者可以使用 SQLite 数据库存储数据，也可以通过文件系统或 SharedPreferences 存储数据，甚至使用网络存储。ContentProvider 提供了一组标准的数据操作接口，使用者只要调用接口中的方法即可完成所有的数据操作。

ContentProvider 的数据集类似于数据库的表，每行是一条记录，每列具有相同的数据类型，如表 6.2 所示。

表 6.2 ContentProvider 数据集

_ID	NAME	AGE	HEIGHT
1	Tom	21	1.81
2	Jim	22	1.78

每条记录都包含一个长整型的字段_ID，用来唯一标识每条记录。可以提供多个数据集，调用者使用 URI 对不同数据集的数据进行操作。

URI 是统一资源标识符，用来定位远程或本地的可用资源，其格式为：

content://<授权者>/<数据路径>/< id>

其中：

- **content://**：表示该 URI 用于 ContentProvider 定位资源。
- **<授权者>**：确定 ContentProvider 资源。一般由类的小写全称组成，以保证唯一性。
- **<数据路径>**：确定数据集。如果 ContentProvider 仅提供一个数据集，数据路径可省略；如果提供多个数据集，则必须用数据路径指明具体是哪一个数据集。数据路径可以写成多段格式。
- **<id>**：数据编号，用来唯一确定数据集中的一条记录，匹配数据集中_ID 字段的值。如果请求的数据不只限于一条，则<id>可以省略。

例如：

请求整个 computer 数据集的 URI 应写为：

content://edu.njone.computerprovider/computer

请求 computer 数据集中第 3 条记录的 URI 则应写为：

content://edu.njone.computerprovider/computer/3

6.4.2 ContentProvider 创建

通过继承 ContentProvider 类可以创建一个新的数据提供者，过程分为 3 步。

1. 继承 ContentProvider 重载方法

新创建的类继承 ContentProvider 后，需要重载 delete()、insert()、query()、update()和 onCreate()、getType()方法。其中，前面 4 个方法分别用于对数据集的删除、添加、查询和更新操作。onCreate()一般用来初始化底层数据集和建立数据连接等工作。getType()用来返回指定 URI 的 MIME 数据类型，如果 URI 是单条数据，则返回的 MIME 数据类型应以 vnd.android.cursor.item/开头；如果 URI 是多条数据，则返回的 MIME 数据类型应以 vnd.android.cursor.dir/开头。

下面是继承 ContentProvider 后需要开发人员重载的代码框架：

```java
import android.content.*;
import android.database.Cursor;
import android.net.Uri;

public class PeopleProvider extends ContentProvider{
    @Override
    public int delete(Uri uri, String selection, String[] selectionArgs) {
        // TODO Auto-generated method stub
        return 0;
    }
    @Override
    public String getType(Uri uri) {
        // TODO Auto-generated method stub
        return null;
    }
    @Override
    public Uri insert(Uri uri, ContentValues values) {
        // TODO Auto-generated method stub
        return null;
    }

    @Override
    public boolean onCreate() {
        // TODO Auto-generated method stub
        return false;
    }
    @Override
    public Cursor query(Uri uri, String[] projection, String selection,
            String[] selectionArgs, String sortOrder) {
        // TODO Auto-generated method stub
        return null;
    }
    @Override
    public int update(Uri uri, ContentValues values, String selection,
            String[] selectionArgs) {
        // TODO Auto-generated method stub
        return 0;
    }
}
```

2. 声明 CONTENT_URI 构造 UriMatcher

在新创建的 ContentProvider 类中，还需构造一个 UriMatcher 判断 URI 是单条还是多条数据。同时，为了便于判断和使用 URI，一般将 URI 的授权者名称和数据路径等内容声明为静态常量，并声明 CONTENT_URI。声明 CONTENT_URI 和构造 UriMatcher 的代码如下：

```java
public static final String AUTHORITY = "edu.hrbeu.peopleprovider";      // (a)
public static final String PATH_SINGLE = "people/#";                     // (b)
public static final String PATH_MULTIPLE = "people";                     // (b)
```

```
        public static final String CONTENT_URI_STRING = "content://" + AUTHORITY + "/" + PATH_MULTIPLE;
                                                                                        // (c)
        public static final Uri CONTENT_URI = Uri.parse(CONTENT_URI_STRING);     // (d)
        private static final int MULTIPLE_PEOPLE = 1;                            // (e)
        private static final int SINGLE_PEOPLE = 2;                              // (e)

        private static final UriMatcher uriMatcher;                              // (f)
        static {
            uriMatcher = new UriMatcher(UriMatcher.NO_MATCH);
            uriMatcher.addURI(AUTHORITY, PATH_SINGLE, MULTIPLE_PEOPLE);
            uriMatcher.addURI(AUTHORITY, PATH_MULTIPLE, SINGLE_PEOPLE);
        }
```

其中：

（a）**public static final String AUTHORITY = "edu.hrbeu.peopleprovider";**：声明 URI 的授权者名称。

（b）**public static final String PATH_SINGLE = "people/#";** 和 **public static final String PATH_MULTIPLE = "people";**：声明单条和多条数据的数据路径。

（c）**public static final String CONTENT_URI_STRING = "content://" + AUTHORITY + "/" + PATH_MULTIPLE;**：声明 CONTENT_URI 的字符串形式。

（d）**public static final Uri CONTENT_URI = Uri.parse(CONTENT_URI_STRING);**：正式声明 CONTENT_URI。

（e）**private static final int MULTIPLE_PEOPLE = 1;** 和 **private static final int SINGLE_PEOPLE = 2;**：声明多条数据和单条数据的返回代码。

（f）**private static final UriMatcher uriMatcher;**：声明 UriMatcher，并在静态构造函数中声明匹配方式和返回代码。

其中，UriMatcher.NO_MATCH 是 URI 无匹配时的返回代码。public void addURI (String authority, String path, int code)方法用来添加新的匹配项，authority 表示匹配的授权者名称，path 表示数据路径，#可以代表任何数字，code 表示返回代码。

使用 UriMatcher 时，可以直接调用 match()方法对指定的 URI 进行判断。例如：

```
switch(uriMatcher.match(uri)) {
    case MULTIPLE_PEOPLE:
        //多条数据的处理过程
        break;
    case SINGLE_PEOPLE:
        //单条数据的处理过程
        break;
    default:
        throw new IllegalArgumentException("不支持的 URI:" + uri);
}
```

3. 注册 ContentProvider

SQLiteProvider 直接继承自 ContentProvider，是一个提供数据共享的 ContentProvider 组件，在使用它之前，必须在 AndroidManifest.xml 中注册，代码如下：

```
<application
    ...
    <provider
```

```
        android:authorities="com.easybooks.android.sqlitereg.SQLiteProvider"
        android:name=".SQLiteProvider"
        android:exported="true"/>
</application>
```

6.4.3 ContentProvider 应用举例

【例6.4】 使用 ContentProvider 共享工程 SqliteReg 的注册用户数据，并另外开发一个应用来访问共享的数据。

实现思路：

（1）编写一个继承自 ContentProvider 的类，其中编写实现共享数据添加、删除、更新和查询的接口方法。

（2）在另一个工程的 LoginActivity 中，通过 ContentResolver 调用 ContentProvider 的查询（query()）方法获取 SQLite 数据库中的用户名和密码信息。

1. 共享 SQLite 数据

打开工程 SqliteReg，在其中添加一个 SQLiteProvider 提供数据共享，代码如下：

```java
public class SQLiteProvider extends ContentProvider {
    private SQLiteAdapter sqLiteAdapter;                          //声明 SQLiteAdapter 对象
    private SQLiteDatabase sqLiteDb;                              //数据库实例
    private static final String SQLITE_NAME = "SqliteUserInfo";   //数据库名称
    private static final String TABLE_NAME = "users";             //记录用户信息的表名

    private static final int USERS = 1;
    private static final int USER = 2;
    private static final UriMatcher MATCHER;

    static {
        MATCHER = new UriMatcher(UriMatcher.NO_MATCH);
        MATCHER.addURI("com.easybooks.android.sqlitereg.SQLiteProvider", "users", USERS);
                                                                  //不带主键编号的 Uri
        MATCHER.addURI("com.easybooks.android.sqlitereg.SQLiteProvider", "users/#", USER);
                                                                  //带主键编号的 Uri
    }

    @Override
    public boolean onCreate() {
        sqLiteAdapter = new SQLiteAdapter(getContext(), SQLITE_NAME);
        sqLiteDb = sqLiteAdapter.getSqliteDb();
        if (sqLiteDb == null) return false;
        else return true;
    }

    /*
     * 返回当前 Uri 所代表数据的 MIME 类型：
     * 如果操作的数据属于集合类型，那么 MIME 类型字符串应该以 vnd.android.cursor.dir/开头
     * 如果操作的数据属于非集合类型，那么 MIME 类型字符串应该以 vnd.android.cursor.item/开头
     */
    @Override
    public String getType(Uri uri) {
```

```java
            switch (MATCHER.match(uri)) {
                case USERS:
                    return "vnd.android.cursor.dir/vnd.easybooks.users";
                case USER:
                    return "vnd.android.cursor.item/vnd.easybooks.users";
                default:
                    throw new IllegalArgumentException("Failed to getType:" + uri.toString());
            }
        }
        //供外部应用从 ContentProvider 添加数据
        @Override
        public Uri insert(Uri uri, ContentValues values) {
            switch (MATCHER.match(uri)) {
                case USERS:
                    long userId = sqLiteDb.insert(TABLE_NAME, null, values);
                    Uri insertUri = ContentUris.withAppendedId(uri, userId);
                    getContext().getContentResolver().notifyChange(insertUri, null);
                    return insertUri;
                default:
                    throw new IllegalArgumentException("Failed to insert:" + uri.toString());
            }
        }

        // 供外部应用从 ContentProvider 删除数据
        @Override
        public int delete(Uri uri, String selection, String[] selectionArgs) {
            int count = 0;
            switch (MATCHER.match(uri)) {
                case USERS:
                    count = sqLiteDb.delete(TABLE_NAME, selection, selectionArgs);
                    break;
                case USER:
                    String segment = uri.getPathSegments().get(1);
                    count = sqLiteDb.delete(TABLE_NAME, "_id=" + segment, selectionArgs);
                    break;
                default:
                    throw new IllegalArgumentException("Failed to delete:" + uri.toString());
            }
            getContext().getContentResolver().notifyChange(uri, null);
            return count;
        }

        // 供外部应用更新 ContentProvider 中的数据
        @Override
        public int update(Uri uri, ContentValues values, String selection, String[] selectionArgs) {
            int count = 0;
            switch (MATCHER.match(uri)) {
                case USERS:
                    count = sqLiteDb.update(TABLE_NAME, values, selection, selectionArgs);
                    break;
                case USER:
```

```
                String segment = uri.getPathSegments().get(1);
                count = sqLiteDb.update(TABLE_NAME, values, "_id=" + segment, selectionArgs);
                break;
            default:
                throw new IllegalArgumentException("Failed to update:" + uri.toString());
        }
        getContext().getContentResolver().notifyChange(uri, null);
        return count;
    }

    // 供外部应用从 ContentProvider 中获取数据
    @Override
    public Cursor query(Uri uri, String[] projection, String selection, String[] selectionArgs, String sortOrder)
    {
        SQLiteQueryBuilder qb = new SQLiteQueryBuilder();
        qb.setTables(TABLE_NAME);
        switch (MATCHER.match(uri)) {
            case USER:
                qb.appendWhere("_id=" + uri.getPathSegments().get(1));
                break;
            default:
                break;
        }
        Cursor cursor = qb.query(sqLiteDb, projection, selection, selectionArgs, null, null, sortOrder);
        cursor.setNotificationUri(getContext().getContentResolver(), uri);
        return cursor;
    }
}
```

2. 创建登录应用

创建 Android Studio 工程 SqliteLog，直接套用【例 6.1】开发好的登录框架，将相关的源文件及资源复制到本工程对应的目录即可。

只需修改 LoginActivity 中的 loadUser()方法并声明对应的全局变量，代码如下：

```
public class LoginActivity extends AppCompatActivity {
    ...
    private static final String CONTENT_URI = "content://com.easybooks.android.sqlitereg.SQLite Provider/users/1";
    private ContentResolver resolver = null;
    private String myusername;                                              //全局变量（用户名）
    private String mypassword;                                              //全局变量（密码）
    ......
    private void loadUser() {
        resolver = getContentResolver();
        Uri uri = Uri.parse(CONTENT_URI);
        Cursor cursor = resolver.query(uri, new String[]{"name", "pwd"}, null, null, null);
        if (cursor != null) {
            cursor.moveToFirst();
            myusername = cursor.getString(0);
            myName.setText(myusername);
            mypassword = cursor.getString(cursor.getColumnIndex("pwd"));
        }
```

```
        }
        ......
}
```

3. 运行效果

运行效果同【例 6.1】的注册框架,如图 6.1 和图 6.2 所示。

第 7 章 Android 数据库和网络编程

这部分包括 Android 直连数据库的 JDBC 编程、通过 Web 服务器访问后台数据库的 HTTP 编程、Android 与 WebService 的交互以及 JSON 在移动互联网数据交互中的应用，下面分别进行介绍。

7.1 数据库准备

1. 数据库环境

本章所有的实例程序一共用到 3 个数据库：MySQL、SQL Server 和 Oracle，笔者将它们分别安装在 3 台不同的笔记本计算机上，每台计算机的主机名、IP、操作系统及所安装数据库的版本如表 7.1 所示。

表 7.1　本章实例所用的数据库

主 机 名	IP	操 作 系 统	数 据 库
SMYSQL	192.168.0.251	Windows Server 2012 R2 标准版 64 位	MySQL 8.0
LAPTOP-8SJBOG5R	192.168.0.183	Windows 10	SQL Server 2019
SMONGO	192.168.0.252	Windows Server 2012 R2 标准版 64 位	Oracle 11g

2. Navicat Premium 工具

为管理数据库方便，我们使用当前较为流行的 Navicat Premium 统一管理和操作以上 3 个数据库。Navicat Premium 是 Navicat 的衍生版本，提供了对数据库领域最广泛的 DBMS 兼容性支持，用它几乎能够操作所有类型的 DBMS 产品。Navicat Premium 在本书编写时的最新版本是 12.1，可从官网 https://www.navicat.com.cn/download/navicat-premium 免费下载获得，笔者使用的是 Windows 版（64 位），下载得到的安装文件名为 "navicat121_premium_cs_x64.exe"，双击启动安装向导，按照向导的指引往下安装，每一步都取默认，安装过程从略。

启动 Navicat Premium，单击主界面左上角工具栏的"连接"按钮，出现下拉菜单，可选择创建到不同数据库的连接，如图 7.1 所示。

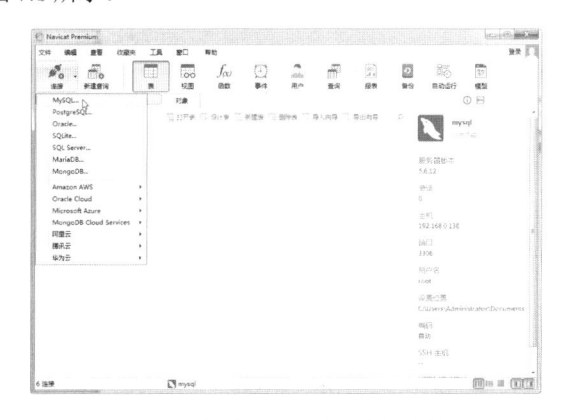

图 7.1　创建到不同数据库的连接

3. 创建数据库及连接

为接下来开发实例做准备，我们在 3 台计算机的 MySQL、SQL Server、Oracle 数据库中分别创建名为 emarket（电子商城）的数据库，其中创建一个 commodity（商品）表，各数据库中的表结构如图 7.2 所示（其中 Oracle 由于默认只支持大写，故数据库名和表名也都使用全大写英文名称），创建完成后分别往 3 个数据库的表中录入 3 条样本数据。

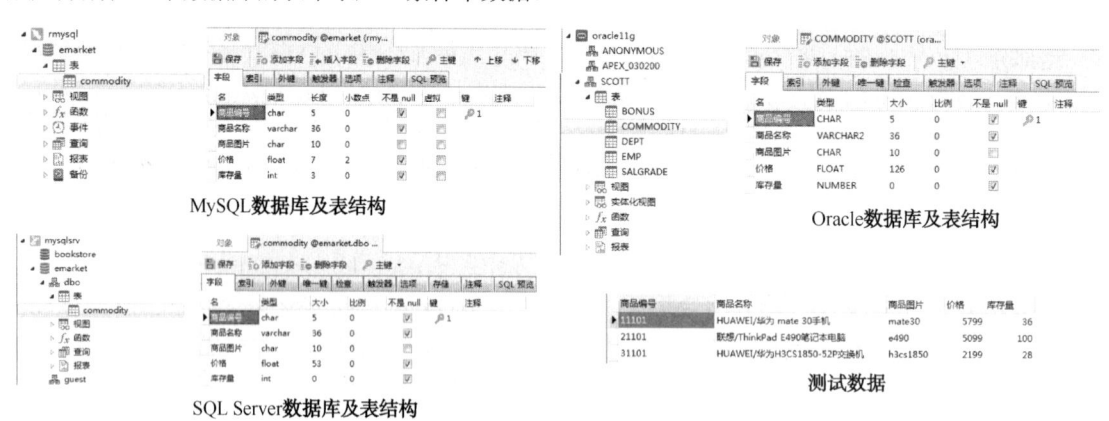

图 7.2 本章用到的各数据库、表及数据

接下来，通过 Navicat Premium 分别创建到这 3 个数据库的连接，连接名分别为 rmysql（MySQL 8.0）、mysqlsrv（SQL Server 2019）和 oracle11g（Oracle 11g），各个连接的具体配置信息如图 7.3 所示，读者请根据自己计算机上安装数据库的实际情况来配置连接。

图 7.3 配置到各数据库的连接

7.2 Android JDBC 编程

7.2.1 基本原理

Android 程序是用 Java 语言编写的，而 JDBC 是 Java 语言访问数据库的通用接口，故 Android 程序可以通过相应数据库的 JDBC 驱动包直连数据库。出于安全性考虑，并不是所有数据库的 DBMS 都

允许 Android 程序直接连接，已有数据库的 JDBC 驱动也未必都支持 Android 程序的连接。例如，Oracle 作为一款重量级 DBMS，它的新版本 Oracle 19c 目前就尚未提供支持 Android 连接的 JDBC 驱动；而另一些大型商用主流数据库，如微软 SQL Server，虽然其官方不提供支持 Android 的 JDBC 驱动，但某些开源组织和第三方却提供了 Android 连接 SQL Server 的驱动；MySQL 数据库支持 Android 的 JDBC 直连，但也要用特定的.jar 包才能做到。基于上述现状，一般 Java 程序能用来操作某个 DBMS 的驱动直接套用在 Android 程序上未必适用，这一点请读者务必知晓。

1. 系统结构

Android 程序直连数据库的系统结构如图 7.4 所示，其中画出了 Android 9.0 程序连接 MySQL、SQL Server 和 Oracle 11g 的情况以及它们所需.jar 包的具体版本信息。

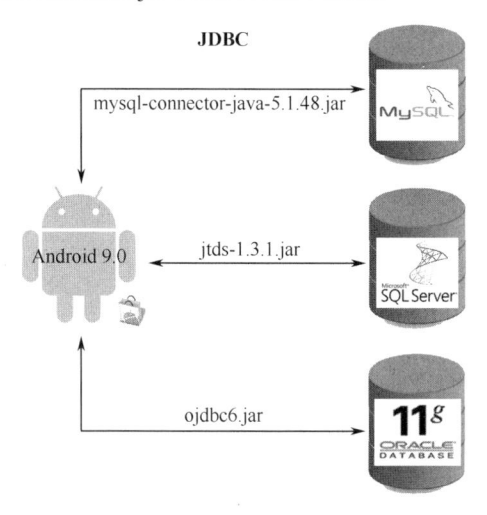

图 7.4 Android 程序直连数据库的系统结构

2. 配置支持直连数据库的工程

当 Android 程序要访问某个 DBMS 时，必须首先在工程中添加该 DBMS 支持 Android 直连的 JDBC 驱动。下面我们来创建和配置一个支持直连数据库的工程，步骤如下：

（1）创建 Android Studio 工程 MySqlTest。

（2）将预先准备好的 JDBC 驱动（.jar 包）拷贝到工程的 app\libs 目录下，刷新。

（3）在 Android Studio 中选择主菜单"File"→"Project Structure..."命令，弹出"Project Structure"窗口，如图 7.5 所示。左侧栏切换至"Dependencies"页，中间"Modules"选"app"，此时右边区域列出了当前工程所依赖的全部 Java 库和包及所在目录。

（4）单击 Declared Dependencies 区域左上角 + 按钮，选择"Jar Dependency"，从弹出的对话框下拉列表中选择需要加载的.jar 包，如图 7.6 所示。

加载完成之后就可以在编程中使用它来连接数据库了。

（5）最后，不要忘了在 AndroidManifest.xml 中添加配置：

```
<uses-permission android:name="android.permission.INTERNET"/>
```

打开 Android 程序的网络连接。

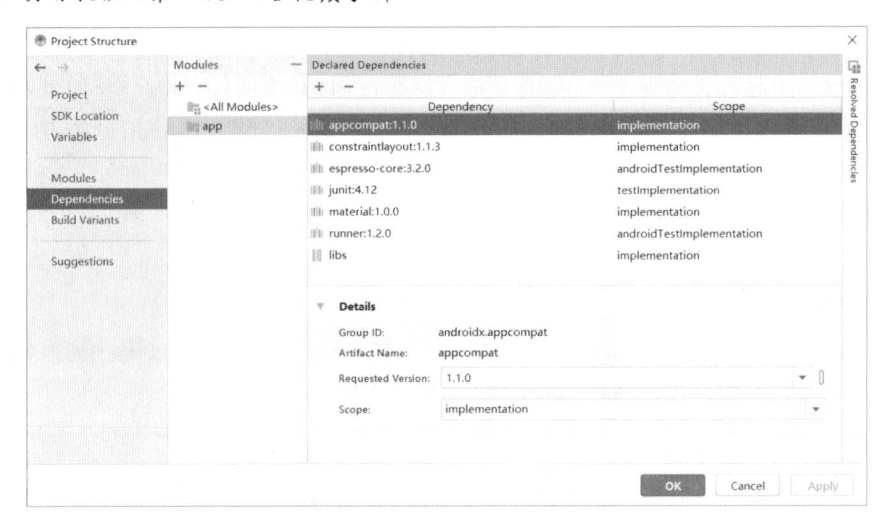

图 7.5　添加 Android Studio 工程的依赖项

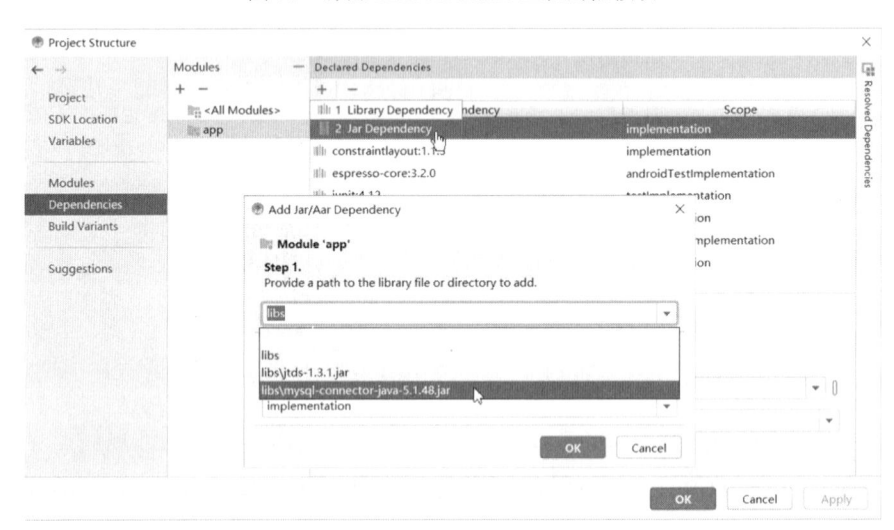

图 7.6　选择需要加载的.jar 包

7.2.2　Android JDBC 直连 MySQL

【例 7.1】 用 JDBC 直连的方式访问 MySQL，读出其中信息。

实现思路：

（1）在程序开头导入 JDBC 驱动相关的类，并在主程序 MainActivity 类中声明数据库连接对象、SQL 语句对象和查询结果集对象。

（2）自定义一个 connToDb()方法专门连接和访问数据库，该方法中开启一个线程，在这个单独的线程中执行连接数据库和查询商品信息的一系列操作。

（3）查询的结果封装在一个 Message 对象中，通过句柄 Handler 发给主程序，刷新 UI 界面。

我们借助的 JDBC 驱动包版本为 mysql-connector-java-5.1.48.jar，这是本书编写时能支持 Android 程序直连 MySQL 的最高版本驱动。更新版本的驱动可以支持一般 Java 程序连 MySQL，但并不支持 Android 程序的直连。

设计 Android 程序界面,为简单起见,界面上仅放置一个文本视图和一个图像视图(代码略),为显示商品图片,将要用到的各商品图片预先置于工程的 drawable 目录下。

MainActivity.java 代码如下:

```java
package com.easybooks.mysqltest;
...
import android.os.Handler;                              //使用 Handler 句柄
import android.os.Message;                              //使用 Message
...
//以下是 JDBC 驱动相关的类
import java.sql.Connection;
import java.sql.DriverManager;
import java.sql.ResultSet;
import java.sql.SQLException;
import java.sql.Statement;

public class MainActivity extends AppCompatActivity {
    private TextView myTextView;                        //用于显示商品信息的文本视图
    private ImageView myImageView;                      //用于显示商品图片的图像视图
    private Connection conn = null;                     //数据库连接对象
    private Statement stmt = null;                      //SQL 语句对象
    private ResultSet rs = null;                        //查询结果集对象

    @Override
    protected void onCreate(Bundle savedInstanceState) {
        super.onCreate(savedInstanceState);
        setContentView(R.layout.activity_main);
        myTextView = findViewById(R.id.myTextView);
        myImageView = findViewById(R.id.myImageView);
        connToDb();                                     //连接访问数据库的方法(自定义)
        ...
    }
    ......
    //此方法通过 JDBC 发起对数据库的连接
    public void connToDb() {
        new Thread(new Runnable() {                     // (a)
            @Override
            public void run() {
                try {
                    Class.forName("com.mysql.jdbc.Driver");  //加载 MySQL 的 JDBC 驱动类库
                    conn = DriverManager.getConnection("jdbc:mysql://192.168.0.251:3306/emarket", "root", "ross123456");
                                                        // (b)
                    stmt = conn.createStatement();      //创建 SQL 语句对象
                    rs = stmt.executeQuery("SELECT * FROM commodity WHERE 商品编号='21101'");
                                                        // (c)
                    Message msg = Message.obtain();     // (d)
                    msg.what = 1000;                    //标识整数随便赋个值即可
                    while (rs.next()) {
                        msg.obj = rs.getString("商品名称") + "   " + rs.getString("价格") + ":" + rs.getString("商品图片");
                                                        //读取商品名称、价格和商品图片等信息
                    }
```

```
                myHandler.sendMessage(msg);           //通过 Handler 句柄发送消息给主线程刷新 UI
            } catch (ClassNotFoundException e) {
            } catch (SQLException e) {
            } finally {
                try {
                    if (rs != null) {
                        rs.close();                    // 关闭 ResultSet 对象
                        rs = null;
                    }
                    if (stmt != null) {
                        stmt.close();                  // 关闭 Statement 对象
                        stmt = null;
                    }
                    if (conn != null) {
                        conn.close();                  // 关闭 Connection 对象
                        conn = null;
                    }
                } catch (SQLException e) {
                }
            }
        }
    }).start();
}

private Handler myHandler = new Handler() {
    public void handleMessage(Message message) {       // （e）
        try {
            myTextView.setText(message.obj.toString().split(":")[0]);
                                                       // ":" 前的部分是可直接显示的商品信息文本
            //下面根据 ":" 前的部分（即商品图片名）定位要显示的商品图片资源
            Class resClass = R.drawable.class;         //Android 项目中的资源类
            Field field = resClass.getField(message.obj.toString().split(":")[1].trim());
                                                       //根据图片名获取资源类中的字段
            int resId = (Integer) field.get(resClass.newInstance());
                                                       //转换为整数型的资源 id
            myImageView.setImageResource(resId);       //显示在界面上的图像视图中
        } catch (NoSuchFieldException e) {
        } catch (IllegalAccessException e) {
        } catch (InstantiationException e) {
        }
    }
};
```

其中：

（a）**new Thread(new Runnable() { ... }).start();**：为保证移动用户的使用体验，Android 规定，程序中一切耗时的操作都必须放在子线程中去完成，数据库操作属于耗时操作，故必须放在一个子线程中来执行。通过 Message 由子线程向主线程 UI 传递消息。

（b）**conn = DriverManager.getConnection("jdbc:mysql://192.168.0.251:3306/emarket", "root", "ross123456");**：通过驱动管理器 DriverManager 获取数据库连接对象，其参数是连接字符串，访问 IP

地址为 192.168.0.251 的机器上的 emarket 数据库，必须提供 root 用户名和密码，且必须**只能**以 IP 地址访问，不能用主机名。

（c）**rs = stmt.executeQuery("SELECT * FROM commodity WHERE 商品编号='21101'");**：执行查询语句返回结果集，这里查询数据库中名为 commodity 的表中编号为"'21101'"的商品信息，表名不区分大小写。

（d）**Message msg = Message.obtain();**：用 Android 系统的 Message 对象来存储要向主线程传递的消息（即从数据库查到的商品名称、价格、图片名等信息）。

（e）**public void handleMessage(Message message):**：句柄的 handleMessage()方法与主线程是共用的，故可以在其中直接操作界面上的 UI 对象，完成对 Android 程序界面的刷新。

程序运行结果查询出 MySQL 数据库中编号为"21101"的商品记录，将该商品的信息和图片在 App 界面上显示出来，结果如图 7.7 所示。

图 7.7 JDBC 直连远程 MySQL 8 运行结果

如果要访问 MySQL 其他版本或者本地 MySQL 数据库，也是用同样的 JDBC 驱动，只需对连接字符串稍做修改。笔者先后试验了远程 MySQL 5 及本地 MySQL 8，它们获取连接的语句分别为：

conn = DriverManager.getConnection("jdbc:mysql://192.168.0.138:3306/emarket", "root", "njnu123456");
　　　　　　　　　　　　　　　　　　　　　　　　　　　　　　　　　　　　　//MySQL 5（远程）
conn = DriverManager.getConnection("jdbc:mysql://192.168.0.183:3306/emarket", "root", "zhou123456");
　　　　　　　　　　　　　　　　　　　　　　　　　　　　　　　　　　　　　//MySQL 8（本地）

但是，无论远程还是本地，都**只能**以 IP 地址访问，不能用主机名访问，运行结果同上。

7.2.3　Android JDBC 直连 SQL Server

SQL Server 是微软公司出品的大型数据库产品，但微软官方并不提供支持 Android 程序访问它的驱动，而且普通 Java 程序访问 SQL Server 的驱动（如 sqljdbc.jar、sqljdbc4.jar 和 mssql-jdbc-7.2.2.jre11.jar 等）均不能用于 Android。好在有第三方开源社区提供的 JTDS 包可以支持 SQL Server 直连，它实质上是 SQL Server 的 JDBC 除微软官方外的另一种实现（加入了 Android 功能），本书编写时的最新版本是 1.3.1，下载地址为 https://sourceforge.net/projects/jtds/files/jtds/，下载得到文件"jtds-1.3.1.jar"。用前面讲的 JDBC 驱动包加载方法将它载入工程中，然后修改程序，只需改动两处（驱动类加载和获取连接）：

Class.forName("net.sourceforge.jtds.jdbc.Driver");
conn = DriverManager.getConnection("jdbc:jtds:sqlserver://192.168.0.183:1433/emarket", "sa", "zhou123456");

我们使用最新的 SQL Server 2019，运行结果同上。

7.2.4 Android JDBC 直连 Oracle

Oracle 的主流版本 Oracle 11g 能够支持 Android 程序直连，但只能使用较早版本的驱动 ojdbc6.jar。同样地，将该驱动包载入工程中，然后修改程序。由于 Oracle 默认只支持大写的表名，故还要修改查询语句（一共要改动 3 处，如下：

```
Class.forName("oracle.jdbc.driver.OracleDriver");                                    //驱动类加载
conn = DriverManager.getConnection("jdbc:oracle:thin:@192.168.0.252:1521:EMARKET", "SCOTT", "Mm123456");
                                    //这里数据库名 EMARKET 必须写在冒号后而不是/后面
rs = stmt.executeQuery("SELECT * FROM COMMODITY WHERE 商品编号='21101'");          //查询语句
```

运行结果同前。

最后，我们总结一下使用 Android 直连数据库时需要注意的几点：

（1）JDBC 驱动 MySQL 最高只能用 mysql-connector-java-5.1.48.jar，虽然 MySQL 8 最新的驱动版本为 mysql-connector-java-8.0.17.jar，但那只能用作**普通** Java 程序访问数据库，Android 程序并不支持；SQL Server 只能使用第三方库 jtds-1.3.1.jar；Oracle 11g 最高只能用 ojdbc6.jar。

（2）仿真器 AVD 中的系统镜像最高只能到 Android 9.0。

（3）不论远程（局域网）还是本地计算机，都只能使用 **IP 地址**访问。本地访问也必须先配置 IP 地址，程序连接字符串中要写出具体的 IP，**不能用** localhost 或 127.0.0.1 替代。

（4）访问数据库的代码**只能**写在线程中，Android 不支持用户将耗时操作放在主线程中去完成！通常的做法是，使用 Handler 机制由子线程刷新界面，用 Handler 的 handleMessage()方法可直接操作主线程 UI 上的控件。

（5）mysql-connector-java-5.1.48.jar、JTDS 包 jtds-1.3.1.jar 与 ojdbc6.jar 可以**同时**加载存在于同一个工程中，程序代码在需要时调用，互不干扰。

7.2.5 以表格形式显示数据库表数据

为了能更清楚地浏览数据库中的信息，可以运用第 4 章介绍过的视图类控件，结合界面布局技术，以表格的形式将数据库中的记录展示出来，一目了然。

【例 7.2】 将 MySQL 中的商品信息通过 JDBC 读取后，以表格形式呈现在界面上，并且当用户长按表格中的条目时显示对应商品的实物图片，如图 7.8 所示。

1. 实现思路

（1）以列表视图（ListView）实现显示商品的表格，整个程序采用 MVC 框架模式，自定义适配器 CommodityAdapter 类和模型 Commodity 类，前端列表项文件 itemlayout.xml 以 View 分隔，结合线性布局（LinearLayout）设计表格样式。

（2）后台用 JDBC 连接 MySQL 读取数据，结合线程和 Message-Handler 句柄机制刷新界面。

（3）所有用到的商品图片存放在工程 res\drawable 目录下，以资源 id 引用。

（4）通过绑定在列表视图上的长按事件监听器（OnItemLongClickListener）响应用户的选择动作，以 Dialog 弹窗的形式显示浮动的商品实物图供用户预览。

创建 Android Studio 工程 MyDbGrid，将 MySQL 的 JDBC 驱动拷贝到工程中并添加依赖项，具体操作见 7.2.1 小节。

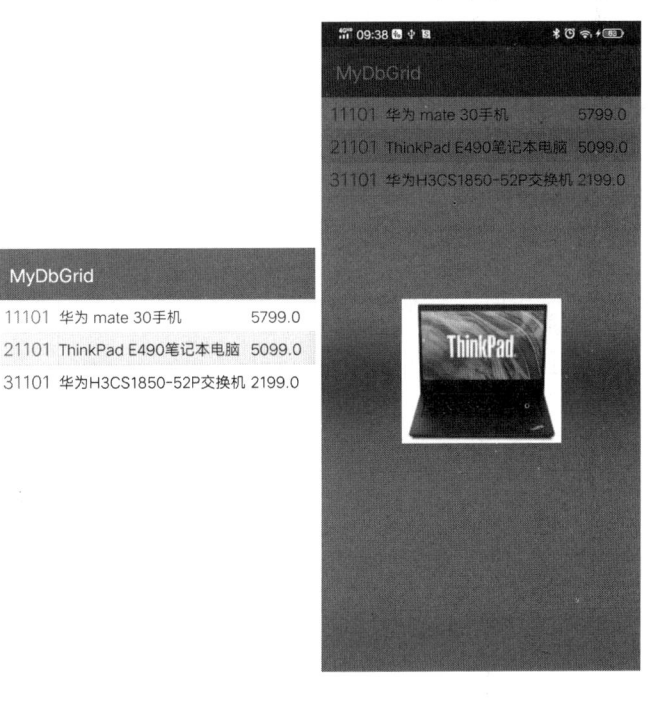

图 7.8　表格形式预览数据运行结果

2. 界面设计

（1）主界面 XML 文件。

在主界面 XML 文件 main.xml 上简单地放置一个 ListView，代码如下：

```
<LinearLayout xmlns:android="http://schemas.android.com/apk/res/android"
    android:layout_width="match_parent"
    android:layout_height="match_parent"
    android:orientation="vertical" >
    <ListView
        android:id="@+id/myListView"
        android:layout_width="match_parent"
        android:layout_height="wrap_content" />
</LinearLayout>
```

（2）列表项 XML 文件。

表格内每一行条目项的具体外观定义在 itemlayout.xml 中，代码如下：

```
<LinearLayout xmlns:android="http://schemas.android.com/apk/res/android"
    android:layout_width="fill_parent"
    android:layout_height="fill_parent"
    android:orientation="horizontal" >                    //最外层是个水平的线性布局

    <View                                                 //分隔线
        android:layout_width="1px"
        android:layout_height="fill_parent"
        android:background="#B8B8B8"
        android:visibility="visible" />

    <TextView                                             //文本视图（显示商品编号）
        android:id="@+id/cid"
```

```xml
        android:layout_width="0dip"
        android:layout_height="35dip"
        android:layout_weight="1"
        android:gravity="center"
        android:textColor="#CD3700"
        android:textSize="20sp" />

    <View                                                    //分隔线
        android:layout_width="1px"
        android:layout_height="fill_parent"
        android:background="#B8B8B8"
        android:visibility="visible" />

    <TextView                                                //文本视图（显示商品名称）
        android:id="@+id/name"
        android:layout_width="0dip"
        android:layout_height="wrap_content"
        android:layout_weight="3"
        android:textColor="#000000"
        android:textSize="17sp" />

    <View                                                    //分隔线
        android:layout_width="1px"
        android:layout_height="fill_parent"
        android:background="#B8B8B8"
        android:visibility="visible" />

    <TextView                                                //文本视图（显示商品价格）
        android:id="@+id/price"
        android:layout_width="0dip"
        android:layout_height="wrap_content"
        android:layout_weight="1"
        android:textColor="#000000"
        android:textSize="17sp" />

    <View                                                    //分隔线
        android:layout_width="1px"
        android:layout_height="fill_parent"
        android:background="#B8B8B8"
        android:visibility="visible" />

</LinearLayout>
```

可以看到，表格每一行都是一个水平的线性布局，行中单元格以文本视图（TextView）实现，单元格之间以 View 作为分隔线，这样就制作出一个表格的效果。

3. 开发数据模型和适配器

（1）定义数据模型。

既然采用 MVC，就要将商品数据封装在模型中，本例定义模型类 Commodity.java 的代码如下：

```java
public class Commodity {
    public String cid;                                       //商品编号
```

```
        public String name;                              //商品名称
        public String image;                             //商品图片
        public float price;                              //商品价格

        public Commodity(String cid, String name, String image, float price) {
            this.cid = cid;
            this.name = name;
            this.image = image;
            this.price = price;
        }
    }
```

（2）开发适配器。

下面来为 ListView 视图开发适配器，适配器类 CommodityAdapter.java 的代码如下：

```
public class CommodityAdapter extends BaseAdapter {        //继承自 BaseAdapter，便于灵活扩展
    private Context myContext;                             //声明上下文
    private LayoutInflater myInflater;                     //布局填充器
    private int myLayoutId;                                //布局 id
    private ArrayList<Commodity> myCommodityList;          //商品类数据列表
    private int[] colors = {Color.WHITE, Color.rgb(219, 238, 244)};    // （a）
    private int myBackground;                              //列表项背景

    public CommodityAdapter(Context ctx, int lid, ArrayList<Commodity> flst, int bg) {
        myContext = ctx;
        myInflater = LayoutInflater.from(ctx);             //从上下文获取布局
        myLayoutId = lid;                                  //设置布局 id
        myCommodityList = flst;                            //获取商品类数据
        myBackground = bg;                                 //设置列表项背景
    }

    @Override
    public int getCount() {                                //获得列表项数目
        return myCommodityList.size();
    }

    @Override
    public Object getItem(int arg0) {
        return myCommodityList.get(arg0);
    }

    @Override
    public long getItemId(int arg0) {
        return arg0;
    }

    @Override
    public View getView(final int position, View convertView, ViewGroup parent) {
        ViewHolder vholder = null;
        if (convertView == null) {
            vholder = new ViewHolder();
            convertView = myInflater.inflate(myLayoutId, null);
```

```
                    vholder.list_cid = convertView.findViewById(R.id.cid);        //获得商品编号
                    vholder.list_name = convertView.findViewById(R.id.name);      //获得商品名称
                    vholder.list_price = convertView.findViewById(R.id.price);    //获得商品价格
                    convertView.setTag(vholder);
                } else {
                    vholder = (ViewHolder) convertView.getTag();
                }
                Commodity commodity = myCommodityList.get(position);
                vholder.list_cid.setText(commodity.cid);                          //显示商品编号
                vholder.list_name.setText(commodity.name);                        //显示商品名称
                vholder.list_price.setText(String.valueOf(commodity.price));      //显示商品价格
                convertView.setBackgroundColor(colors[position % 2]);             //（b）
                return convertView;
            }

            public final class ViewHolder {                                       //（c）
                public TextView list_cid;                                         //商品编号
                public TextView list_name;                                        //商品名称
                public TextView list_price;                                       //商品价格
            }
        }
```

其中：

（a）**private int[] colors = {Color.WHITE, Color.rgb(219, 238, 244)};**：这里我们定义一个颜色值的数组，预先存放两种不同颜色，是为了能给表格中的条目间隔着切换上色，这样可使表格的行与行之间界限分明，数据显示效果更加清晰醒目。

（b）**convertView.setBackgroundColor(colors[position % 2]);**：相邻条目之间颜色不同，以模 2 取余间隔引用颜色数组中的元素进行设定。

（c）**public final class ViewHolder { ... }**：ViewHolder 类对应于 convertView 参数要设定的列表项中各控件元素的引用对象，根据界面上实际需要显示的内容来设计其中的属性项，本例就只包含商品编号、商品名称和商品价格这 3 项，而未包含商品图片。但商品图片信息仍然存储于模型及其对象列表 ArrayList<Commodity>中，需要时可以随时根据索引访问获取。

4. 实现功能

在主程序中使用 ListView 实现显示表格的功能，主程序 GridActivity.java 代码如下：

```
public class GridActivity extends AppCompatActivity {
    private ListView myListView;                                                  //列表视图对象引用
    private ArrayList<Commodity> commodityList;                                   //商品数据列表
    private Context myContext;                                                    //（a）
    //以下是以 JDBC 访问 MySQL 的对象
    private Connection conn = null;
    private Statement stmt = null;
    private ResultSet rs = null;

    @Override
    public void onCreate(Bundle savedInstanceState) {
        super.onCreate(savedInstanceState);
        setContentView(R.layout.main);
        myContext = this.getApplicationContext();                                 //（a）
```

```java
        myListView = findViewById(R.id.myListView);
        connToDb();                                                //连接访问 MySQL 的方法
        //表格上绑定长按事件响应用户操作，显示商品图片预览
        myListView.setOnItemLongClickListener(new AdapterView.OnItemLongClickListener() {
            @Override
            public boolean onItemLongClick(AdapterView<?> parent, View view, int position, long id) {
                try {
                    Class resClass = R.drawable.class;             //获取资源类
                    Field field = resClass.getField(commodityList.get(position).image);
                                                                   //获取资源类中的字段
                    int resId = (Integer) field.get(resClass.newInstance());  //转换为图片资源的 ID
                    //弹窗预览图片
                    Dialog dlg = new Dialog(GridActivity.this, R.style.myDialogStyle);
                                                                   // (b)
                    dlg.setContentView(R.layout.preview_dialog);   // (c)
                    ImageView imageView = dlg.findViewById(R.id.view_img);
                    imageView.setImageResource(resId);             //图片显示在对话框中
                    dlg.show();
                    dlg.setCanceledOnTouchOutside(true);           // (d)
                    Window w = dlg.getWindow();
                    WindowManager.LayoutParams lp = w.getAttributes();
                    lp.x = 0;
                    lp.y = 40;
                    dlg.onWindowAttributesChanged(lp);
                } catch (NoSuchFieldException e) {
                } catch (IllegalAccessException e) {
                } catch (InstantiationException e) {
                }
                return false;
            }
        });
    }

    public void connToDb() {
        new Thread(new Runnable() {
            @Override
            public void run() {
                try {
                    Class.forName("com.mysql.jdbc.Driver");        //加载 MySQL 驱动
                    conn = DriverManager.getConnection("jdbc:mysql://192.168.0.138:3306/emarket", "root", "njnu123456");
                                                                   //获取 MySQL 数据库连接
                    stmt = conn.createStatement();
                    rs = stmt.executeQuery("SELECT * FROM commodity");
                    Message msg = Message.obtain();                //消息传递商品信息
                    msg.what = 1000;
                    msg.obj = "";
                    while (rs.next()) {
                        msg.obj += rs.getString("商品编号") + ":" + rs.getString("商品名称") + ":" + rs.getString("商品图片") + ":" + rs.getString("价格") + ";";
                                                                   // (e)
                    }
                    myHandler.sendMessage(msg);
```

```
                    } catch (ClassNotFoundException e) {
                    } catch (SQLException e) {
                    } finally {
                        try {
                            if (rs != null) {
                                rs.close();                                    // 关闭 ResultSet 对象
                                rs = null;
                            }
                            if (stmt != null) {
                                stmt.close();                                  // 关闭 Statement 对象
                                stmt = null;
                            }
                            if (conn != null) {
                                conn.close();                                  // 关闭 Connection 对象
                                conn = null;
                            }
                        } catch (SQLException e) {
                        }
                    }
                }
            }).start();
        }

        private Handler myHandler = new Handler() {
            public void handleMessage(Message message) {
                try {
                    ArrayList<Commodity> cList = new ArrayList<Commodity>();
                    for (int i = 0; i < 3; i++) {
                        String cInfo = message.obj.toString().split(";")[i];
                        cList.add(new Commodity(cInfo.split(":")[0], cInfo.split(":")[1].split("/")[1], cInfo.split(":")[2], Float.parseFloat(cInfo.split(":")[3])));
                    }
                    commodityList = cList;                                     // (f)
                    CommodityAdapter adapter = new CommodityAdapter(myContext, R.layout.itemlayout, commodityList, Color.WHITE);   // (a)
                    myListView.setAdapter(adapter);
                } catch (Exception e) {
                }
            }
        };
    }
```

其中:

(a) **private Context myContext;**、**myContext = this.getApplicationContext();** 和 **CommodityAdapter adapter = new CommodityAdapter(myContext, R.layout.itemlayout, commodityList, Color.WHITE);**: 由于本例中创建适配器对象的语句是放在线程句柄的 handleMessage() 方法中, 无法直接引用主程序上下文, 故这里要将主程序的上下文声明为一个全局变量, 在初始化 onCreate() 方法中通过 getApplicationContext() 获取后, 就可以在程序中的任何地方引用。

(b) **Dialog dlg = new Dialog(GridActivity.this, R.style.myDialogStyle);**: 创建对话框弹窗, 用来显示商品图片的预览, 这里指定对话框样式为 myDialogStyle。这个样式是我们自定义的, 位于工程 values

目录下的 styles.xml 文件中，代码如下：

```xml
<resources>
    ......
    <!--弹出框样式定义-->
    <style name="myDialogStyle" parent="@android:style/Theme.Dialog">
        <item name="android:windowIsFloating">true</item>
        <item name="android:windowIsTranslucent">true</item>
        <item name="android:windowNoTitle">true</item>                          //取消标题栏
        <item name="android:windowBackground">@android:color/transparent</item>
        <item name="android:background">@android:color/transparent</item>
        <item name="android:backgroundDimEnabled">true</item>                   //设为半透明背景
    </style>
</resources>
```

需要特别强调的是，创建 Dialog 时只能在主程序（GridActivity 类）中直接引用 GridActivity.this 上下文，而**不能**使用全局变量 myContext 中存储的上下文，否则，对话框会因"找不到"调用它的主类而无法成功实例化，进而导致程序崩溃。

（c）**dlg.setContentView(R.layout.preview_dialog);**：设置对话框的显示内容视图。我们设计的视图文件 preview_dialog.xml 位于工程 layout 目录下，内容为：

```xml
<?xml version="1.0" encoding="utf-8"?>
<LinearLayout xmlns:android="http://schemas.android.com/apk/res/android"
    xmlns:app="http://schemas.android.com/apk/res-auto"
    android:orientation="vertical" android:layout_width="match_parent"
    android:layout_height="match_parent">
    <ImageView
        android:id="@+id/view_img"
        android:layout_width="179dp"
        android:layout_height="159dp" />
</LinearLayout>
```

即在线性布局中简单放置一个图像视图用来显示预览的商品图片。

（d）**dlg.setCanceledOnTouchOutside(true);**：设置这个属性的目的是在用户触击图片外部区域时对话框自动消失，回到原来的表格显示主页。

（e）**msg.obj += rs.getString("商品编号") + ":" + rs.getString("商品名称") + ":" + rs.getString("商品图片") + ":" + rs.getString("价格") + ";";**：为了便于主程序解析商品信息，我们对从数据库查询得到的商品信息进行格式化，同一条记录的字段之间以":"分隔，而在记录之间则以";"隔开，解析时通过字符串 split()方法拆分即可。

（f）**commodityList = cList;**：由于本例用了消息句柄机制在线程间异步传递数据对象，故此处赋值的时候只能先创建一个临时的 ArrayList<Commodity>对象，将 Message 中解析出的信息存储在其中，然后再赋值给全局列表对象 commodityList，而**不能**直接对全局的商品数据列表对象调用 add()方法进行赋值。

7.3 Android HTTP 编程

7.3.1 基本原理

Android 程序通过 JDBC 直连网络数据库的方式虽然简便，但在实际开发中并不提倡使用这种方法，基于以下两点原因：

(1)不安全。移动端用户直接操作,数据库连接字符串(含用户名、密码)就会在网络上以明文传输,很容易被截获导致机密信息泄露,且这样的 Android 程序也有很大可能被黑客反编译从中获取数据库用户密码信息。

(2)加重数据库系统负荷。大量用户并发直接操作数据库,对 DBMS 的性能和载荷量是个严峻考验,弄不好有把数据库搞崩溃的风险。

实际的互联网应用系统大多采用如下"移动端—Web 服务器—后台数据库"的 3 层架构方式,如图 7.9 所示,保证安全性的同时又能提高系统的性能和可用性。

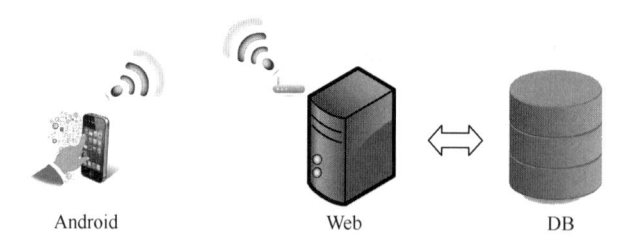

图 7.9 互联网应用的通用架构

在这种架构下,移动端是通过 HTTP 协议,由 Web 服务器间接操作数据库的。Android 为 HTTP 编程提供了 HttpURLConnection 类,它的功能非常强大,具有广泛的通用性,可用它连接 Java/Java EE、.NET、PHP 等几乎所有主流平台的 Web 服务器,如图 7.10 所示。

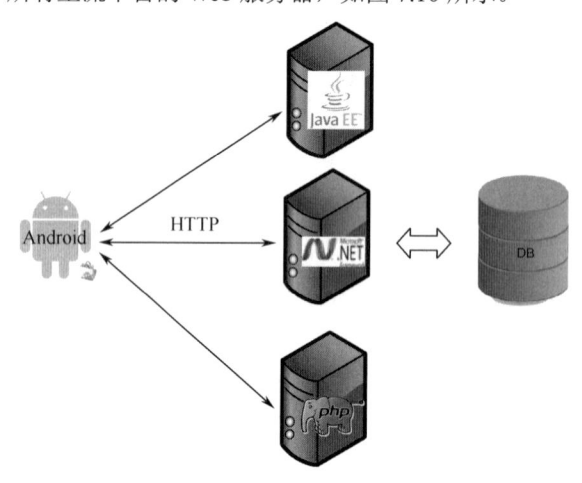

图 7.10 Android 通过 HTTP 与各种主流 Web 平台互动

本节我们来实现一个最简单(也是最典型)的以 Java Servlet 作为 Web 服务程序操作后台数据库并向移动前端返回信息的系统,它一共涉及 3 台计算机,分别为:

- **移动端**:华硕笔记本(192.168.0.183,Windows 10)。安装 Android Studio 3.5 和 Eclipse,开发程序,运行 Android 移动端。
- **Web 服务器**:联想笔记本(192.168.0.138,Windows 7 64 位),其上有 Tomcat 9.0 和 JDK,部署开发好的 Java Servlet 服务器程序。
- **DB 服务器**:ThinkPad 笔记本(192.168.0.251,Windows Server 2012 R2 标准版 64 位),其上有 MySQL 8.0 数据库。

为能更深刻地理解当前互联网应用的架构和运行模式,建议读者最好准备 3 台计算机来做实验。当然,如果条件实在不具备,也可以在一台计算机上模拟完成。

7.3.2 环境安装

客户端上已经安装了 Android Studio 3.5，为了能开发 Java Servlet 服务器程序，还需要安装 Eclipse 和 Tomcat（这个 Tomcat 的主要作用是供 Eclipse 开发过程中测试 Web 程序之用，实际运行时还必须部署到 Web 服务器上才行）。

1. 安装 Eclipse

我们使用最新的 Eclipse IDE 2019-09 作为开发 Web 端 Java 程序的工具，目前官方只提供 Eclipse 安装器的下载，下载地址为 https://www.eclipse.org/downloads/，获取文件名为 "eclipse-inst-win64.exe"。

实际安装时先必须确保计算机处于联网状态，然后启动 eclipse-inst-win64.exe，选择要安装的 Eclipse IDE 类型，我们选择 "Eclipse IDE for Enterprise Java Developers"（即 Jave EE 版），如图 7.11 所示。安装全过程要始终确保联网以实时下载所需的文件。单击 "INSTALL" 开始安装，如图 7.12 所示。

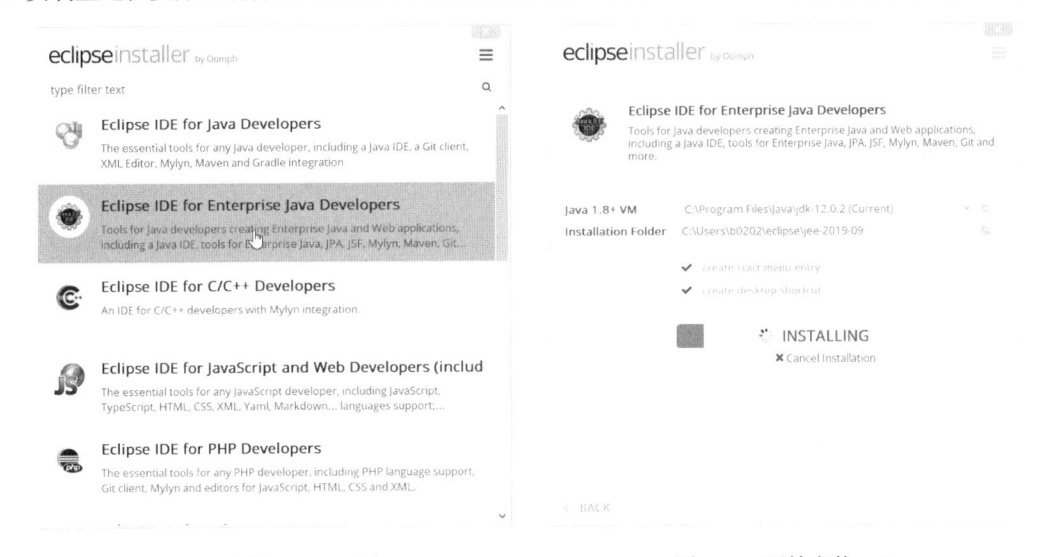

图 7.11　选择安装 Java EE 版 Eclipse　　　　图 7.12　开始安装 Eclipse

安装过程中会出现几次对话框以确认许可协议条款，如图 7.13 所示，分别单击 "Accept"、"Select All" 和 "Accept selected" 按钮，一律接受。安装完成后单击 "LAUNCH" 启动并设置工作区。

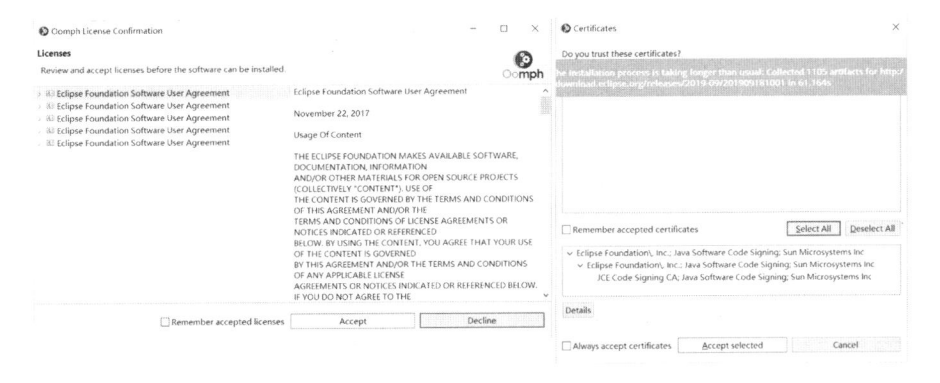

图 7.13　接受全部协议条款

2. 安装 Tomcat

如果是干净系统，须先安装 JDK（为配合 Eclipse 调试程序用，笔者在搭建 Android 开发环境时已经装好了 JDK12）。我们使用最新的 Tomcat 9.0 作为 Web 服务器，下载地址为 https://tomcat.apache.org/download-90.cgi，获得安装包"apache-tomcat-9.0.26.exe"，双击启动安装向导，按照向导的指引往下操作，过程略。

> **注意：**
> 在安装向导 Configuration（配置）页上要修改"Server Shutdown Port"，由-1 改为 8005（或者在安装完成之后由配置文件中改再重启 Tomcat），如图 7.14 所示，只有这样修改之后才能由 Eclipse 环境来启动 Tomcat 实例调试程序。

笔者使用的 JDK12 的 JRE 已经集成，安装后无独立 JRE 目录，故安装 Tomcat 时直接选择 JDK 目录即可，如图 7.15 所示。

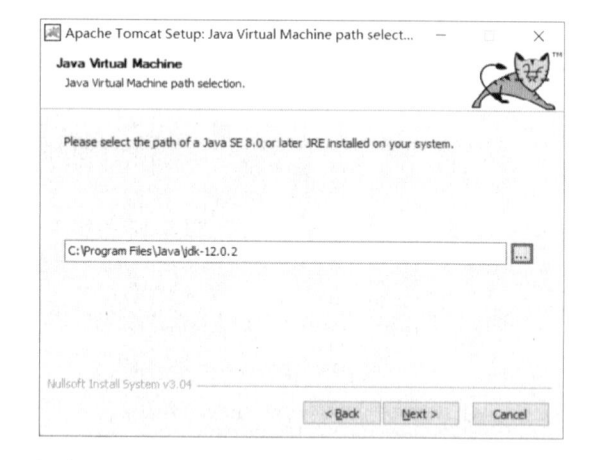

图 7.14 修改"Server Shutdown Port"　　　　图 7.15 直接选择 JDK 目录为 JRE 目录

3. 配置 Eclipse 环境中的 Tomcat

要将 Tomcat 与 Eclipse IDE 整合起来，才能在开发调试程序时使用。下面简述整合的步骤。

（1）在 Eclipse 中选择主菜单"Window"→"Preferences"命令，在出现的"Preferences"对话框中选择左侧项目树中的"Server"→"Runtime Environments"，单击"Add"按钮，如图 7.16 所示。

（2）在打开的"New Server Runtime Environment"对话框中，服务器类型列表中选中"Apache Tomcat v9.0"，勾选"Create a new local server"复选框，如图 7.17 所示，单击"Next"按钮。

（3）接下来选择服务器路径为我们安装的 Tomcat 9.0 所在目录，Tomcat 运行基于 JRE 为我们计算机上安装的 JDK（这里是"jdk-12.0.2"），如图 7.18 所示。

> **注意：**
> Eclipse 环境使用的 Tomcat 与从开发环境外部直接启动的 Tomcat 是两个不同的 Web 服务器实例，但它们共用同一个端口（某一时刻两者只能启动一个）。

经过上述配置后，就可以从 Eclipse 环境启动 Tomcat 来调试运行程序，这样 Android 应用架构中 Web 服务器端程序的开发环境就搭建好了。

接下来我们将用它开发与 Android 程序直接交互的 Web 服务端的 Java Servlet 程序。

第 7 章　Android 数据库和网络编程　217

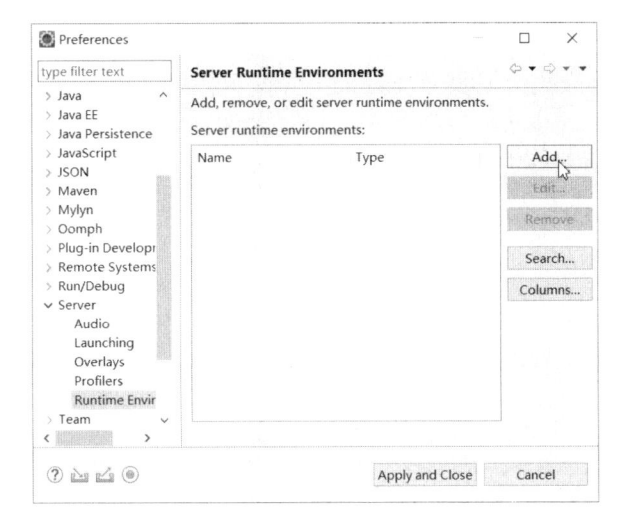

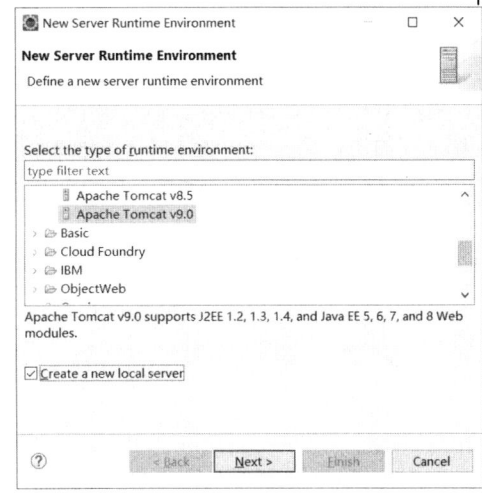

图 7.16　配置服务器运行时环境

图 7.17　选择创建服务器的类型

图 7.18　选择 Tomcat 目录及所使用的 JRE

7.3.3　Web 应用开发和部署

【例 7.3】　开发 Web 服务端的 Java Servlet 程序。

1. 创建动态 Web 项目

我们用 Java 的 Servlet 实现服务端的 Web 程序，在刚刚搭建的 Eclipse IDE 环境下开发，选择主菜单"File"→"New"→"Dynamic Web Project"命令，出现如图 7.19 所示对话框，将项目命名为"MyServlet"。

在"Web Module"页勾选"Generate web.xml deployment descriptor"复选框，单击"Finish"按钮，如图 7.20 所示。

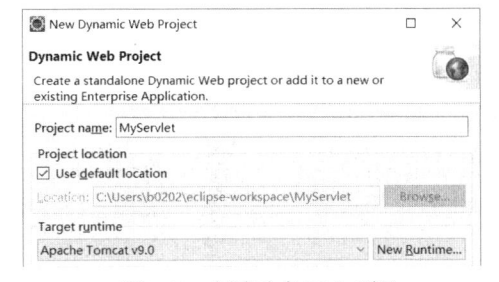

图 7.19　创建动态 Web 项目

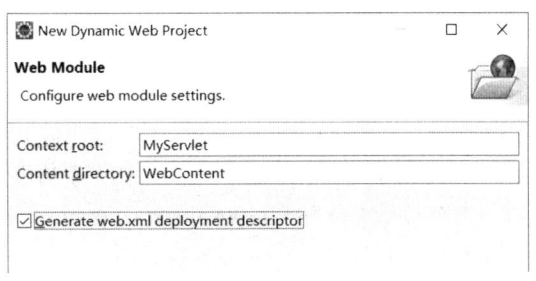

图 7.20　自动生成 web.xml 文件

项目创建完成后，在 Eclipse 开发环境左侧的树状视图中，可看到该项目的组成目录结构。这个运行在 Web 端的程序负责接收 Android 程序发来的请求，再根据 Android 程序的要求去操作后台数据库，故离不开 JDBC 驱动包。这里仍然使用 mysql-connector-java-5.1.48.jar，将它拷贝到项目的 lib 目录直接刷新即可，最终形成的项目目录细节如图 7.21 所示。

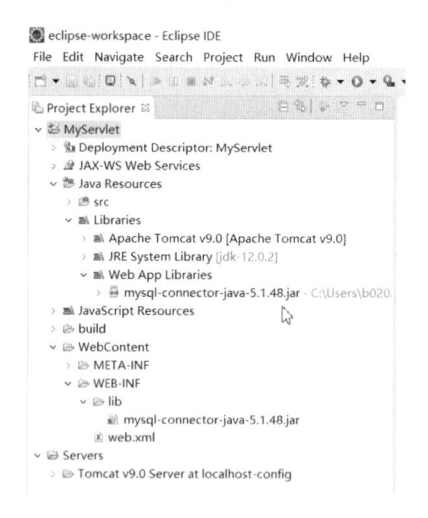

图 7.21　项目目录细节

2. 编写 Servlet 程序

现在 Eclipse IDE 已能支持在 src 下直接创建 Servlet 的源文件模板，自动生成 Servlet 的代码框架即可运行，无须再配置 web.xml。在项目 src 下创建包 org.easybooks.myservlet，右击此包，在弹出的快捷菜单中选择"New"→"Servlet"命令，在弹出的向导对话框中输入 Servlet 类名，于多个页面上根据需要配置 Servlet 的具体属性（我们都使用默认），如图 7.22 所示。

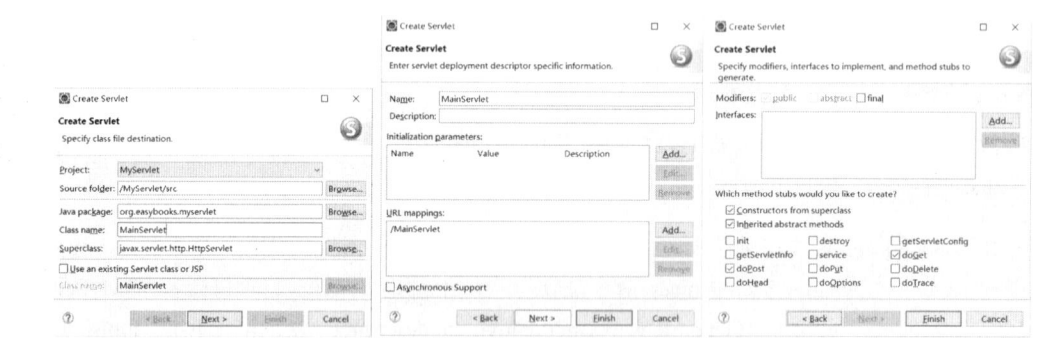

图 7.22　创建 Servlet

单击"Finish"按钮，Eclipse 就会自动生成 Servlet 源文件模板，其中的代码框架都已经生成好了，我们只需修改加入自己的代码即可。

实现思路：

（1）在主 Servlet 类 MainServlet 中声明数据库连接对象、SQL 语句对象和结果集对象。

（2）主要功能代码全部在 doGet()方法中，根据移动端发来的请求 HttpServletRequest 的内容决定要执行的操作。

（3）如果请求中包含新价格，则更新写入数据库，否则读取并返回原来的商品信息。

下面给出本例使用的 Servlet 源代码（加粗语句为添加的内容），内容如下：

```java
package org.easybooks.myservlet;

import java.io.IOException;
import javax.servlet.ServletException;
import javax.servlet.annotation.WebServlet;
import javax.servlet.http.HttpServlet;
import javax.servlet.http.HttpServletRequest;
import javax.servlet.http.HttpServletResponse;
import java.io.*;                                    //IO 操作的库
import java.sql.*;                                   //SQL 操作的库

/**
 * Servlet implementation class MainServlet
 */
@WebServlet("/MainServlet")
public class MainServlet extends HttpServlet {
    private static final long serialVersionUID = 1L;
    private Connection conn = null;                  //数据库连接对象
    private Statement stmt = null;                   //SQL 语句对象
    private ResultSet rs = null;                     //结果集对象

    /**
     * @see HttpServlet#HttpServlet()
     */
    public MainServlet() {
        super();
        // TODO Auto-generated constructor stub
    }

    /**
     * @see HttpServlet#doGet(HttpServletRequest request, HttpServletResponse response)
     */
    protected void doGet(HttpServletRequest request, HttpServletResponse response) throws ServletException, IOException {
        // TODO Auto-generated method stub
        response.setCharacterEncoding("utf-8");       //必须有这句！否则中文显示为???
        response.setContentType("text/html");         //也必须有这句！否则中文显示为乱码
        String result = "";                           //需要返回给移动端（Android 程序）的字符串
        //访问数据库读取内容
        try {
            Class.forName("com.mysql.jdbc.Driver");   //加载 MySQL 驱动库
            conn = DriverManager.getConnection("jdbc:mysql://SMYSQL:3306/emarket", "root", "ross123456");
                                                      //获取 MySQL 连接（支持主机名访问）
            //conn = DriverManager.getConnection("jdbc:mysql://192.168.0.251:3306/emarket", "root", "ross123456");
                                                      //这是用 IP 地址访问的语句，与上一句等价（二选一）
            stmt = conn.createStatement();
            //如果用户发来新的价格，则更新写入数据库
            String id = request.getParameter("id");   //获取请求 URL 中携带的商品编号
            String price = request.getParameter("price"); //获取请求 URL 中携带的新价格
            if(!(price == null||price.length() <= 0)) {
```

```java
                    String sql = "UPDATE commodity SET 价格 = " + Float.parseFloat(price) + " WHERE 商品编号 = '" + id + "'";
                    stmt.executeUpdate(sql);
                }
                //如果用户发来空信息(未输入新价格),则直接读取返回数据库中原来的商品信息
                rs = stmt.executeQuery("SELECT * FROM commodity WHERE 商品编号='" + id + "'");
                while (rs.next()) {
                    result = rs.getString("商品名称") + "   " + rs.getString("价格") + ":" + rs.getString("商品图片");
                }
            } catch (ClassNotFoundException e) {
                result = e.getMessage();
            } catch (SQLException e) {
                //可用于向移动端页面返回 DBMS 本身内置的错误信息,便于用户查找原因
                result = e.getMessage();
            } finally {
                try {
                    if (rs != null) {
                        rs.close();                        // 关闭 ResultSet 对象
                        rs = null;
                    }
                    if (stmt != null) {
                        stmt.close();                      // 关闭 Statement 对象
                        stmt = null;
                    }
                    if (conn != null) {
                        conn.close();                      // 关闭 Connection 对象
                        conn = null;
                    }
                } catch (SQLException e) {
                    result = e.getMessage();
                }
            }
            PrintWriter return_to_client = response.getWriter();
            return_to_client.println(result);
            return_to_client.flush();
            return_to_client.close();
        }

        /**
         * @see HttpServlet#doPost(HttpServletRequest request, HttpServletResponse response)
         */
        protected void doPost(HttpServletRequest request, HttpServletResponse response) throws ServletException, IOException {
            // TODO Auto-generated method stub
            doGet(request, response);
        }
    }
```

可见,操作数据库的过程与之前 JDBC 直连模式下几乎一模一样,只不过这里的操作者换成了 Web 服务器端的 Servlet 程序,对数据库的操作全部集中在服务器端进行。

3. 打包部署 Web 项目

（1）项目打包。

将编写完成的 Servlet 程序打包成.war 文件。用 Eclipse 对项目打包的基本操作为：右击项目 MyServlet，在弹出的快捷菜单中选择"Export"→"WAR file"命令，从弹出对话框中选择打包.war 文件存放的路径，如图 7.23 所示，单击"Finish"按钮即可。

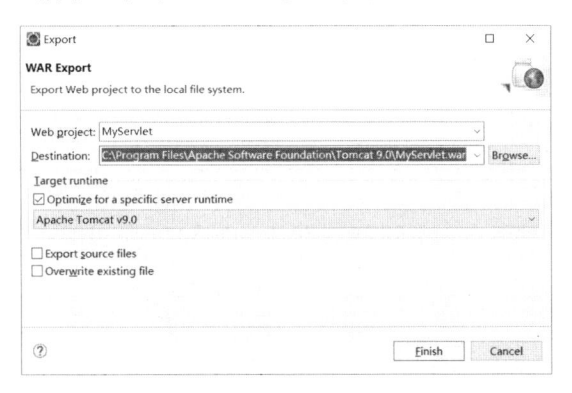

图 7.23　项目打包

将打包形成的.war 文件**直接复制到** Web 服务器上 Tomcat（注意**不是**本地 Eclipse 开发环境的 Tomcat）的 webapps 目录下。

（2）升级 Web 服务器的 Java 运行时环境。

如果 Web 服务器上的 JDK 版本低于客户端开发机器上的 JDK 版本，则需要先升级 Java 运行时环境。笔者 Web 服务器计算机上原来安装的是 JDK11，而 Servlet 是基于最新 JDK12 开发，故要进行升级，方法如下：

① 先卸载重装高版本的 JDK12，然后重设 Tomcat 的 jvm.dll 虚拟机运行时，进入 Web 服务器上 Tomcat 安装的 bin 目录，双击 Tomcat9w.exe，弹出对话框切换至"Java"选项页，单击"Java Virtual Machine"右侧的"…"按钮，如图 7.24 所示。

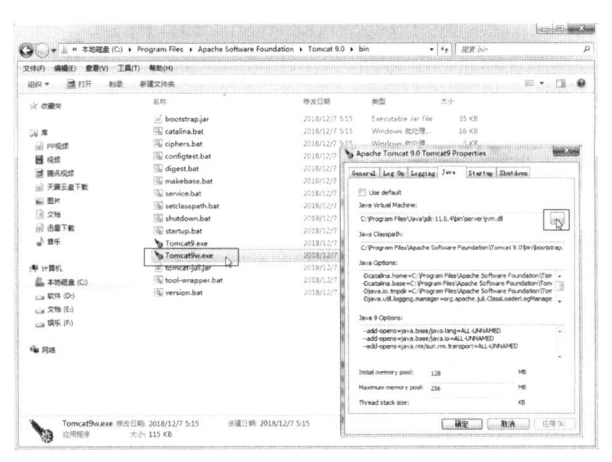

图 7.24　重设 Tomcat 的 jvm.dll 虚拟机运行时

② 然后选中 JDK 安装 bin\server 目录下的 jvm.dll，如图 7.25 所示，单"应用"按钮后重启 Tomcat 即可。

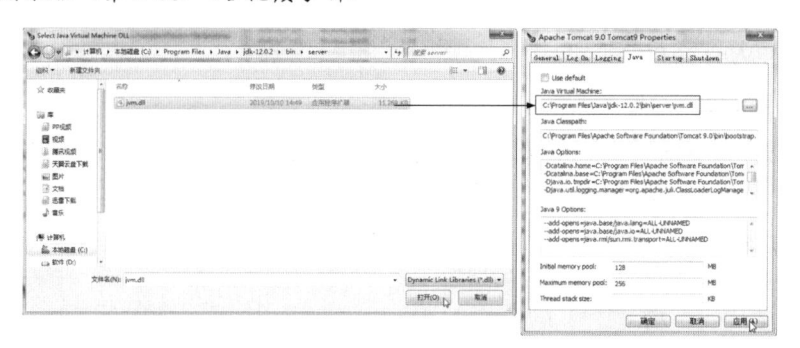

图 7.25 设为新的 jvm.dll

> 👀注意：
> 如果不升级 Java 运行时环境，网页将无法访问；如果只升级 JDK 而不重设 Tomcat 的 jvm，则 Tomcat 将无法启动。

（3）测试 Web 服务器。

打包部署完成，启动 Web 服务器上的 Tomcat，可先在客户端用浏览器访问 http://192.168.0.138:8080/MyServlet/MainServlet?id=31101 测试是否成功，如果能正确查询出特定编号的商品信息，如图 7.26 所示，表示 Web 服务器环境已经搭建成功。

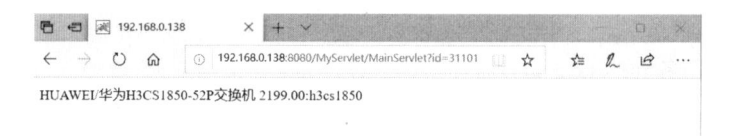

图 7.26 测试 Web 服务器

7.3.4 移动端 Android 程序开发

【例 7.4】 在移动端计算机上开发一个 Android 程序来访问 Web 服务器。

在移动端 Android Studio 中创建名为 HttpServletMySqlDemo 的工程。在界面上拖曳放置一个文本视图、一个图像视图、一个编辑框和一个按钮，如图 7.27 所示。

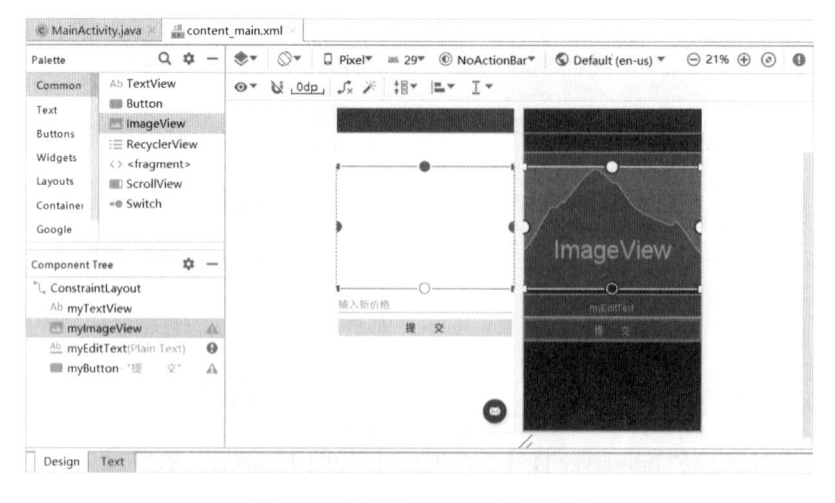

图 7.27 移动端 Android 程序界面

界面完整代码

这个界面很简单，故不再罗列其源码。同样地，将要用到的各商品图片预先置于工程的 drawable 目录下。下面重点来介绍移动端通过 HTTP 向 Web 服务器请求获取数据库中商品信息的过程。

实现思路：

（1）初始启动 Android 程序时通过 connToWeb() 方法连接到 Web 服务器。该方法提交的请求 URL 只带一个 id（商品编号）参数，指明要显示的商品。

（2）若用户输入了新的价格，点击"提交"按钮后，就执行 onSubmitClick() 方法，此方法带两个参数，除了表示商品编号的 id 外，还有一个 price 就是用户输入的新价格。

（3）连接通过 getInputStream() 方法获取到 Web 服务器返回响应的输入流之后，调用 refresh_UI() 方法来刷新前端界面，该方法中用到了 Android 的 Message-Handler 消息句柄刷新机制，将移动端对获取到的输入流的解析及刷新前端 UI 的 Message-Handler 操作全都封装在其中，避免了代码冗余。

代码全部位于主程序类 MainActivity 中，为使读者清楚程序的结构，我们将以上 3 步在程序代码中特别标示出来，内容如下：

```
...
import android.os.Handler;                          //实现 UI 刷新
import android.os.Message;                          //子线程、主线程间传递信息
...
import java.io.BufferedReader;
import java.io.IOException;
import java.io.InputStream;
import java.io.InputStreamReader;
import java.lang.reflect.Field;                     //获取资源对象中的字段域
import java.net.HttpURLConnection;                  //HTTP 请求的关键类
import java.net.URL;                                //存储访问的 URL 地址

public class MainActivity extends AppCompatActivity {
    private TextView myTextView;                    //显示商品名称、价格
    private ImageView myImageView;                  //显示商品图片
    private EditText myEditText;                    //接受用户修改价格值
    private HttpURLConnection conn = null;          //HTTP 连接对象
    private InputStream stream = null;              //输入流（存放获取到的响应内容）
    private String myCommodityId = "31101";         //商品编号（由 URL 请求携带给服务器）

    @Override
    protected void onCreate(Bundle savedInstanceState) {
        super.onCreate(savedInstanceState);
        setContentView(R.layout.activity_main);
        myTextView = findViewById(R.id.myTextView);
        myImageView = findViewById(R.id.myImageView);
        myEditText = findViewById(R.id.myEditText);
        //发起对 Web 服务器的连接（自定义方法）
        connToWeb();                                // （1）
            ...
    }
    ...
    //连接到 Web 服务器的方法
    public void connToWeb() {                       // （1）
        new Thread(new Runnable() {                 //连接服务器是耗时操作，必须放入子线程
            @Override
```

```java
            public void run() {
                try {
                    URL url = new URL("http://192.168.0.138:8080/MyServlet/MainServlet?id=" +
myCommodityId);                                    //Web 端 Servlet 地址（携带商品编号）
                    conn = (HttpURLConnection) url.openConnection();
                                                    //获取 HTTP 连接对象
                    conn.setRequestMethod("GET");   //请求方式为 GET（从指定的资源请求数据）
                    conn.setConnectTimeout(3000);   //连接超时时间
                    conn.setReadTimeout(9000);      //读取数据超时时间
                    conn.connect();                 //开始连接 Web 服务器
                    stream = conn.getInputStream(); //获取服务器的响应（输入）流
                    refresh_UI(stream);             // (3)
                } catch (Exception e) {
                } finally {
                    try {
                        if (stream != null) {
                            stream.close();         //关闭输入流
                            stream = null;
                        }
                        conn.disconnect();          //断开连接
                        conn = null;
                    } catch (Exception e) {
                    }
                }
            }
        }).start();
    }

    //移动端提交价格修改
    public void onSubmitClick(View view) {                  // (2)
        new Thread(new Runnable() {                         //耗时操作，也必须放在线程里
            @Override
            public void run() {
                try {
                    //带两个参数的 GET 请求，参数 id 为操作商品编号，price 为用户提交新的价格
                    URL url = new URL("http://192.168.0.138:8080/MyServlet/MainServlet?id=" +
myCommodityId + "&price=" + myEditText.getText().toString());
                    conn = (HttpURLConnection) url.openConnection();
                    conn.setRequestMethod("GET");   //请求方式为 GET（从指定的资源请求数据）
                    conn.setConnectTimeout(3000);   //连接超时时间
                    conn.setReadTimeout(9000);      //读取数据超时时间
                    conn.connect();                 //开始连接 Web 服务器
                    stream = conn.getInputStream();
                    refresh_UI(stream);             // (3)
                } catch (Exception e) {
                } finally {
                    try {
                        if (stream != null) {
                            stream.close();
                            stream = null;
                        }
```

```java
                    conn.disconnect();
                    conn = null;
                } catch (Exception e) {
                }
            }
        }
    }).start();
}

public void refresh_UI(InputStream in) {                    // (3)
    BufferedReader bufReader = null;
    try {
        bufReader = new BufferedReader(new InputStreamReader(in));
                                                            //输入流数据放入读取缓存
        StringBuilder builder = new StringBuilder();
        String str = "";
        while ((str = bufReader.readLine()) != null) {
            builder.append(str);                            //从缓存对象中读取数据拼接为字符串
        }
        Message msg = Message.obtain();
        msg.what = 1000;
        msg.obj = builder.toString();                       //通过 Message 传递给主线程
        myHandler.sendMessage(msg);                         //通过 Handler 发送
    } catch (IOException e) {
    } finally {
        try {
            if (bufReader != null) {
                bufReader.close();                          //关闭读取缓存
                bufReader = null;
            }
        } catch (IOException e) {
        }
    }
}

private Handler myHandler = new Handler() {
    public void handleMessage(Message message) {
        try {
            myTextView.setText(message.obj.toString().split(":")[0]);
            Class resClass = R.drawable.class;
            Field field = resClass.getField(message.obj.toString().split(":")[1].trim());
            int resId = (Integer) field.get(resClass.newInstance());
            myImageView.setImageResource(resId);
        } catch (NoSuchFieldException e) {
        } catch (IllegalAccessException e) {
        } catch (InstantiationException e) {
        }
    }
};
}
```

最后，不要忘记在 AndroidMainifest.xml 中打开 HTTP 明文传输及互联网访问的权限：

```xml
<manifest xmlns:android="http://schemas.android.com/apk/res/android"
    package="com.easybooks.httpservletmysqldemo">
    <application
        ...
        android:supportsRtl="true"
        android:usesCleartextTraffic="true"                         //允许 HTTP 明文传输
        android:theme="@style/AppTheme">
        ...
    </application>
    <uses-permission android:name="android.permission.INTERNET"/>   //打开互联网访问权限
</manifest>
```

运行 Android 程序，显示数据库中的商品名称、价格和图片，在移动界面上填写新的价格后提交，数据库的信息被修改后回显在移动界面上，如图 7.28 所示。

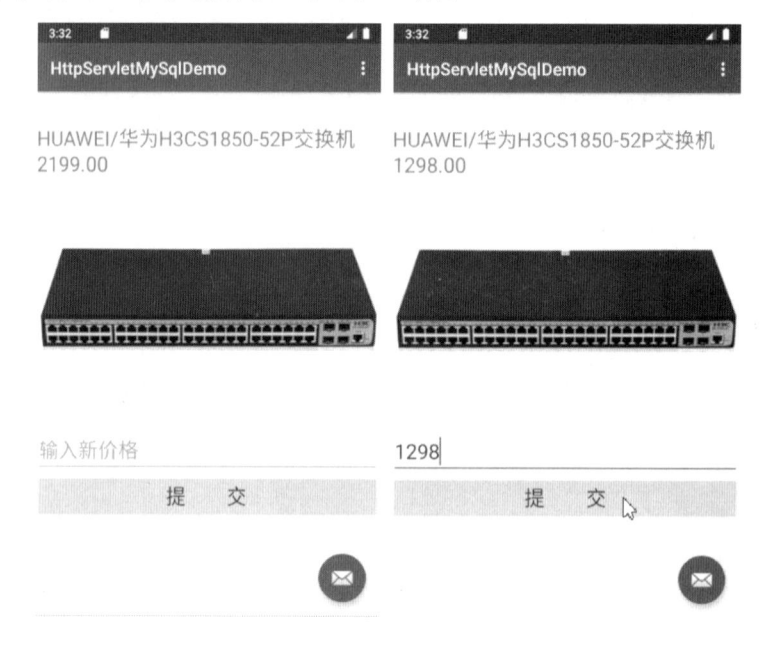

图 7.28 Android 程序访问 Web 服务器运行效果

> **注意：**
> HTTP-Web 方式由于是由 Web 服务端程序来操作后台数据库，故此种方式与 Java 程序直接操作数据库在底层的原理上完全一样。但不同的是，它允许用主机名或 IP 地址**两种方式访问数据库**，而用 JDBC 直连数据库则只能用 IP 地址不能用主机名。

7.4 Android 与 WebService 交互

7.4.1 基本原理

当前实际的互联网应用中，Android 程序访问网络很多情况下只是为了获取所需要的数据和处理数据的服务，并非为了显示网页界面（Android 有自己的一套 UI 系统和规范，不局限于显示桌面 Web 网页），故与它交互的应用程序不一定非要有可视的 UI 界面，只需完成传递数据或者某种处理任务即可。目前通行的做法是，移动端 Android 程序通过 HTTP 访问互联网上的 WebService，由 WebService

提供数据或完成处理功能。仍然沿用前述互联网应用系统的 3 层架构，如图 7.29 所示的是 Android 与 WebService 交互的原理。

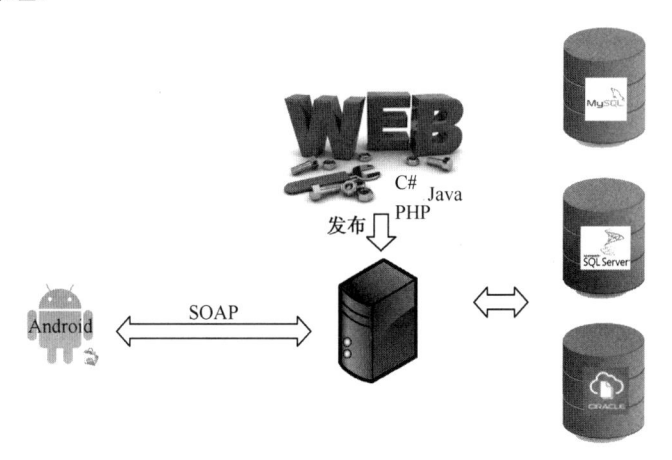

图 7.29　Android 与 WebService 交互的原理

其中，WebService 可以用 Java、C#、PHP 等各种不同编程语言来实现，后台数据库也可以是 MySQL、SQL Server、Oracle 等多种多样异构类型的 DBMS。正因为如此高度的兼容性和扩展性，WebService 目前在互联网中得到了极为广泛的应用。

7.4.2　配置 IIS 服务器

本节我们用基于.NET 平台 C#语言开发的 WebService 与 Android 交互，需要先配置 Windows 上的 IIS 服务器。步骤如下：

（1）在 Windows 控制面板"程序和功能"下单击"打开或关闭 Windows 功能"链接，在出现的"Windows 功能"对话框中补充勾选"Internet 信息服务"下"Web 管理工具"和"万维网服务"中的选项，如图 7.30 所示，确定后出现"Windows 正在更改功能…"的提示进度框，稍候片刻，待进度框消失后重启 Windows。

图 7.30　添加 Windows 功能

（2）在 Windows 控制面板中选择"系统和安全"→"管理工具"→"Internet 信息服务(IIS)管理

器",打开 Windows 的网站管理器界面,如图 7.31 所示。

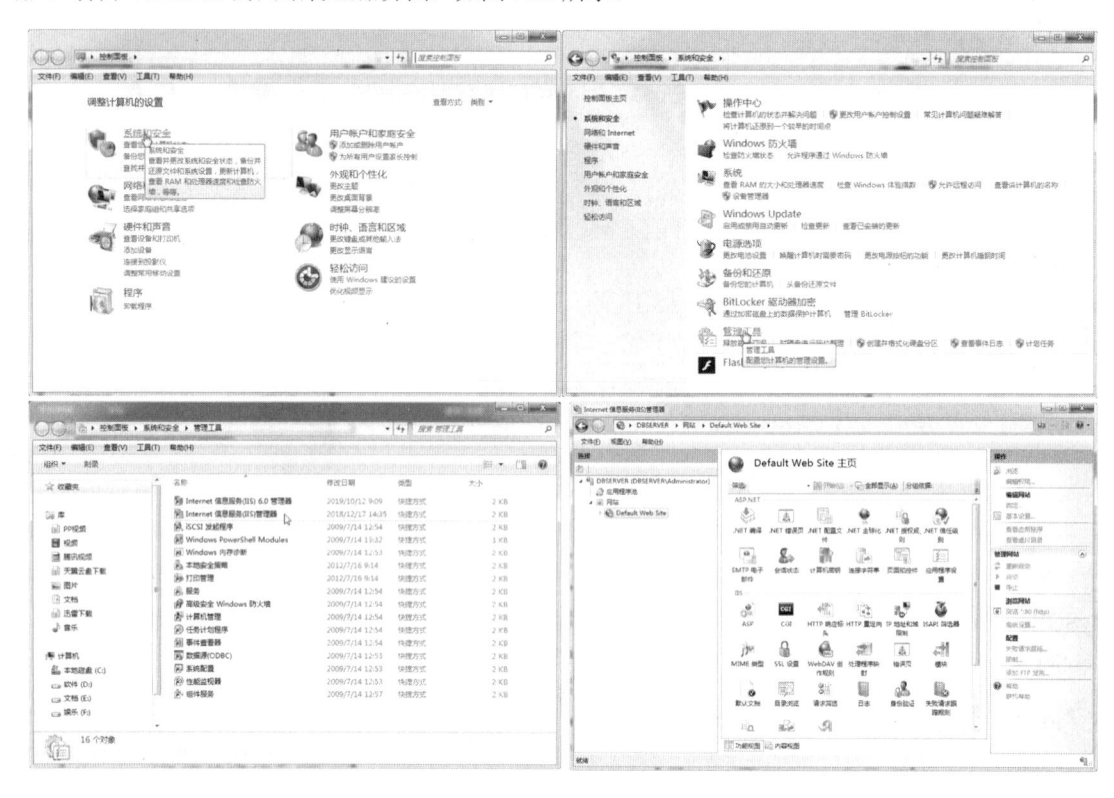

图 7.31 打开 Windows 的网站管理器界面

(3)在计算机 E 盘下创建一个名为 MyServices 的目录,右击 "Internet 信息服务(IIS)管理器" 左侧树状视图下的 "网站"→"添加网站",在出现的 "添加网站" 对话框中填写网站名称为 "EasyService",选择物理路径为 "E:\MyServices",绑定 IP 地址 "192.168.0.138"、端口 "8080"(读者请根据自己 Web 服务器的实际 IP 地址和端口填写),如图 7.32 所示。

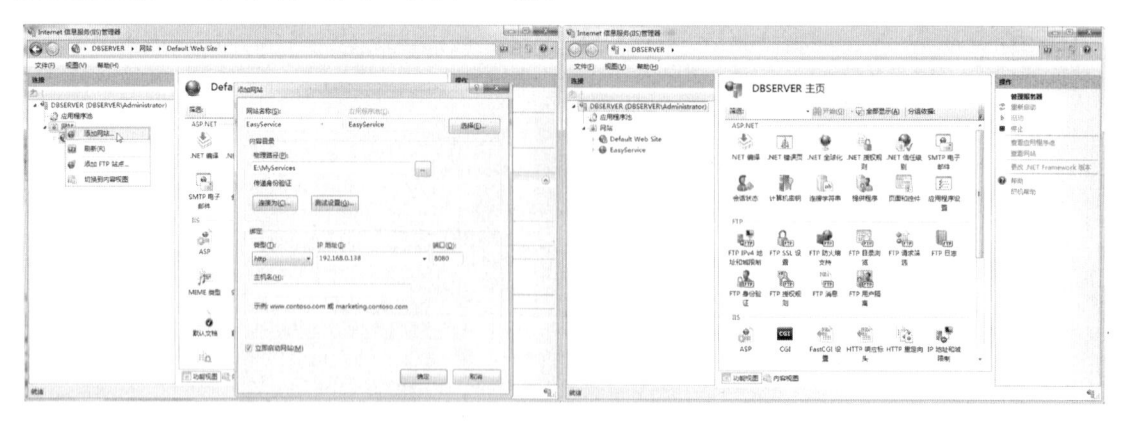

图 7.32 添加网站

(4)将网站 "EasyService" 的应用程序池设为 ASP.NET v4.0,如图 7.33 所示,选中网站 "EasyService",单击右侧 "操作" 下的 "基本设置",出现 "编辑网站" 对话框,单击 "选择" 按钮,从弹出框 "应用程序池" 下拉列表中选择 "ASP.NET v4.0",单击 "确定" 按钮即可。配置应用程序池的目的主要是为了让网站的.NET 框架与 WebService 相兼容。

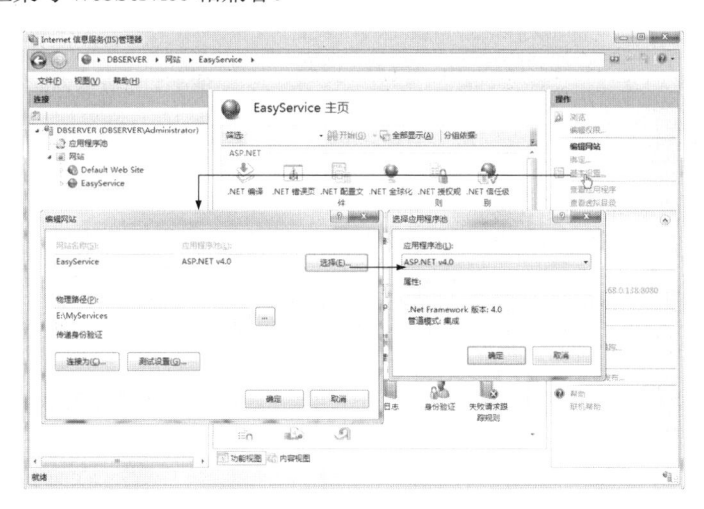

图 7.33 配置应用程序池

这样，我们就配置好了一个 IIS 服务器上运行的 Web 网站，稍后开发的 WebService 将发布在这个网站上。

7.4.3 开发 WebService

【例 7.5】 在服务器计算机上开发一个 WebService。

1. 创建 Web 服务项目

我们使用微软的 Visual Studio 2015（VS2015）来开发 WebService。启动 VS2015，选择主菜单"文件"→"新建"→"项目"命令，弹出"新建项目"对话框，如图 7.34 所示，在左侧选择"已安装"→"模板"→"Visual C#"→"Web"，中间选中"ASP.NET 空 Web 应用程序"项，在上方下拉列表中注意一定要选".NET Framework 4"（图中特别圈出），底部"名称"栏填写项目名称为"MyWebService"，单击"确定"按钮。

图 7.34 创建.NET 4.0 上的空 Web 项目

在"解决方案资源管理器"中右击刚刚创建的项目，在弹出的快捷菜单中选择"添加"→"新建项"命令，弹出"添加新项"对话框，如图 7.35 所示，在左侧选择"已安装"→"Visual C#"→"Web"，中间列表选中"Web 服务（ASMX）"，底部"名称"栏填写 Web 服务名称为"MainWebService.asmx"，单击"添加"按钮，这样就创建好了一个 Web 服务项目。

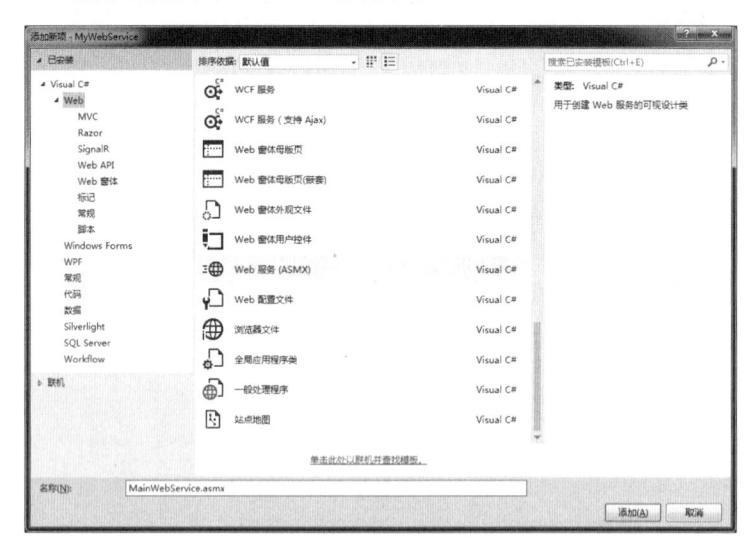

图 7.35　创建 Web 服务项目

2. 编写 WebService 程序

此时，在 VS2015 中打开 MainWebService.asmx 源文件，可以看到 VS 已经自动为我们生成了 WebService 的代码框架，其中有个 HelloWorld 方法（默认的示例方法），如图 7.36 所示，它仅仅简单地返回"Hello World"字符串。用户可改写这个方法，也可以自定义其他方法来扩充此 WebService 的功能。

图 7.36　VS 自动生成的 WebService 代码框架

由于本项目是基于.NET Framework 4 创建的，而 VS2015 默认的.NET 平台版本是 4.5.2，我们希望本项目的 WebService 能够访问后台 MySQL 8 数据库，但 MySQL 8 自带的 C#驱动也是.NET 4.5.2 版的，故我们需要想办法弄到匹配.NET 4.0 的 MySQL 驱动 MySql.Data.dll。这里笔者是从 CSDN 上下载得到与.NET 4.0 兼容的 MySQL 驱动的（附在本书源码中提供给读者），将其添加引用进项目中，如图 7.37 所示。

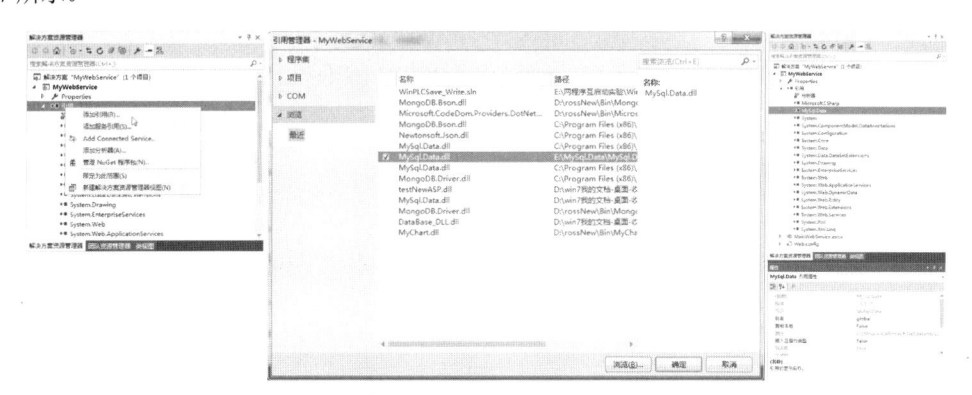

图 7.37 添加引用与.NET 4.0 兼容的 MySQL 驱动

实现思路：

（1）在程序开头 using 引用 MySQL 8 驱动库，并在 WebService 主类 MainWebService 中声明数据库连接对象、MySQL 命令对象和数据读取器。

（2）自定义一个 WebService 方法（以[WebMethod]声明）GetCommodity()连接和访问数据库，该方法有两个参数：id（商品编号）和 price（价格），当 price 不为空时更新商品价格，否则查询出对应 id 编号的商品信息返回。

在 MainWebService.asmx 文件中编写 WebService 代码如下：

```
using System;
using System.Collections.Generic;
using System.Linq;
using System.Web;
using System.Web.Services;
using MySql.Data.MySqlClient;                              //MySQL 8 驱动库
using System.Data;

namespace MyWebService
{
    /// <summary>
    /// MainWebService 的摘要说明
    /// </summary>
    [WebService(Namespace = "http://tempuri.org/")]
    [WebServiceBinding(ConformsTo = WsiProfiles.BasicProfile1_1)]
    [System.ComponentModel.ToolboxItem(false)]
    // 若要允许使用 ASP.NET AJAX 从脚本中调用此 Web 服务，请取消注释以下行
    // [System.Web.Script.Services.ScriptService]
    public class MainWebService : System.Web.Services.WebService
    {
        private MySqlConnection conn = null;               //MySQL 数据库连接对象
        private MySqlCommand cmd = null;                   //MySQL 命令对象
        private MySqlDataReader mdr = null;                //MySQL 数据读取器

        [WebMethod]
        public string GetCommodity(string id, string price)    //（a）
```

```csharp
            {
                string result = "";                                    //WebService 返回的字符串
                try
                {
                    conn = new MySqlConnection("server=192.168.0.251;User Id=root;password=ross123456;database=emarket;Character Set=utf8");    //获取数据库连接对象
                    conn.Open();                                       //打开连接
                    string sqlstr = "";                                //存放 SQL 语句
                    if (!string.IsNullOrEmpty(price))                  //（b）
                    {
                        sqlstr = "UPDATE commodity SET 价格 = " + float.Parse(price) + " WHERE 商品编号 = '" + id + "'";
                                                                       //更新的 SQL 语句
                        cmd = new MySqlCommand(sqlstr, conn);
                        cmd.ExecuteNonQuery();                         //执行更新操作
                    }
                    sqlstr = "SELECT * FROM commodity WHERE 商品编号 = '" + id + "'";
                                                                       //查询的 SQL 语句
                    cmd = new MySqlCommand(sqlstr, conn);
                    mdr = cmd.ExecuteReader();                         //执行查询操作
                    while (mdr.Read())
                    {
                        result = mdr["商品名称"].ToString() + "    " + mdr["价格"].ToString() + ":" + mdr["商品图片"].ToString();
                                                                       //（c）
                    }
                }
                catch (Exception e)
                {
                    result = e.Message.ToString();                     //出错返回异常信息供用户排查
                }
                finally
                {
                    if (mdr != null)
                    {
                        mdr.Close();                                   //关闭数据读取器
                        mdr = null;
                    }
                    if (cmd != null)
                    {
                        cmd.Cancel();                                  //关闭命令对象
                        cmd = null;
                    }
                    if (conn.State == ConnectionState.Open)
                    {
                        conn.Close();                                  //关闭连接
                        conn = null;
                    }
                }
                return result;                                         //返回结果
            }
        }
    }
}
```

以上代码用 C#语言实现了更新、查询 MySQL 数据库的操作，向移动端返回结果。

其中：

（a）**public string GetCommodity(string id, string price)**：GetCommodity()是我们自定义的方法，它的功能就是从数据库中获取商品信息、更改价格。

（b）**if (!string.IsNullOrEmpty(price))**：如果移动端用户发来价格值（参数 price 不为空），则更新数据库。

（c）**result = mdr["商品名称"].ToString() + "   " + mdr["价格"].ToString() + ":" + mdr["商品图片"].ToString();**：将查到的内容按照一定格式组织起来赋给返回字符串。

7.4.4 发布 WebService

WebService 开发完成之后，还要发布到 IIS 服务器上，才能被 Android 程序访问和使用。发布 WebService 的步骤如下：

（1）在解决方案资源管理器中右击项目，在弹出的快捷菜单中单击"发布"命令，出现如图 7.38 所示的"发布 Web"向导对话框。在第一个"配置文件"页选择"选择发布目标"为"自定义"，弹出"新建自定义配置文件"对话框，填写配置文件的名称（可任取），单击"确定"按钮，单击"下一页"按钮继续。

（2）在"连接"页，选择"Publish method"（发布方式）为"File System"，"Target location"（目标路径）为"E:\MyServices"（也就是我们之前创建的 IIS 网站"EasyService"的物理路径），单击"下一页"按钮，如图 7.39 所示。

图 7.38　"发布 Web"向导对话框 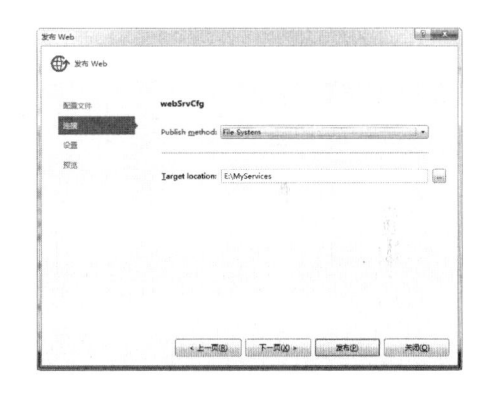　　　　　图 7.39　选择发布的目标路径

（3）在接下来的"设置"页，"Configuration"选为"Debug"（注意一定要选这个！），如图 7.40 所示，单击"发布"按钮，开始生成 WebService。

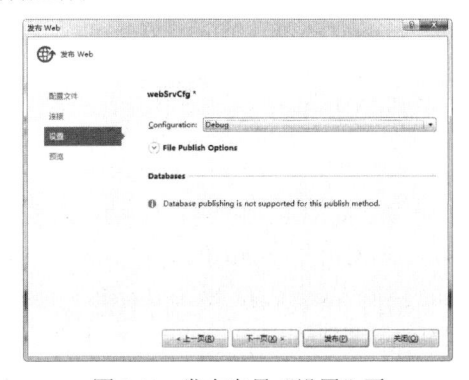

图 7.40　发布向导"设置"页

稍候片刻，在 VS 环境下方的"输出"子窗口出现如图 7.41 所示的信息，表示发布成功。去往"E:\MyServices"目录下可看到发布的文件和目录。

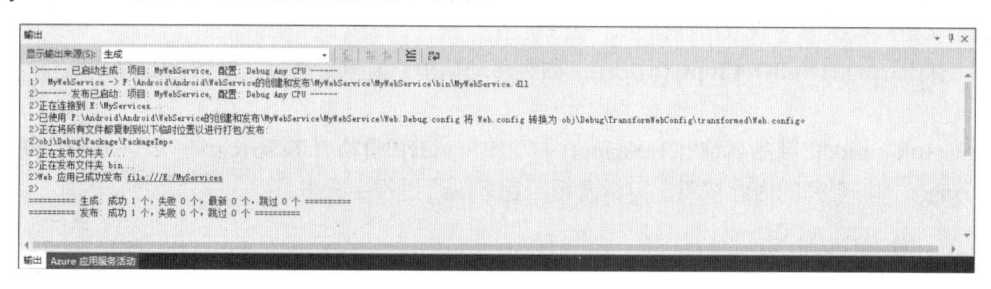

图 7.41　发布成功

读者可以先测试一下这个 WebService 能否正常工作，在移动端浏览器中输入 http://192.168.0.138:8080/MainWebService.asmx 并回车，如果出现如图 7.42 所示的页面就表明该 WebService 已部署成功且能正常提供服务。

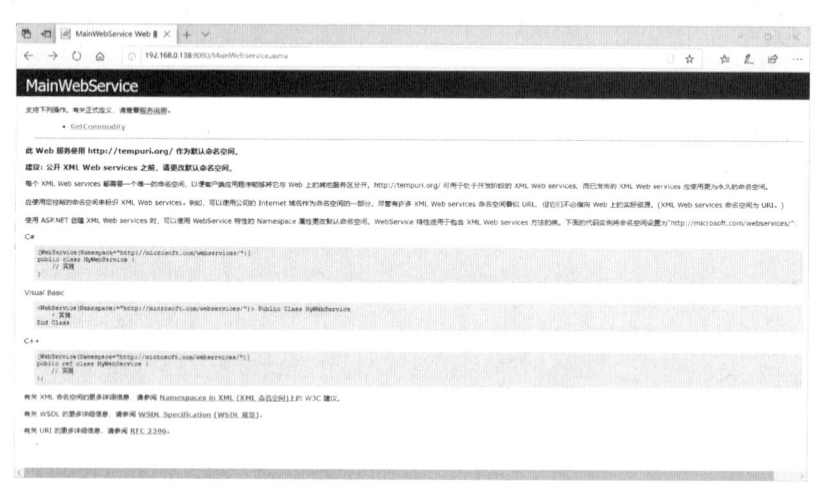

图 7.42　测试 WebService 能否正常工作

接下来我们将开发移动端 Android 程序来访问这个 WebService。

7.4.5　移动端 Android 程序开发

【例 7.6】 在移动端计算机上开发一个 Android 程序访问 WebService。

在移动端 Android Studio 中创建名为 HttpWebServiceDemo 的工程。与前面一样，在界面上拖曳放置一个文本视图、一个图像视图、一个编辑框和一个按钮，代码略。

1. 加载 ksoap2 包

Android 程序与 WebService 通信使用的是 SOAP 协议，谷歌的 ksoap2 库实现了这个协议，需要引入这个库才能实现对 WebService 的访问。到 https://oss.sonatype.org/content/repositories/ksoap2-android-releases/com/google/code/ksoap2-android/ksoap2-android-assembly/上可以下载到 ksoap2，当前最新版本为 3.6.4，下载获得的 JAR 包文件名为"ksoap2-android-assembly-3.6.4-jar-with-dependencies.jar"，用 7.2.1 小节介绍的方法将它加载到工程中。

2. 编写 Android 程序

实现思路：

（1）在程序开头导入 ksoap2 库。

（2）在 MainActivity 中声明 WebService 命名空间、方法名、URL 地址、SOAP 动作等一系列参数。

（3）定义一个 onSubmitClick(null) 方法向 WebService 发起请求，在 onCreate() 中调用该方法。

（4）在 onSubmitClick(null) 方法内另开一个线程实现与 WebService 的交互，先创建一个 SOAP 请求对象，装入信封；然后创建一个 HttpTransportSE 对象，指定 WebService 的 URL；再通过 HttpTransportSE 对象的 call() 方法调用 WebService 方法（GetCommodity()）；最后从信封的 getResponse() 方法得到 WebService 的返回结果。

（5）WebService 返回结果转换为字符串的形式封装于 Message 消息对象中，通过 Handler 句柄发给主程序，刷新 UI 界面。

MainActivity.java 代码如下：

```java
...
//引入 ksoap2 相关类库
import org.ksoap2.SoapEnvelope;
import org.ksoap2.serialization.SoapObject;
import org.ksoap2.serialization.SoapSerializationEnvelope;
import org.ksoap2.transport.HttpTransportSE;
//
import android.os.Handler;
import android.os.Message;
import java.lang.reflect.Field;

public class MainActivity extends AppCompatActivity {
    private TextView myTextView;
    private EditText myEditText;
    private ImageView myImageView;
    private String NAMESPACE = "http://tempuri.org/";              //WebService 默认命名空间
    private String METHOD = "GetCommodity";                        //WebService 中的方法名
    private String URL = "http://192.168.0.138:8080/MainWebService.asmx"; //WebService 的 URL 地址
    private String SOAP_ACTION = "http://tempuri.org/GetCommodity"; //WebService 的 SOAP 动作
    private String myCommodityId = "11101";

    @Override
    protected void onCreate(Bundle savedInstanceState) {
        super.onCreate(savedInstanceState);
        setContentView(R.layout.activity_main);
        myTextView = findViewById(R.id.myTextView);
        myImageView = findViewById(R.id.myImageView);
        myEditText = findViewById(R.id.myEditText);
        onSubmitClick(null);                                        //向 WebService 发起请求
        ...
    }
    ......
    public void onSubmitClick(View view) {
        new Thread(new Runnable() {                                 // （a）
            @Override
            public void run() {
```

```
            try {
                    SoapObject requestObj = new SoapObject(NAMESPACE, METHOD);
                                                                                //创建一个 SOAP 请求对象
                    requestObj.addProperty("id", myCommodityId);              // (b)
                    requestObj.addProperty("price", myEditText.getText().toString());
                    SoapSerializationEnvelope envelope = new SoapSerializationEnvelope(SoapEnvelope.
VER12);                                                                       // (c)
                    envelope.bodyOut = requestObj;                            //将请求对象装入信封
                    envelope.dotNet = true;
                    envelope.setOutputSoapObject(requestObj);
                    HttpTransportSE transportSE = new HttpTransportSE(URL);
                                                                              // (d)
                    transportSE.call(SOAP_ACTION, envelope);                  // (e)
                    Object responseObj = envelope.getResponse();              // (f)
                    Message msg = Message.obtain();
                    msg.what = 1000;
                    msg.obj = responseObj.toString();
                    myHandler.sendMessage(msg);
            } catch (Exception e) {
                    e.printStackTrace();
            }
        }
    }).start();                                                               //开启线程
}

private Handler myHandler = new Handler() {
        public void handleMessage(Message message) {
            try {
                    myTextView.setText(message.obj.toString().split(":")[0]);
                    Class resClass = R.drawable.class;
                    Field field = resClass.getField(message.obj.toString().split(":")[1].trim());
                    int resId = (Integer) field.get(resClass.newInstance());
                    myImageView.setImageResource(resId);
            } catch (NoSuchFieldException e) {
            } catch (IllegalAccessException e) {
            } catch (InstantiationException e) {
            }
        }
};
}
```

其中：

（a）**new Thread(new Runnable() { ... }).start();**：请求 WebService 是耗时操作，要放在线程里。

（b）**requestObj.addProperty("id", myCommodityId);**：这里 id 与 WebService 中的参数名必须完全一样。

（c）**SoapSerializationEnvelope envelope = new SoapSerializationEnvelope(SoapEnvelope.VER12);**：创建 SoapSerializationEnvelope 对象时需要通过 SoapSerializationEnvelope 类的构造方法设置 SOAP 协议的版本号，该版本号需要根据服务端 WebService 的版本号来设置。

（d）**HttpTransportSE transportSE = new HttpTransportSE(URL);**：通过 HttpTransportsSE 类的构造方法可以指定 WebService 的 WSDL 文档的 URL。

(e) **transportSE.call(SOAP_ACTION, envelope);**：使用 call()方法调用 WebService 方法。
(f) **Object responseObj = envelope.getResponse();**：使用 getResponse()方法获得 WebService 方法的返回结果。

最后，同样要记得在 AndroidMainifest.xml 中打开 HTTP 明文传输及互联网访问权限。

运行程序，显示数据库中的商品名称、价格和图片，在移动界面上填写新的价格后提交给 WebService，数据库的信息被修改后回显在移动界面上，如图 7.43 所示。

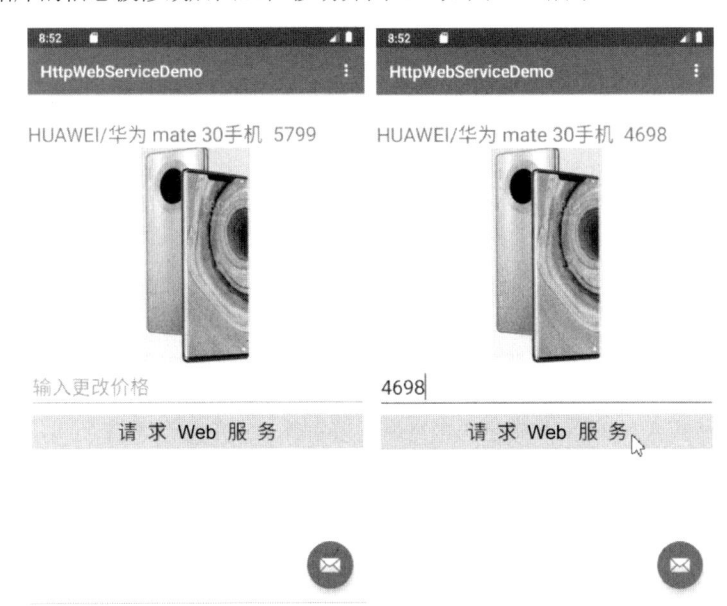

图 7.43 Android 程序访问 WebService 运行效果

7.5 网上商城 JSON 数据操作

7.5.1 基本原理

本系统还是采用 HTTP-Web-DB 的 3 层架构，移动端以 HttpURLConnection 向 Web 服务器发起请求，Web 服务器上的 Java Servlet 接收请求，通过 JDBC 操作后台 Oracle 11g 数据库，将获取的结果数据封装为 JSON 对象返回给移动端。JSON 是目前移动互联网通行的信息传输格式，因此，图 7.44 很好地演示了当前真实移动互联网应用的运作模式。

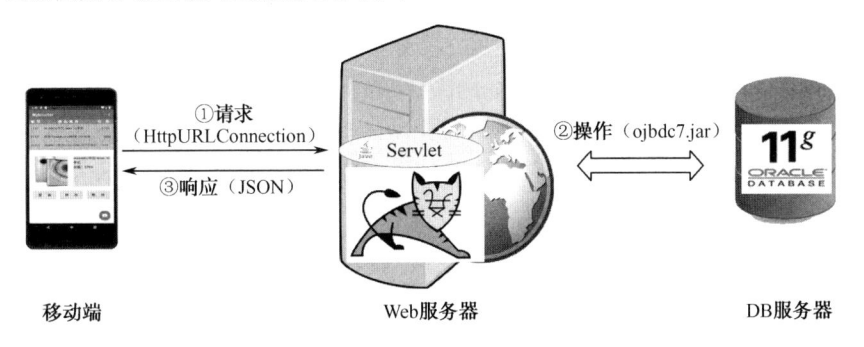

图 7.44 当前真实移动互联网应用的运作模式

7.5.2　Web 端开发

【例 7.7】 开发 Web 端程序，以 JSON 与移动端交互。

Web 端仍然使用 Java Servlet 程序，它负责接收 Android 程序发来的请求，根据 Android 程序的要求去操作后台 Oracle 11g 数据库，故离不开 JDBC 驱动包，这里使用的是 ojdbc7.jar。又由于 Web 服务程序是以 JSON 格式向移动端返回数据的，故还需要使用 JSON 相关的包，从网络下载获得，一共是 6 个.jar 包，具体如下：

- commons-beanutils-1.8.0.jar；
- commons-collections-3.2.1.jar；
- commons-lang-2.5.jar；
- commons-logging-1.1.1.jar；
- ezmorph-1.0.6.jar；
- json-lib-2.3.jar。

将它们连同数据库驱动 ojdbc7.jar 包一起复制到项目的 lib 目录直接刷新即可。

实现思路：

（1）导入 IO、SQL 和 JSON 操作的库。

（2）在主 Servlet 类 MainServlet 中声明数据库连接对象、SQL 语句对象和结果集对象。

（3）本程序中一共创建了 2 个 JSON 数据结构，一个为 JSON 对象 jobj，另一个为 JSON 数组 jarray。

（4）程序从后台 Oracle 11g 数据库中读取的数据会先遍历包装为一个个临时的 JSON 对象（即 jcommodity），将它们存入数组 jarray；然后将数组 jarray 再封装到一个总的 JSON 对象 jobj（"list"）中；最后将这个总的 JSON 对象返回给移动端。

下面给出本例的 Servlet 源代码（略去自动生成的部分），内容如下：

```
import java.io.*;                                          //IO 操作的库
import java.sql.*;                                         //SQL 操作的库
import net.sf.json.*;                                      //JSON 操作的库
...
@WebServlet("/MainServlet")
public class MainServlet extends HttpServlet {
    private static final long serialVersionUID = 1L;
    private Connection conn = null;                        //数据库连接对象
    private Statement stmt = null;                         //SQL 语句对象
    private ResultSet rs = null;                           //结果集对象
    ......
    protected void doGet(HttpServletRequest request, HttpServletResponse response) throws ServletException, IOException {
        // TODO Auto-generated method stub
        response.setCharacterEncoding("utf-8");            //必须有这句！否则中文显示为???
        response.setContentType("application/json");       //设置以 JSON 格式向移动端返回数据
        //创建 JSON 数据结构
        JSONObject jobj = new JSONObject();                //创建 JSON 对象
        JSONArray jarray = new JSONArray();                //创建 JSON 数组对象
        //访问 Oracle 11g 数据库读取内容
        try {
            Class.forName("oracle.jdbc.driver.OracleDriver");   //加载 Oracle 11g 驱动类
            conn = DriverManager.getConnection("jdbc:oracle:thin:@SMONGO:1521:EMARKET", "SCOTT",
```

```java
"Mm123456");                                                              //获取到 Oracle 11g 的连接
            stmt = conn.createStatement();
            //解析移动端请求中的数据项                      //（a）
            String data = request.getParameter("data");   //数据项（商品价格值）
            String id = request.getParameter("id");       //当前要操作（修改、删除）的商品编号
            String opt = request.getParameter("opt");     //所要执行的操作
            if (!(data == null||data.length() <= 0)) {
                if (opt.equals("upt")) {                  //修改商品价格
                    String sql = "UPDATE COMMODITY SET 价格 = " + Float.parseFloat(data) + " WHERE 商品编号 = '" + id + "'";
                    stmt.executeUpdate(sql);
                }
                if (opt.equals("del")) {                  //删除商品记录
                    String sql = "DELETE FROM COMMODITY WHERE 商品编号 = '" + id + "'";
                    stmt.executeUpdate(sql);
                }
            }
            if (opt != null && opt.equals("que") && !(data == null||data.length() <= 0))
                rs = stmt.executeQuery("SELECT * FROM COMMODITY WHERE 商品编号 = '" + id + "'");
                                                          //查询某个编号的商品信息记录
            else
                rs = stmt.executeQuery("SELECT * FROM COMMODITY");
                                                          //（b）
            int i = 0;
            while (rs.next()) {                           //遍历查询结果
                JSONObject jcommodity = new JSONObject(); //临时 JSON，存储结果集中一条记录
                //将需要返回的结果字段封装进 JSON
                jcommodity.put("cid", rs.getString("商品编号").toString().trim());
                jcommodity.put("name", rs.getString("商品名称").toString());
                jcommodity.put("price", rs.getInt("价格"));
                jcommodity.put("image", rs.getString("商品图片").toString().trim());
                jarray.add(i, jcommodity);                //将单个 JSON 对象添加进数组
                i++;
            }
            jobj.put("list", jarray);                     //将 JSON 数组再封装进 JSON 对象
        } catch (ClassNotFoundException e) {
            jobj.put("err", e.getMessage());
        } catch (SQLException e) {
            jobj.put("err", e.getMessage());
        } finally {
            try {
                if (rs != null) {
                    rs.close();                           // 关闭 ResultSet 对象
                    rs = null;
                }
                if (stmt != null) {
                    stmt.close();                         // 关闭 Statement 对象
                    stmt = null;
                }
                if (conn != null) {
                    conn.close();                         // 关闭 Connection 对象
```

```
                        conn = null;
                    }
                } catch (SQLException e) {
                    jobj.put("err", e.getMessage());
                }
            }
            PrintWriter return_to_client = response.getWriter();    //将最终封装好的JSON对象返回移动端
            return_to_client.println(jobj);
            return_to_client.flush();
            return_to_client.close();
        }
        ...
    }
```

其中：

（a）**String data = request.getParameter("data");**、**String id = request.getParameter("id");** 和 **String opt = request.getParameter("opt");**：移动端请求中有 3 个数据项（在请求的 URL 地址后携带，以 & 分隔），其中 data 是要修改的商品价格数值，id 是要对其执行操作（如修改、删除）的商品编号，opt 则表示所要执行的具体操作类型，有 upt（修改）、del（删除）和 que（查询）三个选项，服务器程序就是根据以上 3 个数据项的取值来得知移动端所要执行的操作的。例如：

    data=4799&id='11101'&opt='upt'

表示将数据库中编号为 "11101" 的商品价格修改为 4799。

（b）**rs = stmt.executeQuery("SELECT * FROM COMMODITY");**：如果用户发来空信息（未修改任何内容），则直接读取并返回数据库中所有商品的信息。

将程序部署到 Web 服务器计算机上，启动 Tomcat，可先在移动端计算机上用浏览器访问 http://192.168.0.138:8080/MyServlet/MainServlet 测试是否连通，如果出现如图 7.45 所示页面，上面以 JSON 格式的字符串显示出 Oracle 11g 数据库中存储的商品信息记录，就表示 Web 服务器环境没有问题。

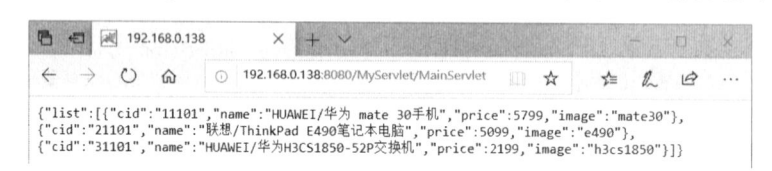

图 7.45　测试 Web 服务器

7.5.3　移动端开发

【例 7.8】　在移动端计算机上开发一个 Android 程序，与 Web 服务器之间以 JSON 格式交互数据，实现基本的商品信息管理操作。

1. 设计界面

创建 Android Studio 工程，名为 MyEmarket。在工程 content_main.xml 文件的设计（Design）模式下拖曳设计 Android 程序界面，如图 7.46 所示。

这里我们在界面顶部以一个列表视图（ListView）控件来显示 Oracle 11g 数据库中存放的商品信息记录，背景设为绿色，列表顶部用一个文本视图（TextView）显示 "编号　商品名称　价格" 列表标题。列表下方并列放置两个控件：一个图像视图（ImageView）用来显示当前用户所选中商品的图片，图片作为资源预先存放于工程 drawable 目录下；一个编辑框（EditText）是用来显示对应商品名称和价格信息的（加灰色打底），用户可修改其中价格值并提交给后台数据库；点击底部的 3 个按钮分别执行查询、修改和删除操作。界面设计的详细代码略。

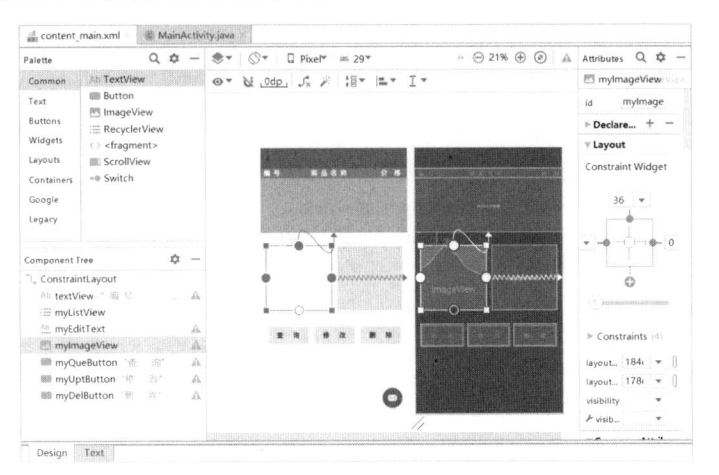

图 7.46 设计 Android 程序界面

2. 实现功能

实现思路：

（1）初始启动 Android 程序时默认就会连接到 Web 服务器，而程序运行起来后，任何时刻用户点击界面按钮也会向服务器发出请求。

（2）为简化代码、避免冗余，本例中我们将初始化和用户点击按钮时所要执行的功能封装于同一个 onSubmitClick() 方法中，通过向其中传递一个字符串参数来"告知"程序具体要做什么。

移动端的功能实现代码全部位于 MainActivity.java 源文件中，内容如下：

```java
...
//导入 Android 内置的 JSON 库
import org.json.JSONArray;
import org.json.JSONException;
import org.json.JSONObject;
......
public class MainActivity extends AppCompatActivity implements AdapterView.OnItemClickListener {
    private ListView myListView;                    //列表视图（显示 Oracle 数据库的商品信息）
    private List<String> list;                      //存储商品信息的 List 结构（与列表视图绑定）
    private List<String> imagelist;                 //存储商品图片名的 List 结构（与 list 对应）
    private ArrayAdapter<String> adapter;           //Array 适配器（给列表视图绑定数据源）
    private EditText myEditText;                    //编辑框（显示当前选中商品的名称和价格）
    private ImageView myImageView;                  //图像视图（显示当前选中商品的图片）
    private HttpURLConnection conn = null;          //HTTP 连接对象（与服务器交互的工具）
    private InputStream stream = null;              //输入流（存放获取的响应数据内容）
    private Button myQueButton;                     //"查询"按钮
    private Button myUptButton;                     //"修改"按钮
    private Button myDelButton;                     //"删除"按钮
    private String commodityId;                     //用户所操作商品的编号（点选列表项确定）

    @Override
    protected void onCreate(Bundle savedInstanceState) {
        super.onCreate(savedInstanceState);
        setContentView(R.layout.activity_main);
```

```java
        myListView = findViewById(R.id.myListView);
        myListView.setOnItemClickListener(this);              //绑定列表项点击事件监听器
        list = new ArrayList<>();                             //创建 List 结构
        adapter = new ArrayAdapter<String>(this, R.layout.support_simple_spinner_dropdown_item, list);
                                                              //创建数据适配器
        imagelist = new ArrayList<>();
        myEditText = findViewById(R.id.myEditText);
        myImageView = findViewById(R.id.myImageView);
        myQueButton = findViewById(R.id.myQueButton);
        myQueButton.setOnClickListener(new View.OnClickListener() {
            @Override
            public void onClick(View view) {
                onSubmitClick("que");                         //点击"查询"按钮时执行
            }
        });
        myUptButton = findViewById(R.id.myUptButton);
        myUptButton.setOnClickListener(new View.OnClickListener() {
            @Override
            public void onClick(View view) {
                onSubmitClick("upt");                         //点击"修改"按钮时执行
            }
        });
        myDelButton = findViewById(R.id.myDelButton);
        myDelButton.setOnClickListener(new View.OnClickListener() {
            @Override
            public void onClick(View view) {
                onSubmitClick("del");                         //点击"删除"按钮时执行
            }
        });
        onSubmitClick("con");                                 //（a）
        ...
    }
    ......
    public void onSubmitClick(final String opt) {
        new Thread(new Runnable() {                           //耗时操作，必须放入线程
            @Override
            public void run() {
                try {
                    URL url;
                    if (opt.equals("con"))                    //传入参数"con"表示是初始连接服务器
                        url = new URL("http://192.168.0.138:8080/MyServlet/MainServlet");
                                                              //URL 不带参数
                    else
                        url = new URL("http://192.168.0.138:8080/MyServlet/MainServlet?data=" +
myEditText.getText().toString().split(": ")[1] + "&id=" + commodityId + "&opt=" + opt);
                                                              //URL 带参数
                    conn = (HttpURLConnection) url.openConnection();
                                                              //获取 HTTP 连接对象
                    conn.setRequestMethod("GET");             //请求方式为 GET
                    conn.setConnectTimeout(3000);             //连接超时时间
                    conn.setReadTimeout(9000);                //读取数据超时时间
```

```java
                    conn.connect();                              //开始连接 Web 服务器
                    stream = conn.getInputStream();              //获取服务器的响应（输入）流
                    refresh_UI(stream);
                } catch (Exception e) {
                } finally {
                    try {
                        if (stream != null) {
                            stream.close();                      //关闭响应流
                            stream = null;
                        }
                        conn.disconnect();                       //断开连接
                        conn = null;
                    } catch (Exception e) {
                    }
                }
            }
        }).start();                                              //开启线程
    }

    public void refresh_UI(InputStream in) {
        BufferedReader bufReader = null;
        try {
            bufReader = new BufferedReader(new InputStreamReader(in));
                                                                 //响应流数据放入缓存
            StringBuilder builder = new StringBuilder();
            String str = "";
            while ((str = bufReader.readLine()) != null) {
                builder.append(str);                             //从缓存对象中读取数据并拼接为字符串
            }
            Message msg = Message.obtain();
            msg.what = 1000;
            msg.obj = builder.toString();                        //通过 Message 传递给主线程
            myHandler.sendMessage(msg);                          //通过 Handler 发送
        } catch (IOException e) {
        } finally {
            try {
                if (bufReader != null) {
                    bufReader.close();                           //关闭缓存
                    bufReader = null;
                }
            } catch (IOException e) {
            }
        }
    }

    private Handler myHandler = new Handler() {
        public void handleMessage(Message message) {
            try {
                JSONObject jObj = new JSONObject(message.obj.toString());
                                                                 //获取返回消息中的 JSON 对象
                JSONArray jArray = jObj.getJSONArray("list");
```

```java
                        list.clear();
                        imagelist.clear();
                        for (int i = 0; i < jArray.length(); i++) {      //遍历、逐条解析商品信息
                            //当前商品信息存储在临时 JSON 中
                            JSONObject jCommodity = jArray.getJSONObject(i);
                            String cid = jCommodity.getString("cid");           //商品编号
                            String name = jCommodity.getString("name");         //商品名称
                            String price = jCommodity.getString("price");       //价格
                            String image = jCommodity.getString("image");       //商品图片（图片文件名）
                            if (cid.length() == 5) {               //商品编号是长度为 5 的定长字符串
                                String item = cid + "    " + name;
                                //以下这段处理是为了使列表中的数据项对齐显示
                                for (int j = 0; j < 24 - name.length(); j++) item += "\t\t";
                                item += "    " + price;
                                list.add(item);                       //数据项添加进 List
                                imagelist.add(image);                 //图片文件名加入另一个 List
                            }
                        }
                        myListView.setAdapter(adapter);           //将界面列表视图绑定适配器（数据源）
                        myEditText.setText(list.get(0).split("    ")[1] + "\n 价格： " + list.get(0).split("    ")[2]);
                                                                  //编辑框设置为多行显示（可换行）
                        commodityId = list.get(0).split("    ")[0];       //当前选中商品的编号
                        loadPic(imagelist.get(0));                // （b）
                    } catch (JSONException e) {
                        myEditText.setText(e.getMessage());
                    }
                }
            };

    @Override                                               //用户选中列表项时触发
    public void onItemClick(AdapterView<?> adapterView, View view, int pos, long id) {
        myEditText.setText(list.get(pos).split("    ")[1] + "\n 价格： " + list.get(pos).split("    ")[2]);
                                                            //商品名称和价格信息显示在编辑框中
        commodityId = list.get(pos).split("    ")[0];       //获取当前选中商品的编号
        loadPic(imagelist.get(pos));                        // （b）
    }

    // 加载商品图片
    public void loadPic(String picname) {                   // （b）
        try {
            Class resClass = R.drawable.class;
            Field field = resClass.getField(picname);
            int resId = (Integer) field.get(resClass.newInstance());
            myImageView.setImageResource(resId);
        } catch (NoSuchFieldException e) {
        } catch (IllegalAccessException e) {
        } catch (InstantiationException e) {
        }
    }
}
```
//取出 JSON 对象中封装的 JSON 数组

其中：

（a）**onSubmitClick("con");**：当传入参数"con"时表示是初始连接服务器，其请求的 URL 只是纯粹的地址，不带任何参数，服务器默认会将后台 Oracle 11g 数据库中所有的商品信息都查询出来包装进 JSON 并返回给移动端显示；而如果传入的是别的字符串，则表示是用户点击了界面按钮所发出的请求，请求 URL 中会携带几个参数来表示用户要求服务器执行的具体操作类型、对象及数据。

（b）**loadPic(imagelist.get(0));、loadPic(imagelist.get(pos));和 public void loadPic(String picname)**：本例通过 List 数据结构中的元素获取商品图片对应的资源文件名，同样地，由于在初始化及用户点选列表项操作后都要加载商品图片，故这里也将图片资源解析的代码封装在 loadPic()方法中，需要时调用，向其中传入要显示的商品图片文件名，程序就会根据文件名解析出对应的资源 id 并将图片加载进界面上的图像视图中显示。

最后，同样要记得在 AndroidMainifest.xml 中打开 HTTP 明文传输及互联网访问权限。

最终移动端 App 运行的界面效果如图 7.47 所示，用户可以通过前端 App 界面查询、修改或删除后台 Oracle 11g 中的商品信息。

图 7.47　移动端 App 运行的界面效果

第 8 章　Android 多媒体和图形图像编程

本章系统介绍 Android 播放视频/音频媒体文件、处理图形图像以及制作手机相册等功能，同时简单介绍 OpenGL 库的入门使用。运用这些知识可开发出丰富多彩的多媒体形式的 App，让用户充分享受移动互联时代的视听盛宴。

8.1　媒体播放器的开发

8.1.1　视频播放

Android 的视频播放器既可以播放 App 内置的视频文件，也可以让用户选择手机存储器上的视频进行播放，还可以定制或绑定媒体控制条，实现更为高级的视频播放控制功能。借助手机自身的能力，还可以制作出画中画特效。

1. 视频的两种播放方式

Android 支持直接播放 App 内置视频和打开手机存储器上的视频进行播放这两种方式。

【例 8.1】　视频放在本项目 raw 中固定路径播放，或从真实手机上选择视频文件进行播放。

创建 Android Studio 工程 MyVideoView。

（1）设计界面。其上放一个 VideoView 控件、两个按钮和一个文本视图，如图 8.1 所示，代码略。

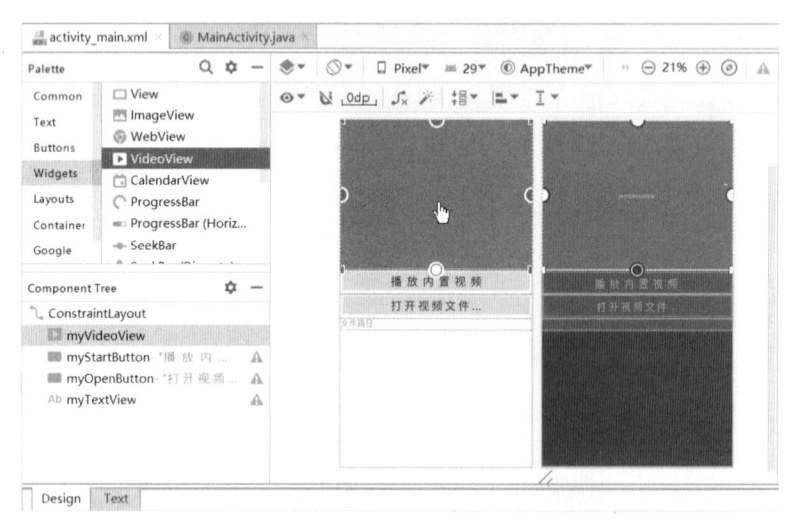

图 8.1　设计界面

（2）预置视频资源。在工程 res 下创建 raw 资源目录，有两种方法：

① 右击 res，在弹出的快捷菜单上选择 "New" → "Folder" → "Raw Resources Folder" 命令，弹出 "New Android Component" 对话框，直接单击 "Finish" 按钮，如图 8.2 所示。

第 8 章　Android 多媒体和图形图像编程

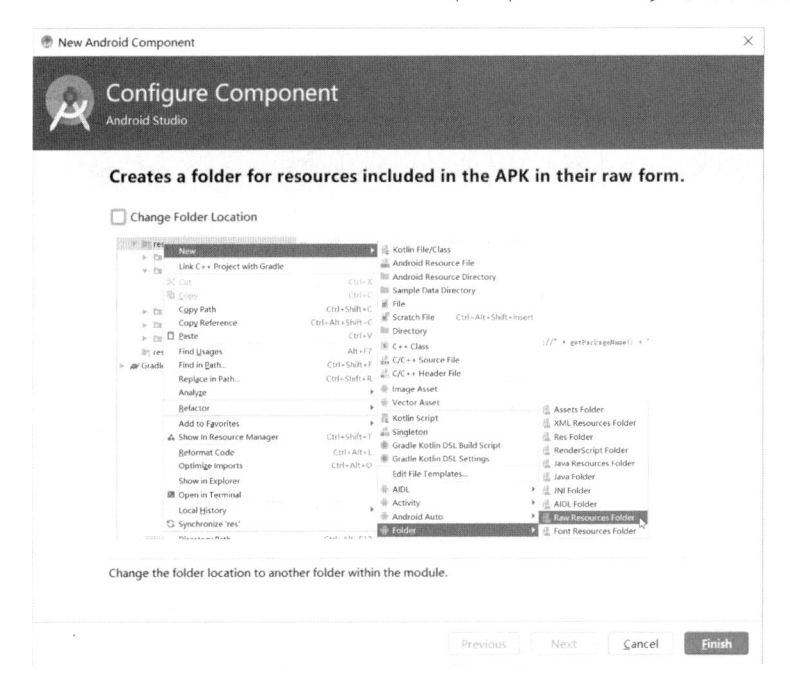

图 8.2　创建 raw 资源目录（方法①）

② 右击 res，在弹出的快捷菜单上选择"New"→"Android Resource Directory"命令，弹出"New Resource Directory"对话框，在"Resource type"栏的下拉列表中选择"raw"，单击"OK"按钮，如图 8.3 所示。

图 8.3　创建 raw 资源目录（方法②）

(3) 功能实现。

实现思路:

① 界面上两个按钮分别演示播放内置视频和打开手机存储器上的视频这两种播放方式；MainActivity 主程序类实现点击事件监听接口 View.OnClickListener，在 onClick()方法中以视图 id 标识用户点击的按钮，分别执行不同播放方式的操作。

② 打开手机存储器上视频播放的方式下，通过 startActivityForResult()启动活动页面，在 onActivityResult()响应方法中获取返回的视频 URI，播放视频。

③ 两种方式的视频播放共用界面上的 VideoView 控件作为放映区。

功能实现在 MainActivity.java 中，代码如下：

```java
public class MainActivity extends AppCompatActivity implements View.OnClickListener {
    private VideoView myVideoView;                    //视频视图（视频播放窗口区域）
    private Button myStartButton;                     //按钮（播放 App 内置视频）
    private Button myOpenButton;                      //按钮（打开手机上存储的视频文件）
    private TextView myTextView;                      //文本视图（显示视频文件的路径）
    private Uri myUri;                                //手机上视频文件的 URI
    @Override
    protected void onCreate(Bundle savedInstanceState) {
        super.onCreate(savedInstanceState);
        setContentView(R.layout.activity_main);
        myVideoView = findViewById(R.id.myVideoView);
        myStartButton = findViewById(R.id.myStartButton);
        myStartButton.setOnClickListener(this);
        myOpenButton = findViewById(R.id.myOpenButton);
        myOpenButton.setOnClickListener(this);
        myTextView = findViewById(R.id.myTextView);
    }

    @Override
    public void onClick(View v) {
        if (v.getId() == R.id.myStartButton) {                      //点击"播放内置视频"按钮
            Uri uri = Uri.parse("android.resource://" + getPackageName() + "/" + R.raw.boydiver);
                                                                     // (a)
            myVideoView.setVideoURI(uri);                            //设置视频 URI
            myVideoView.start();                                     //开始播放
        }
        if (v.getId() == R.id.myOpenButton) {                       //点击"打开视频文件…"按钮
            Intent intent = new Intent((Intent.ACTION_GET_CONTENT));
                                                                     // (b)
            intent.setType("video/*");                               // (c)
            startActivityForResult(intent, 101);                     //启动播放视频的活动页
        }
    }

    protected void onActivityResult(int requestCode, int resultCode, Intent data) {
                                                                     // (d)
        super.onActivityResult(requestCode, resultCode, data);
        myUri = data.getData();                                      // (d)
        myTextView.setText(myUri.getPath());                         // (d)
        myVideoView.setVideoURI(myUri);                              //设置视频 URI
```

```
            myVideoView.start();                              //开始播放
        }
}
```

其中:

(a) **Uri uri = Uri.parse("android.resource://" + getPackageName() + "/" + R.raw.boydiver);**:
Android 程序内置的视频一律要放在工程 res\raw 目录下,编译的时候,Android 编译器会将此目录下的视频文件原封不动地加入 App 可执行文件内。raw 内的文件以 URI 引用,URI 中要带完整的包名,这里通过 getPackageName()方法获得。本例我们播放一段海洋馆潜水员表演视频,视频文件名为 boydiver.mp4。

> **注意**:
> ① raw 目录一定要在 Android Studio 中用右键菜单功能创建,而不能直接进到工程目录文件夹中创建,那样 Android 程序是无法识别到 raw 资源目录的。
> ② 视频文件名必须全是小写的英文字母,不然 Android 程序识别不到这个资源。
> ③ 以 R.raw 引用视频的时候,其后的文件名表示的是资源名称,不用加后缀名(.mp4)。

(b) **Intent intent = new Intent((Intent.ACTION_GET_CONTENT));**:用 Intent 机制选取视频文件,设定 Intent 的动作类型为 ACTION_GET_CONTENT,表示获取所有的资源内容。

(c) **intent.setType("video/*");**:setType()方法指定要获取的资源的具体类型,可用通配符过滤,这里设为"video/*",表示要获取的是所有视频类型的资源;若改为获取音频资源,则类型设定为"audio/*"。

(d) **protected void onActivityResult(int requestCode, int resultCode, Intent data) { ... }**、**myUri = data.getData();**和 **myTextView.setText(myUri.getPath());**:打开视频时重启活动页,在 onActivityResult()方法中获得 Intent 传递过来的视频 URI,getData()从 Intent 数据中解析出 URI,getPath()将其转化为路径名字符串。

运行程序,直接点击"播放内置视频"按钮,程序默认播放的是 App 内置视频(海洋馆潜水员表演),如图 8.4 所示。

图 8.4 播放 App 内置视频

点击"打开视频文件…"按钮,屏幕弹出手机内部存储器上的视频列表页,选中其中一个,程序又回到原来的页面开始播放选中的视频,同时下方显示该视频文件的路径,如图 8.5 所示。

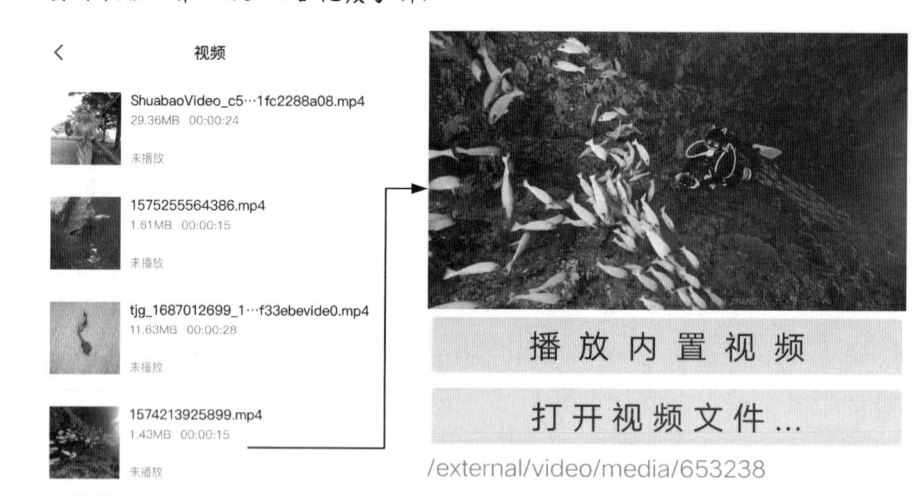

图 8.5　打开手机上的视频运行效果

2. 给画面增加进度条

【例 8.2】　修改【例 8.1】程序，给视频画面增加进度条，实现控制拖曳功能。

有两种方法能实现画面的进度条控制功能。

（1）SeekBar 进度条。在设计页面上加入 SeekBar 控件，重写其方法来定制进度条功能。设计页面上进度条放置在 VideoView 的下方，代码如下：

```
......
    <VideoView
        android:id="@+id/myVideoView"
        android:layout_width="0dp"
        android:layout_height="315dp"
        app:layout_constraintEnd_toEndOf="parent"
        app:layout_constraintStart_toStartOf="parent"
        app:layout_constraintTop_toTopOf="parent" />

    <SeekBar
        android:id="@+id/mySeekBar"
        android:layout_width="0dp"
        android:layout_height="14dp"
        app:layout_constraintEnd_toEndOf="parent"
        app:layout_constraintStart_toStartOf="parent"
        app:layout_constraintTop_toBottomOf="@+id/myVideoView" />
......
```

修改 MainActivity.java 代码（加粗处），增加进度条功能，代码如下：

```
public      class     MainActivity     extends    AppCompatActivity    implements    View.OnClickListener,
SeekBar.OnSeekBarChangeListener {                                       //（a）
    ...
    private SeekBar mySeekBar;                              //声明进度条对象
    @Override
    protected void onCreate(Bundle savedInstanceState) {
        ...
        mySeekBar = findViewById(R.id.mySeekBar);
        mySeekBar.setOnSeekBarChangeListener(this);         //（a）为进度条设置监听器
```

```
        mySeekBar.setEnabled(true);                         //启用进度条
    }

    @Override
    public void onClick(View v) {
        if (v.getId() == R.id.myStartButton) {
            ...
            myVideoView.start();
            myHandler.post(myRefresh);                      //（b）启动刷新任务
        }
        if (v.getId() == R.id.myOpenButton) {...}
    }

    protected void onActivityResult(int requestCode, int resultCode, Intent data) {
        ...
        myVideoView.start();
        myHandler.post(myRefresh);                          //（b）启动刷新任务
    }

    //以下 3 个都是 OnSeekBarChangeListener 接口中需要重写的方法
    @Override
    public void onProgressChanged(SeekBar seekBar, int progress, boolean fromUser) {
    }

    @Override
    public void onStartTrackingTouch(SeekBar seekBar) {
    }

    @Override
    public void onStopTrackingTouch(SeekBar seekBar) {
        int pos = seekBar.getProgress() * myVideoView.getDuration() / 100;
                                                            //（c）
        myVideoView.seekTo(pos);                            //将 VideoView 调整到指定位置继续播放
    }

    private Handler myHandler = new Handler();
    private Runnable myRefresh = new Runnable() {
        @Override
        public void run() {
            mySeekBar.setProgress(100 * myVideoView.getCurrentPosition()/myVideoView.getDuration());
                                                            //（d）
            myHandler.postDelayed(this, 200);               //（b）延迟 200ms 再次启动刷新任务
        }
    };
}
```

其中：

（a）**public class MainActivity...implements...、SeekBar.OnSeekBarChangeListener{...}和 mySeekBar.setOnSeekBarChangeListener(this);**：要想实现程序对进度条的控制，就必须实现 OnSeekBarChangeListener 监听接口。为简单起见，本例采用 1.3.7 小节所讲的范式四，直接在主程序 MainActivity 类中实现监听接口，需要显式地声明重写接口中的 3 个方法：

- **onProgressChanged()**：进度条进度变更时触发。其参数 fromUser 指示进度变更的原因，为 true 表示是由于用户拖曳进度条，为 false 表示是程序代码设置更改进度。
- **onStartTrackingTouch()**：用户开始拖曳进度条时触发。
- **onStopTrackingTouch()**：用户停止拖曳进度条时触发。

本例仅重写了第 3 个 onStopTrackingTouch() 方法，即当用户拖曳进度条停止后在其释放的位置处继续播放视频。其余两个方法虽然暂时用不上，但也必须写出它们的方法体声明。

（b）**myHandler.post(myRefresh);** 和 **myHandler.postDelayed(this, 200);**：用 Handler 机制创建了一个进度条的刷新任务，在其中以线程实时刷新进度条当前进度。创建任务句柄后，还必须在 VideoView 启动后，调用 post() 方法启动刷新任务，才能在视频播放过程中看到进度条的动态变化。

（c）**int pos = seekBar.getProgress() * myVideoView.getDuration() / 100;**：计算视频当前播放到的位置。这里用 VideoView 的 getDuration() 方法获得视频总时长，再通过进度百分比来计算播放位置。

（d）**mySeekBar.setProgress(100 * myVideoView.getCurrentPosition()/myVideoView.getDuration());**：设置进度条的当前进度。这里用 VideoView 的 getCurrentPosition() 方法获得视频已播放的时长，再与总时长相比即得到进度条的当前进度值。

运行程序，画面底部出现进度条且随着视频的播放而动态往前移动，用户也可以拖曳进度条至任一特定位置观看视频，效果如图 8.6 所示。

图 8.6 SeekBar 进度条显示效果

（2）与 MediaController 绑定显示进度条。除了自定义进度条，Android 还支持与媒体控制器 MediaController 绑定显示进度条。

在设计文件中将原来的进度条和 VideoView 控件删去，添加一个线性布局，代码如下：

```
......
    <LinearLayout
        android:id="@+id/myLinearLayout"
        android:layout_width="409dp"
        android:layout_height="294dp"
        android:orientation="vertical"
        app:layout_constraintEnd_toEndOf="parent"
        app:layout_constraintStart_toStartOf="parent"
        app:layout_constraintTop_toTopOf="parent" />
......
```

MainActivity.java 代码如下：

```
public class MainActivity extends AppCompatActivity implements View.OnClickListener {
    private LinearLayout myLinearLayout;            //线性布局（内含 VideoView 与 MediaController）
```

```
        private VideoView myVideoView;                    //视频视图（显示视频画面）
        private MediaController myMediaController;        //媒体控制条（实现进度条播放控制功能）
        ...
        @Override
        protected void onCreate(Bundle savedInstanceState) {
            super.onCreate(savedInstanceState);
            setContentView(R.layout.activity_main);
            myLinearLayout = findViewById(R.id.myLinearLayout);
            myVideoView = new VideoView(this);                       // (a)
            myMediaController = new MediaController(this);
            myVideoView.setMediaController(myMediaController);       // (b)
            myMediaController.setAnchorView(myVideoView);            // (b)
            myLinearLayout.addView(myVideoView);                     // (a)
                ...
        }
            ......
}
```

其中：

（a）**myVideoView = new VideoView(this);** 和 **myLinearLayout.addView(myVideoView);**：创建视频视图对象，并用 addView()方法将其添加进页面上的线性布局中。

（b）**myVideoView.setMediaController(myMediaController);** 和 **myMediaController.setAnchorView(myVideoView);**：创建媒体控制条对象，并将其与视频视图绑定。绑定是相互的，故要分两步：首先调用视频视图的 setMediaController()方法设置与之相关联的媒体控制条；然后调用媒体控制条的 setAnchorView()方法设置与之绑定的视频视图。这里也可以用 setMediaPlayer()取代 setAnchorView()方法，两者作用完全一样。

运行程序，效果如图 8.7 所示。

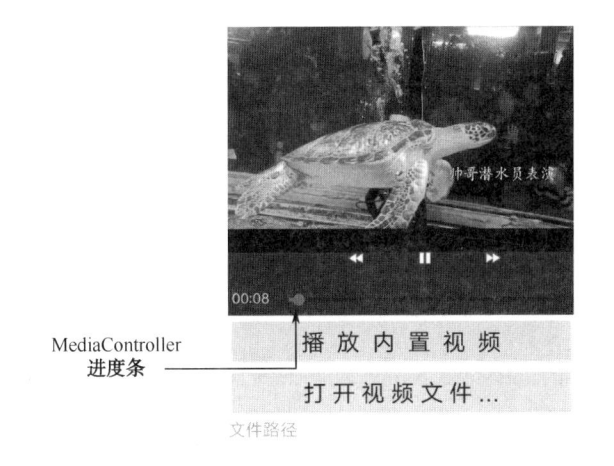

图 8.7　MediaController 进度条显示效果

与 SeekBar 相比，MediaController 的功能更为强大，支持播放、暂停、快进、快退以及显示当前播放到第几分几秒等诸多功能。所以，在一般的开发中多采用将 VideoView 与 MediaController 绑定的方式开发进度条功能。

> **注意:**
> VideoView 要通过代码放在线性视图中（而**不是在设计模式拖曳到界面中**），与 MediaController 绑定后才能紧贴在一起显示。

3. 画中画特效

现在的智能手机上运行的 App 种类繁多，其功能也越来越强大，用户在使用手机时往往会同时运行多个 App（如一边聊微信一边追剧），这就要求媒体播放器窗口不能始终独占着全屏，应当允许用户在需要时将视频画面缩小，只在屏幕上占一小块区域，以便同时做其他的事情。画中画功能就是为了满足这样的需求应运而生的，当前市面上几乎所有的手机都支持画中画模式。

【例 8.3】 修改【例 8.2】程序，增加实现画中画功能。

（1）在设计界面添加一个"进入画中画"按钮，代码略。

（2）修改 MainActivity.java 代码，实现画中画功能，代码如下：

界面完整代码

```java
public class MainActivity extends AppCompatActivity implements View.OnClickListener {
    ...
    private Button myPicInPicButton;                         // "进入画中画"按钮
    @Override
    protected void onCreate(Bundle savedInstanceState) {
        ...
        myPicInPicButton = findViewById(R.id.MyPicInPicButton);
        myPicInPicButton.setOnClickListener(this);
    }

    @Override
    public void onClick(View v) {
        if (v.getId() == R.id.myStartButton) {...}
        if (v.getId() == R.id.myOpenButton) {...}
        if (v.getId() == R.id.MyPicInPicButton) {            //用户点击"进入画中画"按钮
            PictureInPictureParams.Builder builder = new PictureInPictureParams.Builder();
                                                             // (a)
            Rational rational = new Rational(40, 30);        // (b)
            builder.setAspectRatio(rational);                //设置画中画窗口的宽高比
            enterPictureInPictureMode(builder.build());      //进入画中画模式
        }
    }
    ......
    public void onPictureInPictureModeChanged(boolean isInPicInPicMode, Configuration newConfig) {
                                                             // (c)
        super.onPictureInPictureModeChanged(isInPicInPicMode, newConfig);
        if (isInPicInPicMode) {                              //在画中画模式下，隐藏其他控件
            myStartButton.setVisibility(View.INVISIBLE);     // (d)
            myOpenButton.setVisibility(View.GONE);           // (d)
            myTextView.setVisibility(View.INVISIBLE);
            myPicInPicButton.setVisibility(View.GONE);
        } else {                                             //退出画中画模式，显示其他控件
            myStartButton.setVisibility(View.VISIBLE);
            myOpenButton.setVisibility(View.VISIBLE);
            myTextView.setVisibility(View.VISIBLE);
            myPicInPicButton.setVisibility(View.VISIBLE);
```

            }
        }
}
其中：
（a）**PictureInPictureParams.Builder builder = new PictureInPictureParams.Builder();**：PictureInPictureParams.Builder 是画中画模式的参数构建器，通过它的一系列 set()方法为画中画模式设定显示参数。

（b）**Rational rational = new Rational(40, 30);**：Rational 对象存储画中画窗口的宽高比例值，这里的 new Rational(40, 30)表示画中画窗口的宽度与高度之比为 40∶30。

（c）**public void onPictureInPictureModeChanged(boolean isInPicInPicMode, ConfigurationnewConfig)**：当 App 进入或退出画中画时，都会触发页面的 onPictureInPictureModeChanged()方法，通过重写该方法，可以从中获取布尔变量 isInPicInPicMode 的值，为 true 表示处于画中画模式，为 false 表示退出画中画模式，程序根据这个值来控制页面上其他控件的显隐。

（d）**myStartButton.setVisibility(View.INVISIBLE);**和 **myOpenButton.setVisibility(View.GONE);**：View.INVISIBLE 与 View.GONE 效果相同，都表示隐藏控件，编程时可以互换。

（3）画中画模式的配置。除了编写程序代码，还要在工程中进行一些额外配置才能实现画中画效果，具体如下：

① 将工程的最低 SDK 版本设为 26。打开工程 app\build.gradle，修改（加粗处）如下：

```
......
android {
    compileSdkVersion 29
    buildToolsVersion "29.0.2"
    defaultConfig {
        applicationId "com.easybooks.myvideoview"
        minSdkVersion 26                                    //工程的最低 SDK 版本
        targetSdkVersion 29
        versionCode 1
        versionName "1.0"
        testInstrumentationRunner "androidx.test.runner.AndroidJUnitRunner"
    }
    ...
}
......
```

② 设置 Activity 支持画中画模式。Android 的 Activity 默认是不支持画中画显示模式的，要让它支持画中画，必须在工程的 AndroidManifest.xml 中添加如下属性设置：

```
<application
    ...
    android:supportsRtl="true"
    android:theme="@style/AppTheme.NoActionBar">                //（a）
    <activity android:name=".MainActivity"
        android:supportsPictureInPicture="true"                 //设置让 Activity 支持画中画模式
        android:configChanges="screenLayout|orientation">       //（b）
        <intent-filter>
            ...
        </intent-filter>
    </activity>
</application>
```

其中：

（a）**android:theme="@style/AppTheme.NoActionBar"**：在画中画模式下隐藏页面标题栏。无标题栏的"NoActionBar"主题需要在工程 values 目录下的 styles.xml 中定义如下：

```
<style name="AppTheme.NoActionBar">
    <item name="windowActionBar">false</item>
    <item name="windowNoTitle">true</item>
</style>
```

（b）**android:configChanges="screenLayout|orientation"**：当页面从全屏转换到画中画模式时，其 Activity 会被销毁后再重建。为了使转换自如而不影响用户的使用体验，这里设置 configChanges="screenLayout|orientation"是为了不让 Activity 销毁重建，以保持页面的稳定性和性能。

运行程序，打开一个视频播放，然后点击"进入画中画"按钮，效果如图 8.8 所示。

图 8.8 画中画特效

8.1.2 音频播放

Android 音频播放的原理与视频播放的一样，同样支持播放 App 内置音频和从手机存储器上打开音频文件进行播放。

【例 8.4】 以两种方式播放音频文件。

创建 Android Studio 工程 MyMediaPlayer。

实现思路：

（1）由于单纯音频播放是不需要显示画面的，故本例程序只在界面上放置"播放内置音频"和"打开音频文件"两个按钮以及一个用于显示音频文件路径的文本视图，控件的命名同前面视频的例子。

（2）MainActivity 主程序类实现点击事件监听接口 View.OnClickListener，在 onClick() 方法中以视图 id 标识用户点击的按钮，分别执行不同音频播放方式的操作。

（3）播放内置音频时，要先根据 MediaPlayer 播放器对象是否为空来判断其当前是否正在播放中的状态。

界面完整代码

（4）打开音频文件播放的方式下，还是通过 startActivityForResult()启动活动页面，在 onActivityResult()响应方法中获取返回的音频 URI，播放音频。

在 MainActivity.java 中实现功能，代码如下：

```java
public class MainActivity extends AppCompatActivity implements View.OnClickListener {
    private MediaPlayer myMediaPlayer;                      //媒体播放器（播放音频）
    private Button myStartButton;                           //按钮（播放 App 内置音频）
    private Button myOpenButton;                            //按钮（打开手机上存储的音频文件）
    private TextView myTextView;                            //文本视图（显示音频文件的路径）
    private Uri myUri;                                      //手机上音频文件的 URI
    @Override
    protected void onCreate(Bundle savedInstanceState) {
        super.onCreate(savedInstanceState);
        setContentView(R.layout.activity_main);
        myMediaPlayer = null;                               //媒体播放器对象初始置空
        myStartButton = findViewById(R.id.myStartButton);
        myStartButton.setOnClickListener(this);
        myOpenButton = findViewById(R.id.myOpenButton);
        myOpenButton.setOnClickListener(this);
        myTextView = findViewById(R.id.myTextView);
    }

    @Override
    public void onClick(View v) {
        if (v.getId() == R.id.myStartButton) {              //点击"播放内置音频"按钮
            if (myMediaPlayer == null) {                    //播放器无播放的情况
                myMediaPlayer = MediaPlayer.create(this, R.raw.apple);
                                                            //直接由 raw 资源引用创建播放器对象
                myMediaPlayer.start();                      //开始播放
                myStartButton.setText("播 放 中... 停 止");
            } else {                                        //播放器正在播放的情况
                myMediaPlayer.stop();                       //停止当前音频播放
                myMediaPlayer.release();                    //释放播放器资源
                myMediaPlayer = null;                       //播放器置空
                myStartButton.setText("播 放 内 置 音 频");
            }
        }
        if (v.getId() == R.id.myOpenButton) {               //点击"打开音频文件..."按钮
            Intent intent = new Intent((Intent.ACTION_GET_CONTENT));
            intent.setType("audio/*");                      //类型"audio/*"表示获取音频资源
            startActivityForResult(intent, 100);            //启动播放音频的活动页
        }
    }

    protected void onActivityResult(int requestCode, int resultCode, Intent data) {
        super.onActivityResult(requestCode, resultCode, data);
        myUri = data.getData();
        myTextView.setText(myUri.getPath());
        if (myMediaPlayer != null && myMediaPlayer.isPlaying()) {
                                                            //若有音频正在播放，要先将播放器复位
            myMediaPlayer.stop();
```

```
                    myMediaPlayer.reset();
            }
            myMediaPlayer = MediaPlayer.create(this, myUri);    //创建对应特定 URI 的播放器对象
            myMediaPlayer.start();                              //开始播放
        }
    }
```

从上面的代码可见，播放音频与播放视频的原理及代码的逻辑结构几乎完全一样，只是将视频视图 VideoView 换成了媒体播放器 MediaPlayer，并增加了一些释放和复位资源的操作，其他的都不变。

将事先准备好的音乐文件放在工程 raw 目录下，点击"播放内置音频"按钮即可听到音乐；点击"打开音频文件…"按钮，从手机上选择存储的音乐文件打开播放，其操作方式与播放视频一样，不再重复演示。

8.1.3 录像功能

除了播放现成的视频和音频文件，Android 程序还支持获取并操控手机摄像头，实现录像功能。下面的实例演示用手机录像（同时录音）。

【例 8.5】 编写 Android 程序实现手机录像功能。

创建 Android Studio 工程 MyMediaRecorder。

界面设计非常简单，在其上放一个表面视图（SurfaceView）作为录像控件，再放一个"开始录制"按钮和文本视图（显示录像视频文件的保存路径），代码如下：

```
……
    <SurfaceView
        android:id="@+id/mySurfaceView"
        ... />
    <Button
        android:id="@+id/myButton"
        ...
        android:text="开 始 录 制"
        ...
        app:layout_constraintTop_toBottomOf="@+id/mySurfaceView" />
    <TextView
        android:id="@+id/myTextView"
        ...
        app:layout_constraintTop_toBottomOf="@+id/myButton" />
……
```

实现思路：

（1）本例涉及 3 个对象：SurfaceHolder（表面视图的回调接口类）、Camera（照相机）和 MediaRecorder（录像机）。

（2）录像功能是通过手机照相机进行的，故在表面视图回调接口的 surfaceCreated() 方法中执行 initCamera() 初始化照相机，surfaceDestroyed() 方法中执行 closeCamera() 关闭照相机。

（3）自定义 recordStart() 方法完成录制前的准备工作，主要包括：

- 重启照相机；
- 创建 MediaRecorder（录像机）对象并设置参数；
- 获取（或生成）将要录制视频文件的存储路径；
- 申请相关权限；
- 打开录像机。

（4）自定义 recordStop() 方法停止录像。

在 MainActivity.java 中开发录像功能，代码如下：

```
public class MainActivity extends AppCompatActivity implements View.OnClickListener {
    private SurfaceView mySurfaceView;                        //表面视图（也就是录像区域）
    private SurfaceHolder mySurfaceHolder;                    // （a）
    private Button myButton;                                  // "开始录制"按钮
    private TextView myTextView;                              //文本视图（显示录像文件的保存路径）
    private Camera myCamera = null;                           //照相机对象
    private MediaRecorder myMediaRecorder = null;             //录像机对象
    private String myVideoFilePath;                           //录像文件的保存路径
    private boolean started = false;                          //是否处于录制中状态

    @Override
    protected void onCreate(Bundle savedInstanceState) {
        super.onCreate(savedInstanceState);
        setContentView(R.layout.activity_main);
        mySurfaceView = findViewById(R.id.mySurfaceView);
        mySurfaceHolder = mySurfaceView.getHolder();          // （a）
        mySurfaceHolder.addCallback(myCallback);              // （b）
        myButton = findViewById(R.id.myButton);
        myButton.setOnClickListener(this);
        myTextView = findViewById(R.id.myTextView);
    }

    private SurfaceHolder.Callback myCallback = new SurfaceHolder.Callback() {
                                                              // （b）
        @Override
        public void surfaceCreated(SurfaceHolder surfaceHolder) {
            initCamera();                                     //初始化照相机
        }

        @Override
        public void surfaceChanged(SurfaceHolder surfaceHolder, int i, int i1, int i2) {
        }

        @Override
        public void surfaceDestroyed(SurfaceHolder surfaceHolder) {
            closeCamera();                                    //关闭照相机
        }
    };

    private void initCamera() {
        try {
            myCamera = Camera.open();
            myCamera.setDisplayOrientation(90);               // （c）
            myCamera.setPreviewDisplay(mySurfaceHolder);
            myCamera.startPreview();                          // （d）打开相机预览
            myCamera.unlock();                                // （d）解锁相机
        } catch (IOException e) {
            myCamera.release();
```

```java
            myCamera = null;
        }
    }

    private void closeCamera() {
        if (myCamera != null) {
            myCamera.release();                              //释放相机
            myCamera = null;
        }
    }

    @Override
    public void onClick(View v) {
        if (v.getId() == R.id.myButton) {
            if ((!started)) {
                recordStart();                               //开始录制
            } else {
                recordStop();                                //停止录制
            }
        }
    }

    public void recordStart() {
        try {
            closeCamera();                                   // (e)
            initCamera();                                    // (e)
            if (myMediaRecorder == null) {
                myMediaRecorder = new MediaRecorder();       //新建录像机
            } else {
                myMediaRecorder.reset();                     //重置录像机
            }
            myMediaRecorder.setCamera(myCamera);             //将录像机与照相机绑定
            // (f) 设置录像机参数
            myMediaRecorder.setVideoSource(MediaRecorder.VideoSource.CAMERA);
                                                             //影像源自照相机
            myMediaRecorder.setAudioSource(MediaRecorder.AudioSource.MIC);
                                                             //声音源自麦克风
            myMediaRecorder.setProfile(CamcorderProfile.get(CamcorderProfile.QUALITY_720P));
                                                             //录制质量标准
            myMediaRecorder.setVideoEncodingBitRate(5 * 1024 * 1024);
                                                             //编码率
            myMediaRecorder.setOrientationHint(90);          // (c)
            String path = Environment.getExternalStorageDirectory() + "/DCIM/Camera/";
                                                             //获取手机存储器照相机的目录路径
            File file = new File(path);
            if (!file.exists()) file.mkdirs();               //不存在则新建
            myVideoFilePath = path + "MY_VIDEO_" + System.currentTimeMillis() + ".mp4";
                                                             //生成录像视频文件名及路径
            myMediaRecorder.setOutputFile(myVideoFilePath);  //将录像视频文件输出到手机存储
            myMediaRecorder.setPreviewDisplay(mySurfaceHolder.getSurface());
            if (Build.VERSION.SDK_INT >= 23) {               // (g)
```

```
                int REQUEST_CODE_CONTACT = 101;
                String[] permissions = {Manifest.permission.WRITE_EXTERNAL_STORAGE};
                                                               // (g)
                for (String str : permissions) {               //验证是否许可权限
                    if (this.checkSelfPermission(str) != PackageManager.PERMISSION_GRANTED) {
                        this.requestPermissions(permissions, REQUEST_CODE_CONTACT);
                                                               //申请权限
                        return;
                    }
                }
                myMediaRecorder.prepare();                     //录像机准备好了
                myMediaRecorder.start();                       //开始录像
                myTextView.setText("录制进行中...");
                myButton.setText("结　束　录　制");
                started = true;
            } catch (IOException e) {
                Toast.makeText(this, e.toString(), Toast.LENGTH_SHORT).show();
                myMediaRecorder.release();                     //释放录像机
                myMediaRecorder = null;
            }
        }

        public void recordStop() {
            if (myMediaRecorder != null) {
                myMediaRecorder.stop();                        //停止录像
                myMediaRecorder.release();
                myMediaRecorder = null;
                this.sendBroadcast(new            Intent(Intent.ACTION_MEDIA_SCANNER_SCAN_FILE,
Uri.parse(myVideoFilePath)));                                  //广播发出录像保存的路径
                String result = "录像已保存到:" + myVideoFilePath;
                myTextView.setText(result);                    //显示路径
                Toast.makeText(this, result, Toast.LENGTH_SHORT).show();
                myButton.setText("开　始　录　制");
                started = false;
            }
        }
    }
```

其中：

（a）**private SurfaceHolder mySurfaceHolder;** 和 **mySurfaceHolder = mySurfaceView.getHolder();**：SurfaceHolder 是一个接口，通过它可以编辑和控制 SurfaceView 录像视图，用于获取和设置 Surface 参数的 API 就是通过 SurfaceHolder 实现的。一个 SurfaceView 包含一个 SurfaceHolder。

（b）**mySurfaceHolder.addCallback(myCallback);** 和 **private SurfaceHolder.Callback myCallback = new SurfaceHolder.Callback() { ... };**：为 SurfaceHolder 添加一个 SurfaceHolder.Callback 回调接口。SurfaceHolder.Callback 主要是当底层的 Surface 被创建、销毁或者改变时提供回调通知，由于绘制必须在 Surface 被创建后才能进行，因此 SurfaceHolder.Callback 中的 surfaceCreated 和 surfaceDestroyed 就成了绘图处理代码的边界。

SurfaceHolder.Callback 中定义了三个接口方法：

- **abstract void surfaceChanged(SurfaceHolder holder, int format, int width, int height)**：当 surface 发生任何结构性（格式或者大小）的变化时，该方法就会被立即调用。
- **abstract void surfaceCreated(SurfaceHolder holder)**：当 surface 对象创建后，该方法就会被立即调用。
- **abstract void surfaceDestroyed(SurfaceHolder holder)**：当 surface 对象将要销毁前，该方法会被立即调用。

（c）**myCamera.setDisplayOrientation(90);** 和 **myMediaRecorder.setOrientationHint(90);**：必须设置成这样才能保证摄像竖屏影像是正的。

（d）**myCamera.startPreview();** 和 **myCamera.unlock();**：只能放在这里解锁相机，不能放在录像之前设置的时候解锁。

（e）**closeCamera();** 和 **initCamera();**：每次开始录像前都要先关闭再初始化（相当于重启）Camera，不然就只能录一次，第二次再录的时候程序就崩溃了。

（f）**myMediaRecorder.setVideoSource(MediaRecorder.VideoSource.CAMERA);**、**myMediaRecorder.setAudioSource(MediaRecorder.AudioSource.MIC);**、**myMediaRecorder.setProfile(CamcorderProfile.get(CamcorderProfile.QUALITY_720P));** 和 **myMediaRecorder.setVideoEncodingBitRate(5 * 1024 * 1024);**：只能设置这几项。还有其他很多选项。由于存在与具体手机硬件驱动的兼容性问题，如果设置了反而会影响录像机 MediaRecorder 的正常启动，故此处不予设置。

（g）**if (Build.VERSION.SDK_INT >= 23) { ... } String[] permissions = {Manifest.permission.WRITE_EXTERNAL_STORAGE};**：对于相机、麦克风，可以先在 AndroidManifest.xml 中配置打开权限，此时在手机"设置"→"应用管理"下打开此工程的 App，看到出现（只有在 xml 中配置了才会出现）对应权限的设置开关项，如图 8.9 所示，在手机中设置开启即可，不必再在程序中书写烦琐的验证允许权限代码，但此处存储权限的开通则还必须通过编写代码来实现。

图 8.9　打开权限操作演示

最后不要忘了在 AndroidManifest.xml 中打开权限：

```
<manifest xmlns:android="http://schemas.android.com/apk/res/android"
    package="com.easybooks.mymediarecorder">
    <uses-permission android:name="android.permission.CAMERA" />
    <uses-permission android:name="android.permission.RECORD_AUDIO" />
    <uses-permission android:name="android.permission.READ_EXTERNAL_STORAGE" />
    <uses-permission android:name="android.permission.WRITE_EXTERNAL_STORAGE" />
    <application
        ...
    </application>
</manifest>
```

运行程序，点击"开始录制"按钮，进入录像模式，效果如图 8.10 所示。点击"结束录制"按钮，退出录像模式，底部文本视图显示出录像视频文件保存的路径，读者可打开观看。

图 8.10　进入和退出录像模式

8.2　图形图像处理

Android 用像素矩阵的方式实现对静态图形图像的处理，图片在内存中存放的就是一个个像素点，而对于图片的变换实质上就是处理图片的每个像素点，对每个像素点进行相应的变换，即可完成对整个图像的变换。为此，Android 专门提供了一个矩阵工具类 Matrix 来辅助用户实现图形图像处理功能，该类位于 android.graphics.Matrix 包下。

8.2.1　图像倾斜缩放

【例 8.6】　实现图像倾斜及放大、缩小功能。

创建 Android Studio 工程 MyScale。

设计界面，在其上放置一个自定义视图 ScaleView，外加 4 个命令按钮，代码如下：

```
<?xml version="1.0" encoding="utf-8"?>
<LinearLayout xmlns:android="http://schemas.android.com/apk/res/android"
    android:orientation="vertical"
    android:layout_width="fill_parent"
    android:layout_height="fill_parent"
    android:background="#f0f">

    <com.easybooks.myscale.ScaleView                       //自定义的视图
        android:id="@+id/myScaleView"
        android:layout_width="fill_parent"
        android:layout_height="548dp"
        android:layout_gravity="center" />

    <LinearLayout
        android:layout_width="match_parent"
        android:layout_height="48dp"
        android:layout_gravity="center"
```

```xml
            android:orientation="horizontal">

    <Button
        android:id="@+id/myLeftButton"
        android:layout_width="90dp"
        android:layout_height="wrap_content"
        android:text="左倾"
        android:textSize="18sp" />
    //省略其余 3 个按钮元素的代码
    ...
    </LinearLayout>
</LinearLayout>
```

用一个可爱小男孩的照片作为素材，变换前的程序界面如图 8.11 所示。

图 8.11 变换前的程序界面

实现思路：

（1）将图像倾斜缩放功能封装在一个自定义视图 ScaleView 中实现，在其构造方法中获取原始图片并转换为适于处理的位图对象。

（2）ScaleView 视图对外提供 public void optBitmap(String opt)方法，主程序调用时给它传递一个参数 opt，"告诉"视图需要对图像进行什么样的处理。

（3）在 ScaleView 视图的 onDraw()方法中由全局变量 isScale 决定是倾斜还是缩放。若是倾斜，按全局变量 sx 设定的倾斜度调用 Matrix 类的 setSkew()方法重绘图像；若是缩放，则按全局变量 scale 设定的比例调用 Matrix 类的 setScale()方法重绘图像。

（4）用 postInvalidate()机制刷新界面。

编写实现自定义视图的类 ScaleView.java，代码如下：

```java
public class ScaleView extends View {
    private Bitmap myBitmap;                          //位图对象（存储图片资源）
    private Matrix myMatrix = new Matrix();           //矩阵对象
    private float sx = 0.0f;                          //倾斜度
```

```java
    private int width, height;                          //位图宽和高
    private float scale = 1.0f;                         //缩放比例
    private boolean isScale = false;                    //判断是缩放还是倾斜

    public ScaleView(Context ctx, AttributeSet set) {
        super(ctx, set);
        myBitmap = ((BitmapDrawable) ctx.getResources().getDrawable(R.drawable.p1)).getBitmap();
                                                        //获取图片资源并转为位图对象
        width = myBitmap.getWidth();                    //获取图片宽度
        height = myBitmap.getHeight();                  //获取图片高度
        this.setFocusable(true);                        //设置当前视图获得焦点
    }

    @Override
    protected void onDraw(Canvas canvas) {
        super.onDraw(canvas);
        myMatrix.reset();                               //重置矩阵
        if (!isScale) {
            myMatrix.setSkew(sx, 0);                    //（a）
        } else {
            myMatrix.setScale(scale, scale);            //（b）
        }
        Bitmap bitmap = Bitmap.createBitmap(myBitmap, 0, 0, width, height, myMatrix, true);
                                                        //根据原位图和变换后的矩阵创建新图片
        canvas.drawBitmap(bitmap, myMatrix, null);      //绘制新图
    }

    public void optBitmap(String opt) {                 //根据用户点击不同按钮传入参数执行不同的操作
        switch (opt) {
            case "left":                                //左倾
                isScale = false;
                sx += 0.2;
                postInvalidate();                       //（c）
                break;
            case "right":                               //右倾
                isScale = false;
                sx -= 0.2;
                postInvalidate();
                break;
            case "in":                                  //放大
                isScale = true;
                if (scale < 2.0)
                    scale += 0.2;
                postInvalidate();
                break;
            case "out":                                 //缩小
                isScale = true;
                if (scale > 0.5)
                    scale -= 0.2;
                postInvalidate();
                break;
```

            }
        }
}
其中：
（a）**myMatrix.setSkew(sx, 0);**：控制矩阵倾斜的方法。有两个重载的版本：
① **setSkew(float kx,float ky)**：控制 Matrix 进行倾斜，kx、ky 为 X、Y 方向上的比例。
② **setSkew(float kx,float ky,float px,float py)**：控制 Matrix 以(px,py)为轴心进行倾斜，kx、ky 为 X、Y 方向上的倾斜比例。
（b）**myMatrix.setScale(scale, scale);**：控制矩阵缩放的方法。也有两个重载的版本：
① **setScale(float sx,float sy)**：设置 Matrix 进行缩放，sx、sy 为 X、Y 方向上的缩放比例。
② **setScale(float sx,float sy,float px,float py)**：设置 Matrix 以(px,py)为轴心进行缩放，sx、sy 为 X、Y 方向上的缩放比例。
（c）**postInvalidate();**：该方法能够在非 UI 线程中去调用刷新 UI，那么它是如何做到的呢？原来，调用 postInvalidate()方法最后会调用视图中的句柄发送一个 MSG_INVALIDATE 消息，而句柄是 ViewRootHandler 的一个实例，通过 ViewRootHandler 的 handleMessage()方法向 UI 发送消息，实现刷新。

最后，在 MainActivity.java 中使用自定义视图来实现功能，代码如下：

```java
public class MainActivity extends AppCompatActivity implements View.OnClickListener {
    private Button myLeftButton;                              //"左倾"按钮
    private Button myRightButton;                             //"右倾"按钮
    private Button myZoomInButton;                            //"放大"按钮
    private Button myZoomOutButton;                           //"缩小"按钮
    private ScaleView myScaleView;                            //自定义 ScaleView 视图的引用
    @Override
    protected void onCreate(Bundle savedInstanceState) {
        super.onCreate(savedInstanceState);
        setContentView(R.layout.activity_main);
        myLeftButton = findViewById(R.id.myLeftButton);
        myLeftButton.setOnClickListener(this);
        myRightButton = findViewById(R.id.myRightButton);
        myRightButton.setOnClickListener(this);
        myZoomInButton = findViewById(R.id.myZoomInButton);
        myZoomInButton.setOnClickListener(this);
        myZoomOutButton = findViewById(R.id.myZoomOutButton);
        myZoomOutButton.setOnClickListener(this);
        myScaleView = findViewById(R.id.myScaleView);
    }

    @Override
    public void onClick(View view) {
        switch (view.getId()) {
            case R.id.myLeftButton:                           //点击"左倾"按钮
                myScaleView.optBitmap("left");
                break;
            case R.id.myRightButton:                          //点击"右倾"按钮
                myScaleView.optBitmap("right");
                break;
```

```
            case R.id.myZoomInButton:                    //点击"放大"按钮
                myScaleView.optBitmap("in");
                break;
            case R.id.myZoomOutButton:                   //点击"缩小"按钮
                myScaleView.optBitmap("out");
                break;
        }
    }
}
```

运行程序，点击4个命令按钮，分别实现图片左倾、右倾、放大及缩小，如图8.12所示。

(a) 左倾　　　　　　　　　　　　(b) 右倾

(c) 放大　　　　　　　　　　　　(d) 缩小

图 8.12　图像变换

8.2.2　图像扭曲

图像扭曲实际就是对图像一定区域内的像素点进行处理，按照特定的算法进行统一变换，得到新的像素值，再按照新的像素值来重绘图像即可呈现出扭曲的效果。

【例 8.7】　触击图像，使得触点及其周边小范围内的区域出现扭曲效果。

创建 Android Studio 工程 MyWarp。

实现思路：

（1）将图像扭曲功能封装在一个自定义视图 WarpView 中实现。

（2）将图片分别沿横向和纵向等分为 30 格（等分数由常量 WIDTH 和 HEIGHT 指定），这样图片总共包含 961（31×31，存于常量 COUNT 中）个关键点，只要对这些关键点的坐标进行变换即可实现对图像的扭曲。所有关键点扭曲前后的坐标分别存放于两个数组中。

（3）在 wrap() 方法中执行算法，实现扭曲功能。

（4）用户触击屏幕时产生 MotionEvent 事件，在事件响应的 onTouchEvent() 方法中调用 wrap() 方法实现对图像的扭曲。

（5）用 invalidate() 机制刷新界面。

定义视图 WarpView.java，代码如下：

```java
public class WarpView extends View {
    private Bitmap myBitmap;                                        //存储图片资源对象
    private float width, height;                                    //保存图片宽和高
    private final int WIDTH = 30;
    private final int HEIGHT = 30;
    private final int COUNT = (WIDTH + 1) * (HEIGHT + 1);
    private final float[] cdvals = new float[COUNT * 2];            //（a）扭曲前的坐标数组
    private final float[] cdvals_warp = new float[COUNT * 2];       //（a）扭曲后的坐标数组

    public WarpView(Context context) {
        super(context);
        setFocusable(true);                                         //设置当前视图获得焦点
        myBitmap = BitmapFactory.decodeResource(getResources(), R.drawable.p8);
                                                                    //加载图片
        width = myBitmap.getWidth();                                //获取图片宽度
        height = myBitmap.getHeight();                              //获取图片高度
        int index = 0;
        for (int y = 0; y <= HEIGHT; y++) {                         // (b)
            float fy = height * y / HEIGHT;
            for (int x = 0; x <= WIDTH; x++) {
                float fx = width * x / WIDTH;
                cdvals[index * 2 + 0] = cdvals_warp[index * 2 + 0] = fx;
                cdvals[index * 2 + 1] = cdvals_warp[index * 2 + 1] = fy;
                index += 1;
            }
        }
        setBackgroundColor(Color.WHITE);                            //设置背景色
    }

    private void wrap(float cx, float cy) {                         // (c)
        for (int i = 0; i < COUNT * 2; i += 2) {
            float dx = cx - cdvals[i + 0];
            float dy = cy - cdvals[i + 1];
            float dd = dx * dx + dy * dy;
            float d = (float) Math.sqrt(dd);                        //计算与触点的距离
            float pull = 60000 / ((float) (dd * d));                //计算扭曲度
            //保存扭曲后的坐标
            if (pull >= 1) {
```

```
                    cdvals_warp[i + 1] = cx;
                    cdvals_warp[i + 1] = cy;
                } else {
                    cdvals_warp[i + 0] = cdvals[i + 0] + dx * pull;
                    cdvals_warp[i + 1] = cdvals[i + 1] + dx * pull;
                }
            }
            invalidate();                                          // （d）
        }

        public void onDraw(Canvas canvas) {
            super.onDraw(canvas);
            canvas.drawBitmapMesh(myBitmap, WIDTH, HEIGHT, cdvals_warp, 0, null, 0, null);
                                                                   // （e）
        }

        public boolean onTouchEvent(MotionEvent event) {           //响应触击事件
            wrap(event.getX(), event.getY());
            return true;
        }
    }
}
```

其中：

（a）**private final float[] cdvals = new float[COUNT * 2];** 和 **private final float[] cdvals_warp = new float[COUNT * 2];**：定义两个数组，cdvals[]存放原坐标，cdvals_warp[]存放扭曲变换后的坐标。

（b）**for (int y = 0; y <= HEIGHT; y++) { ... }**：初始化数组 cdvals[]和 cdvals_warp[]，最开始时它们中都均匀地存放着 961 个关键点的原坐标。

（c）**private void wrap(float cx, float cy)**：该方法是执行扭曲算法的核心方法，它接收的两个参数 cx、cy 分别是用户触击屏幕位置的 X、Y 坐标值。在算法中，先计算图像上每个关键点与触点之间的距离 d，然后计算关键点的扭曲度 pull，距触点越远的地方扭曲度越小，这样就能使各关键点向触点方向偏移，造成整体上扭曲的效果。最后将扭曲后的新坐标存于数组 cdvals_warp[]中。

（d）**invalidate();**：该方法用来刷新视图，作用与【例 8.6】中的 postInvalidate()方法相同，唯一的区别是 postInvalidate()支持跨线程刷新视图，而 invalidate()仅用于本 UI 线程内部的刷新。其实 postInvalidate()在底层调用的也是 invalidate()方法，只不过在其基础上封装了跨线程的消息机制。如果我们自定义的视图本身就是在 UI 线程，没有用到多线程的话，直接用 invalidate()方法来刷新视图就可以了。

（e）**canvas.drawBitmapMesh(myBitmap, WIDTH, HEIGHT, cdvals_warp, 0, null, 0, null);**：对 myBitmap 位图对象按照数组 cdvals_warp[]的内容进行重绘，即进行扭曲操作。该方法带有 8 个参数，其中第 5 个参数 0 表示从第 1 个点开始扭曲。

在 MainActivity.java 中引用定义的视图 WarpView，代码如下：

```
public class MainActivity extends AppCompatActivity {
    private WarpView myWarpView;                                   //视图 WarpView 对象
    @Override
    protected void onCreate(Bundle savedInstanceState) {
        super.onCreate(savedInstanceState);
        myWarpView = new WarpView(this);                           //创建视图对象
        setContentView(myWarpView);                                //引用自定义视图
```

        }
    }

运行程序，出现一个可爱的小男孩，手指触击手机屏上他的小嘴处，Android 执行像素处理将图像扭曲，呈现出小男孩撅起小嘴的效果，如图 8.13 所示。

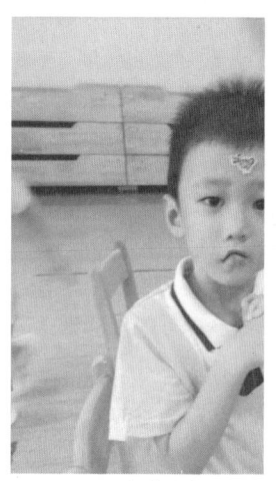

（a）扭曲前　　　　　　　（b）扭曲后

图 8.13　图像扭曲

8.3　手机相册功能

运用 Android 的 ImageSwitcher（图像切换器）控件，结合 Gallery（画廊）控件，可以制作手机相册，将一组照片组织在一起滚动翻页查看。本节就来演示一个制作手机相册的例子。

【例 8.8】　制作手机相册，内放一组潜水员和海洋生物的照片，底部相框中的小照片可以翻动浏览，点击小照片在屏幕主显示区显示对应的照片大图，效果如图 8.14 所示。

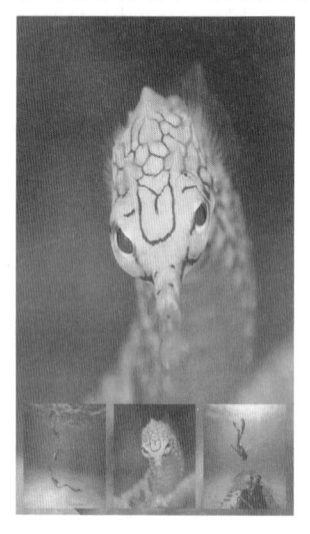

图 8.14　手机相册

创建 Android Studio 工程 MyGallery，在 drawable 目录下预先放入要显示的一组照片资源。
自定义界面文件 activity_gallery.xml，设计相册的布局，代码如下：

```
<RelativeLayout xmlns:android="http://schemas.android.com/apk/res/android"
```

```
        android:layout_width="match_parent"
        android:layout_height="match_parent" >
    <ImageSwitcher                                      //图像切换器（实现照片大图预览）
        android:id="@+id/myImageSwitcher"
        android:layout_width="match_parent"
        android:layout_height="match_parent" />
    <Gallery                                            //画廊（实现底部小图相框）
        android:id="@+id/myGallery"
        android:layout_width="match_parent"
        android:layout_height="wrap_content"
        android:layout_alignParentBottom="true" />
</RelativeLayout>
```

Gallery 作为一种视图类控件，同样要先为其开发适配器类才能使用。开发一个相框适配器类 GalleryAdapter.java，代码如下：

```
public class GalleryAdapter extends BaseAdapter {        //（a）
    private Context myContext;                           //声明上下文
    private int[] myImages;                              //照片资源引用 id 数组

    public GalleryAdapter(Context context, int[] imageArray) {
        myContext = context;
        myImages = imageArray;
    }

    @Override
    public int getCount() {                              //获得画廊照片张数
        return myImages.length;
    }

    @Override
    public long getItemId(int pos) {
        return pos;
    }

    @Override
    public Object getItem(int pos) {
        return myImages[pos];
    }

    @Override
    public View getView(int position, View convertView, ViewGroup parent) {
                                                         //（b）
        ImageView view = new ImageView(myContext);
        view.setImageResource(myImages[position]);
        view.setLayoutParams(new Gallery.LayoutParams(320, 400));
                                                         //设定底部相框中每张照片的尺寸
        view.setScaleType(ImageView.ScaleType.FIT_XY);
        return view;
    }
}
```

其中：

（a）**public class GalleryAdapter extends BaseAdapter**：这个适配器继承自 BaseAdapter 类，这与第 4 章介绍的 ListView、GridView 等视图类控件适配器的基本原理是一样的。由此可见，Android 所有的视图显示类控件都具有共同的本质。

（b）**public View getView(int position, View convertView, ViewGroup parent)**：凡是继承自 BaseAdapter 类的视图适配器都是通过重写 getView()方法，在其中获取内容项的视图的，这里就是获取要显示的相框照片。

实现思路：

（1）相片集预先存放于工程 drawable 目录下，全局变量 imageArray[]数组引用其资源 id。

（2）ImageSwitcher 上绑定触摸监听器，在其 onTouch()方法中响应手势检测器的触摸事件，在手势检测器所绑定的 GestureListener 类中，通过 onFling()方法回调 GalleryActivity 主程序类中的 showPre()和 showNext()两个方法实现相册照片滑动显示功能。

（3）相册底部小图相框的属性由 Gallery 类设定，该类与相框适配器 GalleryAdapter 配合一起显示底部相框中的相片集，其上同时绑定点击事件监听器 OnItemClickListener，在点击事件监听器的 onItemClick()方法中设定 ImageSwitcher 上当前要显示的照片。

GalleryActivity.java 是相册主体的程序类，代码如下：

```
public class GalleryActivity extends AppCompatActivity implements GestureCallback,OnTouchListener,OnItemClickListener {
    //（a）
    private ImageSwitcher myImageSwitcher;            //图像切换器（照片大图预览）
    private Gallery myGallery;                        //相框（底部滚动小照片）
    private GestureDetector gDetector;                //手势检测器
    private int[] imageArray = {R.drawable.p1, R.drawable.p2, R.drawable.p3, R.drawable.p4, R.drawable.p5, R.drawable.p6, R.drawable.p7, R.drawable.p8, R.drawable.p9, R.drawable.p10};
                                                      //照片资源 id 数组
    @Override
    protected void onCreate(Bundle savedInstanceState) {
        super.onCreate(savedInstanceState);
        setContentView(R.layout.activity_gallery);
        myImageSwitcher = findViewById(R.id.myImageSwitcher);
        myImageSwitcher.setFactory(new myViewFactory());           //（b）
        myImageSwitcher.setImageResource(imageArray[0]);           //设置初始显示的照片
        GestureListener gListener = new GestureListener(this);     //创建手势监听器
        gDetector = new GestureDetector(this, gListener);          //（c）
        gListener.setGestureCallback(this);                        //（c）设置手势监听器的手势回调对象
        myImageSwitcher.setOnTouchListener(this);                  //设置触摸监听器

        myGallery = findViewById(R.id.myGallery);
        myGallery.setPadding(10, 25, 10, 25);
        myGallery.setSpacing(30);                                  //设置照片之间的间距
        myGallery.setUnselectedAlpha(0.8f);                        //设置未选中照片的透明度
        myGallery.setOnItemClickListener(this);                    //设置项目点击事件监听器
        myGallery.setAdapter(new GalleryAdapter(this, imageArray));
                                                                   //设置相框的适配器
    }

    @Override
    public void showPre() {                                        //（c）显示前一张照片
```

```
        myImageSwitcher.setOutAnimation(AnimationUtils.loadAnimation(this, R.anim.right_out));
        myImageSwitcher.setInAnimation(AnimationUtils.loadAnimation(this, R.anim.right_in));
        int pre = (int) (myGallery.getSelectedItemId() - 1);
                                                        //getSelectedItemId()方法获取当前选中的视图序号
        if (pre < 0) {
            pre = imageArray.length - 1;
        }
        myImageSwitcher.setImageResource(imageArray[pre]);
        myGallery.setSelection(pre);                    //设置选中前一个视图
    }

    @Override
    public void showNext() {                            // (c) 显示后一张照片
        myImageSwitcher.setOutAnimation(AnimationUtils.loadAnimation(this, R.anim.left_out));
        myImageSwitcher.setInAnimation(AnimationUtils.loadAnimation(this, R.anim.left_in));
        int next = (int) (myGallery.getSelectedItemId() + 1);
        if (next >= imageArray.length) {
            next = 0;
        }
        myImageSwitcher.setImageResource(imageArray[next]);
        myGallery.setSelection(next);                   //设置选中下一个视图
    }

    @Override
    public boolean onTouch(View view, MotionEvent event) {
        gDetector.onTouchEvent(event);
        return true;
    }

    @Override
    public void onItemClick(AdapterView<?> parent, View view, int pos, long id) {
        myImageSwitcher.setOutAnimation(AnimationUtils.loadAnimation(this, R.anim.out));
                                                        // (d) 设置前一张照片的退出动画
        myImageSwitcher.setInAnimation(AnimationUtils.loadAnimation(this, R.anim.in));
                                                        // (d) 设置下一张照片的进入动画
        myImageSwitcher.setImageResource(imageArray[pos]);
    }

    public class myViewFactory implements ViewFactory {     // (b)
        @Override
        public View makeView() {
            ImageView view = new ImageView(GalleryActivity.this);
            view.setBackgroundColor(0xFFFFFFFF);
            view.setScaleType(ScaleType.FIT_XY);
            view.setLayoutParams(new     ImageSwitcher.LayoutParams(LayoutParams.MATCH_PARENT,
LayoutParams.MATCH_PARENT));
            return view;
        }
    }
}
```

其中:

(a) **public class GalleryActivity extends AppCompatActivity implements GestureCallback,**

OnTouchListener,OnItemClickListener：本例程序需要实现 3 种监听器对象。
- **GestureCallback**：手势监听器的回调对象，执行回调接口中的方法来实现当用户手指滑动界面上方图像切换器中的照片大图时，照片能够左右滚动切换的功能。
- **OnTouchListener**：触摸监听器，重写其 onTouch()方法使得屏幕能响应手势检测器的触摸事件，才能进而执行回调接口。
- **OnItemClickListener**：项目点击事件监听器，用于实现当用户点选底部相框中的小照片时，图像切换器能响应操作自动切换显示对应的照片大图。

（b）**myImageSwitcher.setFactory(new myViewFactory());** 和 **public class myViewFactory implements ViewFactory**：设置图像切换器的视图工厂，视图工厂的作用是生成新视图动态添加到图像切换器中，这样就能实现照片大图的轮换显示。视图工厂实现 Android 的 ViewFactory 接口，通过重写其中的 makeView()方法来创建要添加的新视图。

（c）**gDetector = new GestureDetector(this, gListener);**、**gListener.setGestureCallback(this);**、**public void showPre() { ... }** 和 **public void showNext() { ... }**：手势检测器 GestureDetector 与一个手势监听器 GestureListener 关联，监听器检查用户左右滑动手势，再通过调用回调对象 GestureCallback 接口中的方法来轮放照片大图的预览。而 showPre()和 showNext()都是回调对象 GestureCallback 接口中定义的抽象方法，稍后会看到。

（d）**myImageSwitcher.setOutAnimation(AnimationUtils.loadAnimation(this, R.anim.out));** 和 **myImageSwitcher.setInAnimation(AnimationUtils.loadAnimation(this, R.anim.in));**：设置图像切换器中照片进入和退出时需要的动画效果，动画内容预先定义在工程 res\anim 下的 XML 文件中，通过 AnimationUtils.loadAnimation()方法载入。以这里点击相框中小照片触发的动画为例，图像切换器上大照片的进入动画定义在 in.xml 中，代码如下：

```xml
<?xml version="1.0" encoding="UTF-8"?>
<alpha xmlns:android="http://schemas.android.com/apk/res/android"
        android:interpolator="@android:anim/decelerate_interpolator"
        android:fromAlpha="0.0"                              //初始照片背景透明
        android:toAlpha="1.0"                                //进入以后变为不透明
        android:duration="@android:integer/config_longAnimTime" />
```

这样设置之后，当用户手指滑动进入下一张照片时，就会看到照片背景由完全透明状态渐变为不透明状态（淡入）的效果。

在本例中，当用户手指滑动切换照片时的效果主要是通过手势监听器实现的，该类 GestureListener.java 的代码如下：

```java
public class GestureListener implements GestureDetector.OnGestureListener {
        private Context myContext;                           //声明上下文
        private float myFlipGap = 30f;                       // （a）定义滑动间隔

        public GestureListener(Context ctx) {
            myContext = ctx;
        }

        @Override
        public final boolean onFling(MotionEvent me1, MotionEvent me2, float velX, float velY) {
                                                             // （a）
            if (me1.getX() - me2.getX() > myFlipGap) {
                if (myCallback != null) {
                    myCallback.showNext();                   //当向后滑动幅度大于间隔时进入下一张照片
```

```
            }
        }
        if (me1.getX() - me2.getX() < -myFlipGap) {
            if (myCallback != null) {
                myCallback.showPre();            //当向前滑动幅度大于间隔时显示前一张照片
            }
        }
        return true;
    }

    private GestureCallback myCallback;

    public void setGestureCallback(GestureCallback callback) {
        myCallback = callback;
    }

    public static interface GestureCallback {           // (b)
        public abstract void showNext();

        public abstract void showPre();
    }

    /**以下都是按照编程规范实现手势监听器时需要声明重写的方法*/
    @Override
    public final boolean onScroll(MotionEvent me1, MotionEvent me2, float disX, float disY) {
        return false;
    }

    @Override
    public boolean onSingleTapUp(MotionEvent event) {
        return false;
    }

    @Override
    public final boolean onDown(MotionEvent event) {
        return true;
    }

    @Override
    public final void onLongPress(MotionEvent event) {
    }

    @Override
    public final void onShowPress(MotionEvent event) {
    }
}
```

其中:

(a) **private float myFlipGap = 30f;和 public final boolean onFling(MotionEvent me1, MotionEvent me2, float velX, float velY)**: 本例中控制手指滑动特效的主要就是 onFling()这个方法,一旦用户滑动的幅度达到了变量 myFlipGap 所定义的距离,就实现照片的切换。

（b）**public static interface GestureCallback**：前面所提到的回调对象 GestureCallback 正是定义在这里，其中声明的方法 showNext()和 showPre()的实现则位于主程序类 GalleryActivity.java 中。

8.4 OpenGL 图形库

Android 处理复杂的图形可以使用 OpenGL 库，这个库是专业的图像处理库，需要系统的专业知识才能真正用好。本节仅介绍它的引入和基本使用方法，更高级的用法请读者参考这方面的专业书籍，本书不予展开。

8.4.1 OpenGL 简介

OpenGL 的前身是 SGI 公司为其图形工作站开发的 IRIS GL，后来因为 IRIS GL 的移植性不好，所以在其基础上开发出了 OpenGL。完整的 OpenGL 是重量级的图形库，多用于专业图形工作站或高性能 PC，出于性能各方面原因，在移动端是无法使用完全的 OpenGL 库的。为此，Khronos 公司对原 OpenGL 进行了裁剪，分离出它的一个子集 OpenGL ES（OpenGL for Embedded System）来作为 Android 使用的版本，故下面要讲的 OpenGL 实际指的是对应 Android 系统的 OpenGL ES 版本。

目前 OpenGL ES 的 API 已更新至 2.0 版，在 Android 中使用 OpenGL 绘图主要就是调用 GLES20 中的各个方法，其一般流程如图 8.15 所示。

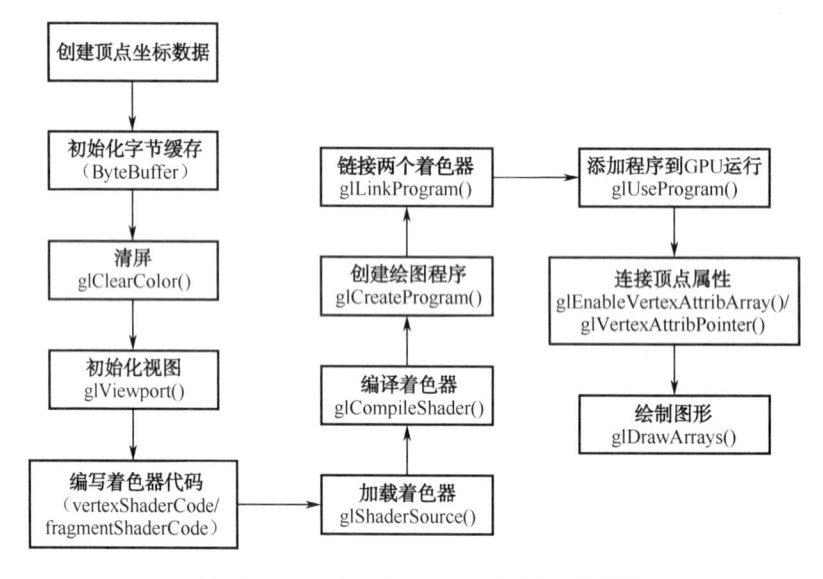

图 8.15 Android 使用 OpenGL 绘图的一般流程

【例 8.9】 创建一个 Android Studio 工程 MyOpenGL，使用 OpenGL 绘制一个三角形。

8.4.2 构建 OpenGL 环境

1. 配置 AndroidManifest.xml

为了让我们的程序可以使用 OpenGL ES 2.0 API，必须在工程 AndroidManifest.xml 文件中添加配置（加粗处），代码如下：

```
<manifest xmlns:android="http://schemas.android.com/apk/res/android"
    package="com.easybooks.myopengl">
    <uses-feature
```

```
            android:glEsVersion="0x00020000"                                            // (a)
            android:required="true" />                                                  // (a)
        <supports-gl-texture android:name="GL_OES_compressed_ETC1_RGB8_texture" />      // (b)
        <supports-gl-texture android:name="GL_OES_compressed_paletted_texture" />       // (b)
        <application
            ...
        </application>
</manifest>
```

其中：

（a）**android:glEsVersion="0x00020000"** 和 **android:required="true"**：指定所要使用的 OpenGL 版本及打开 API 使用权限。

（b）**<supports-gl-texture android:name="GL_OES_compressed_ETC1_RGB8_texture" />** 和 **<supports-gl-texture android:name="GL_OES_compressed_paletted_texture" />**：声明 App 支持的纹理压缩格式，只允许它安装在兼容这些格式的设备上。

2. 创建 GLSurfaceView 视图

OpenGL 库所绘制的图形元素都必须呈现在 GLSurfaceView 类型的视图中，因此，我们要在主程序中定义一个派生自 GLSurfaceView 的视图类，并将它设定为主页面 Activity 的内容视图。本例在主程序类 OpenGLActivity.java 中声明和创建视图，代码如下：

```java
public class OpenGLActivity extends AppCompatActivity {
    private GLSurfaceView myGLSView;                    //声明 GLSurfaceView 类型的视图对象
    @Override
    protected void onCreate(Bundle savedInstanceState) {
        super.onCreate(savedInstanceState);
        myGLSView = new MyGLSurfaceView(this);          //创建一个 GLSurfaceView 实例
        setContentView(myGLSView);                      //将它设定为当前页 Activity 的内容视图
    }

    class MyGLSurfaceView extends GLSurfaceView {       // (a)
        private final MyGLRenderer renderer;            //声明渲染器类
        public MyGLSurfaceView(Context context) {
            super(context);
            setEGLContextClientVersion(2);              //创建一个 OpenGL ES 2.0 上下文
            renderer = new MyGLRenderer();              //创建渲染器对象
            setRenderer(renderer);                      //设置在 GLSurfaceView 上绘图的渲染器
        }
    }
}
```

其中：

（a）**class MyGLSurfaceView extends GLSurfaceView**：MyGLSurfaceView 是我们自定义的视图类，它继承了 OpenGL 库的 GLSurfaceView 视图，GLSurfaceView 在 OpenGL 中实际上仅仅作为显示视图使用，而图形的具体绘制和操控则主要是通过渲染器 Renderer 进行的，故 GLSurfaceView 要扩展的代码很少，一般都会将它定义成主程序 Activity 的内部类（当然也可以在另外一个单独的源文件中定义）。之所以要自定义 MyGLSurfaceView 类而非直接使用 GLSurfaceView，是为了在有需要的时候让视图能够响应用户触摸事件。

3. 实现 Renderer 渲染器

OpenGL 的渲染器与 GLSurfaceView 视图绑定，可完全控制和绘制在视图上要显示的图形内容。本例的渲染器定义在 MyGLRenderer.java 中，代码如下：

```java
public class MyGLRenderer implements GLSurfaceView.Renderer {         //（a）
    private Triangle myTriangle;                                      //要绘制的图形对象（三角形）
    @Override
    public void onSurfaceCreated(GL10 gl10, EGLConfig eglConfig) {    //（a）
        GLES20.glClearColor(0.0f, 0.0f, 1.0f, 0.0f);                  //设置框架背景色
        myTriangle = new Triangle();                                  //初始化三角形
    }

    @Override
    public void onDrawFrame(GL10 gl10) {                              //（a）
        GLES20.glClear(GLES20.GL_COLOR_BUFFER_BIT);                   //重绘背景色
        myTriangle.draw();                                            //绘制三角形
    }

    @Override
    public void onSurfaceChanged(GL10 gl10, int width, int height) {  //（a）
        GLES20.glViewport(0, 0, width, height);
    }

    public static int loadShader(int type, String shaderCode) {       //（b）
        int shader = GLES20.glCreateShader(type);                     //创建着色器类型
        //将源码添加到着色器并编译
        GLES20.glShaderSource(shader, shaderCode);                    //加载着色器
        GLES20.glCompileShader(shader);
        return shader;
    }
}
```

其中：

（a）**public class MyGLRenderer implements GLSurfaceView.Renderer**：实现渲染器需要重写 GLSurfaceView.Renderer 接口中的 3 个方法。

- **public void onSurfaceCreated(GL10 gl10, EGLConfig eglConfig)**：设置视图的绘图环境，该方法只需调用一次，一般在其中完成设置视图框架的背景色、初始化创建要绘制的图形对象等工作。
- **public void onDrawFrame(GL10 gl10)**：每当需要重绘视图时调用，在其中绘制背景和图形元素。
- **public void onSurfaceChanged(GL10 gl10, int width, int height)**：当视图的几何图形发生变化（例如当设备的屏幕方向发生变化）时调用。

以上每个方法都带一个 GL10 类型的参数，是为了使它们的方法签名可以简单地重用于 OpenGL ES 2.0 API，以简化 Android 框架代码。

（b）**public static int loadShader(int type, String shaderCode)**：绘制图形时使用的着色程序包含 OpenGL 着色语言（GLSL）代码，必须先对其进行编译，然后才能在 OpenGL ES 环境中使用，该方法就是用来对 GLSL 语言代码进行预编译的，将在接下来定义和绘制图形的 Triangle 类中调用。

8.4.3 定义和绘制图形

在 OpenGL 中使用三维空间中的坐标描述要绘制的图形对象，因此要绘制图形必须先定义其坐标。在 OpenGL 中，通常采用的方式是为坐标定义浮点型的顶点数组。本例要绘制的三角形定义在 Triangle.java 中，代码如下：

```java
public class Triangle {
    private FloatBuffer vertexBuffer;
    private final int myProgram;
    private final String vertexShaderCode =                    //（a）顶点着色器代码
            "attribute vec4 vPosition;" +
                    "void main() {" +
                    "  gl_Position = vPosition;" +
                    "}";

    private final String fragmentShaderCode =                  //（a）片段着色器代码
            "precision mediump float;" +
                    "uniform vec4 vColor;" +
                    "void main() {" +
                    "  gl_FragColor = vColor;" +
                    "}";

    private int positionHandle;
    private int colorHandle;

    private final int vertexCount = triangleCoords.length / COORDS_PER_VERTEX;
    private final int vertexStride = COORDS_PER_VERTEX * 4;    //每个顶点 4 个字节

    static final int COORDS_PER_VERTEX = 3;                    //每个顶点在矩阵中的坐标数
    static float triangleCoords[] = {                          //（b）
            0.0f, 0.622008459f, 0.0f,                          //顶部
            -0.5f, -0.311004243f, 0.0f,                        //左下角
            0.5f, -0.311004243f, 0.0f                          //右下角
    };

    //用红、绿、蓝和 alpha（透明度）值设置颜色
    float color[] = {0.63671875f, 0.76953125f, 0.22265625f, 1.0f};

    public Triangle() {
        ByteBuffer bb = ByteBuffer.allocateDirect(triangleCoords.length * 4);
                                                               //（c）
        bb.order(ByteOrder.nativeOrder());                     //采用设备硬件的本地字节顺序
        vertexBuffer = bb.asFloatBuffer();                     //由字节缓存创建一个浮点缓存
        vertexBuffer.put(triangleCoords);                      //将坐标添加到浮点缓存
        vertexBuffer.position(0);                              //设置缓存读取第一个坐标
        int vertexShader = MyGLRenderer.loadShader(GLES20.GL_VERTEX_SHADER, vertexShaderCode);
        int fragmentShader = MyGLRenderer.loadShader(GLES20.GL_FRAGMENT_SHADER, fragmentShaderCode);
        myProgram = GLES20.glCreateProgram();                  //创建空 OpenGL ES 程序
        GLES20.glAttachShader(myProgram, vertexShader);        //将顶点着色器添加到程序
        GLES20.glAttachShader(myProgram, fragmentShader);      //将片段着色器添加到程序
```

```
            GLES20.glLinkProgram(myProgram);                    //创建 OpenGL ES 程序可执行文件
        }

        public void draw() {                                    // (d)
            GLES20.glUseProgram(myProgram);                     //添加程序到 OpenGL ES 环境
            positionHandle = GLES20.glGetAttribLocation(myProgram, "vPosition");
                                                                //获取顶点着色器 vPosition 成员的句柄
            GLES20.glEnableVertexAttribArray(positionHandle);   //启用三角形顶点的控制句柄
            GLES20.glVertexAttribPointer(positionHandle, COORDS_PER_VERTEX, GLES20.GL_FLOAT, false,
vertexStride, vertexBuffer);                                    //准备三角形坐标数据
            colorHandle = GLES20.glGetUniformLocation(myProgram, "vColor");
                                                                //获取片段着色器 vColor 成员的句柄
            GLES20.glUniform4fv(colorHandle, 1, color, 0);      //设置所绘三角形的颜色
            GLES20.glDrawArrays(GLES20.GL_TRIANGLES, 0, vertexCount);
                                                                //绘制三角形
            GLES20.glDisableVertexAttribArray(positionHandle);  //禁用顶点数组
        }
    }
```

由上述代码可见，在 OpenGL 中定义和绘制图形需要比较多的代码，这是因为 OpenGL 库是一个比较开放的 API，为了做到让用户能够完全地控制绘图过程的每一个环节，要求用户必须向图形渲染管道提供大量详细信息，这样一来就不得不将很多图形渲染的细节暴露给用户，封装性就差了。但对于精通 OpenGL 的用户来说，这么做可以更灵活地操控复杂的高端图形对象。

其中：

（a）**private final String vertexShaderCode =...** 和 **private final String fragmentShaderCode =...**：这两段代码都是用 GLSL 语言描述的，定义了两个着色器。

- **顶点着色器**：即 vertexShaderCode 的代码，用于渲染图形的每个顶点。
- **片段着色器**：即 fragmentShaderCode 的代码，使用颜色或纹理渲染图形的表面，也就是为图形着色。

使用刚刚渲染器中的 loadShader()方法对以上这两个着色器的代码执行预编译，代码如下：

```
int vertexShader = MyGLRenderer.loadShader(GLES20.GL_VERTEX_SHADER, vertexShaderCode);
                                                                //编译顶点着色器
int fragmentShader = MyGLRenderer.loadShader(GLES20.GL_FRAGMENT_SHADER, fragmentShaderCode);
                                                                //编译片段着色器
```

然后用 glAttachShader()方法将它们添加到程序中使用：

```
GLES20.glAttachShader(myProgram, vertexShader);                 //将顶点着色器添加到程序中
GLES20.glAttachShader(myProgram, fragmentShader);               //将片段着色器添加到程序中
```

预编译的语句需要写在图形对象的构造方法（本例的 Triangle()）中，且只执行一次。

> ◎◎注意：
> 就 CPU 周期和处理时间而言，编译着色器及关联程序的开销非常大，因此应尽量避免多次执行预编译操作。故通常将编译及添加语句写在图形对象的构造方法中，保证其仅被执行一次，然后缓存以备后用。

（b）**static float triangleCoords[] = { ... };**：默认情况下，OpenGL 假定一个坐标系，它按照逆时针顺序定义，其中[0,0,0](X,Y,Z)指定 GLSurfaceView 视图帧的中心，[1,1,0]指定帧的右上角，而[-1,-1,0]指定帧的左下角。

（c）**ByteBuffer bb = ByteBuffer.allocateDirect(triangleCoords.length * 4);**：ByteBuffer 是字节缓存，为了最大限度地提高绘图效率，通常都会将图形的顶点坐标写入字节缓存，然后再传递给 OpenGL 图形管道做进一步处理。本例中为图形顶点坐标初始化分配的字节缓存，其中的每个坐标值是以一个 4 字节的浮点数存储的。

（d）**public void draw() { ... }**：这个是用于绘制图形的 draw()方法。它将位置和颜色值设置为图形的顶点着色器和片段着色器，然后执行绘制功能。

运行程序，绘制出三角形，如图 8.16 所示。

图 8.16　用 OpenGL 绘制的三角形

第 9 章　Android 第三方开发与设备操作

所谓 Android 第三方开发，也就是借助 Android 官方之外的第三方组织所提供的库和 API 来进行二次开发和扩展。由于 Android 是开源的操作系统，任何组织和个人在遵守 GPL 协议的前提下都可以开发针对某一类特定应用功能的库，然后将其发布出去供更多的人使用。本章将介绍一些应用较为广泛的第三方库，重点讲它们的获取、集成和基本的使用方法，至于更深入的内容，请读者参考各个库的第三方文档。另外，作为当前主流的智能手机操作系统，Android 提供了对手机上硬件设备较为全面的支持，几乎能够实现对各种类型设备的操作，本章对此也将分别举例说明。

9.1　生成和扫描二维码（ZXing 库）

9.1.1　ZXing 概述

1. ZXing 简介

ZXing 是谷歌自己推出的一个开放源码的二维码框架，可以实现使用手机的摄像头完成条形码的扫描及解码。它内置了用 Java 实现的多种格式的 1D/2D 条码图像处理库，还包含了连接到其他语言的接口，支持 UPC、EAN、39 码、93 码、代码 128、Codabar、RSS-14（所有变体）、RSS 扩展（大多数变体）、QR 码、DataMatrix、Aztec 码、PDF417 等诸多二维码格式，目前市面上差不多一半以上的公司都在使用这个框架开发二维码相关的 App 功能。

2. ZXing 的获取

ZXing 官方提供了 Maven 库，让我们可以根据自己的需要选择想要的.jar 包版本进行下载，Maven 库的地址为 https://repo1.maven.org/maven2/com/google/zxing/core/。本书使用 ZXing 3.4.0 版，需要下载一个名为 "core-3.4.0.jar" 的包，单击相应版本的链接下载即可。

然后从网上获取二维码 Demo，Demo 的工程名称为 QrScan（.jar 包和 Demo 将放在本书源码资源中一起免费提供给读者）。

接下来就要将.jar 包和 Demo 整合进自己的 Android Studio 工程，才能够使用 ZXing 框架的二维码功能。

9.1.2　整合 ZXing 框架

整合 ZXing 框架的过程比较烦琐，请读者严格按照以下步骤来进行。

（1）创建一个 Android Studio 工程，名为 MyQRCodeDemo。

（2）加载.jar 包。将预先获取的 core-3.4.0.jar 包复制到工程的 app\libs 目录，刷新。然后在 Android Studio 中选择主菜单 "File" → "Project Structure..." 命令，在打开的 "Project Structure" 窗口中完成添加依赖（具体操作见 7.2.1 小节）。

（3）集成 Java 源码。将 Demo 工程 QrScan 中 app\src\main\java 目录下 com.example.administrator. qrscan 包中的 zxing、util 复制到本工程对应的 app\src\main\java 目录下。

然后，要对集成进来的源码进行一系列修改才能将其整合进自己的项目工程，修改的大致顺序为：

修改 package 包名→修改 import 路径→修改类包名

① 修改 package 包名。依次修改每个.java 文件头的起始 package 包名为本工程实际的全路径包名，如图 9.1 所示。

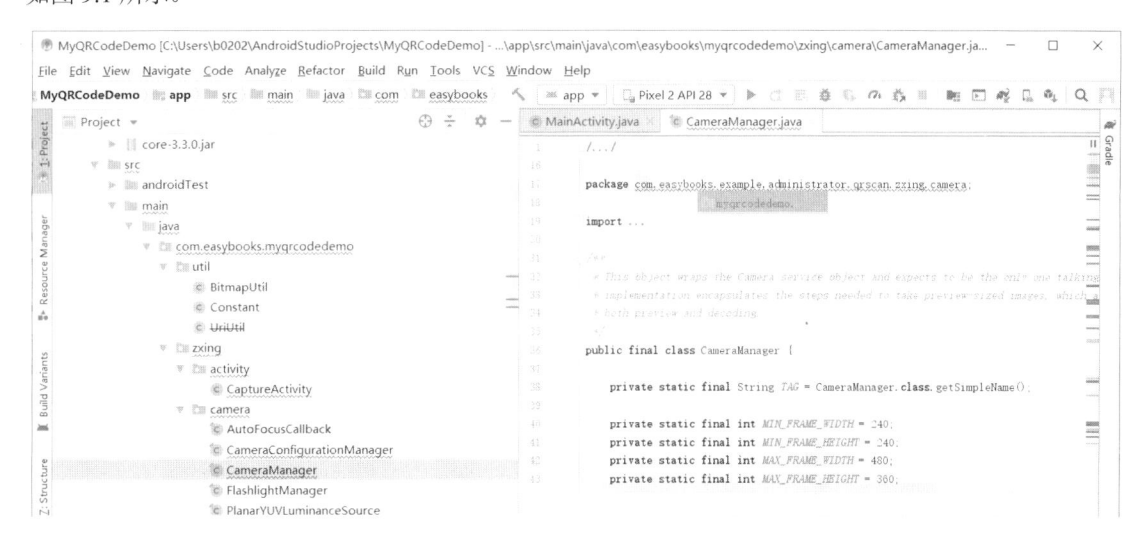

图 9.1　修改 package 包名

② 修改 import 路径。在某些.java 文件中，用 import 语句导入的包路径名为原 Demo 中路径的，要改为本工程实际的路径包名，依次检查每个.java 文件，在所有需要修改的地方做出更正，如图 9.2 所示。

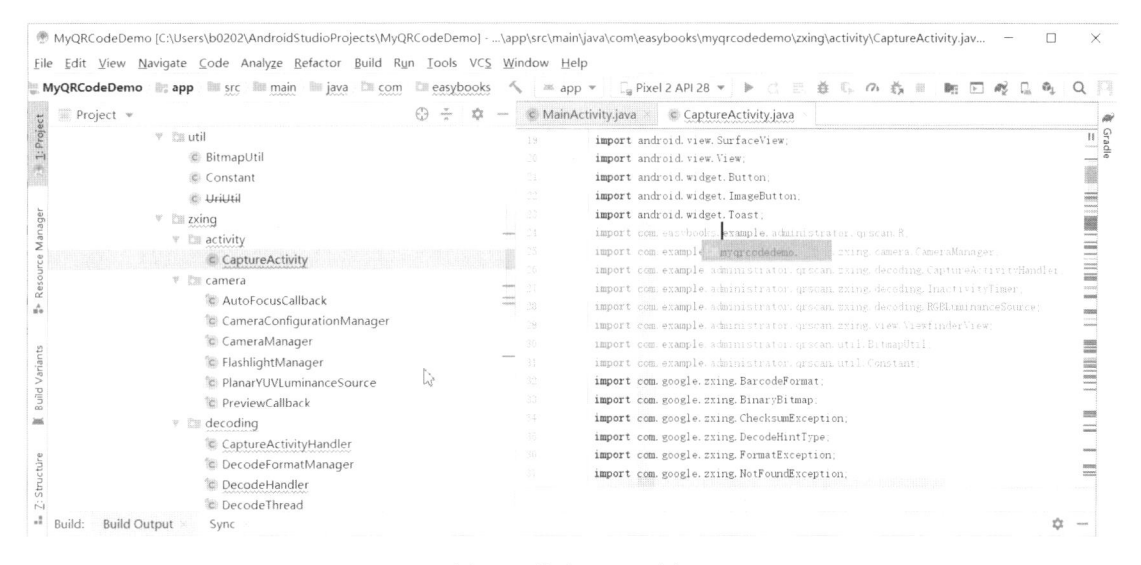

图 9.2　修改 import 路径

③ 修改类包名。由于 Demo 开发所用的 Android Studio 环境与用户自身所用环境存在差异，源码中的某些类所在的包名有变动，例如，CaptureActivity.java 中的 AppCompatActivity 类在原 Demo 中的包名是"android.support.v7.app.AppCompatActivity"，移植到新环境则变成"androidx.appcompat.app.AppCompatActivity"，如图 9.3 所示。在 IDE 加红标注的该类处重新输入一遍类名（往往在输入前几个

字符时 IDE 就会自动给出下拉列表提示），IDE 会将该类所在的新包自动导入源文件开头的 import 语句序列中，将该类包原有 import 语句删除（也可不删）即可。

图 9.3　修改类包名

请读者依次检查各个 .java 文件，看到有红色醒目标注的类名就重新输入一遍，将其更改为新环境下的类包名。

（4）同步资源。在完成以上对 .java 源文件的修改后，.java 源码中仍然有一些标红的错误，如图 9.4 所示，这些都是资源引用上的错误，所以接下来就要同步资源。

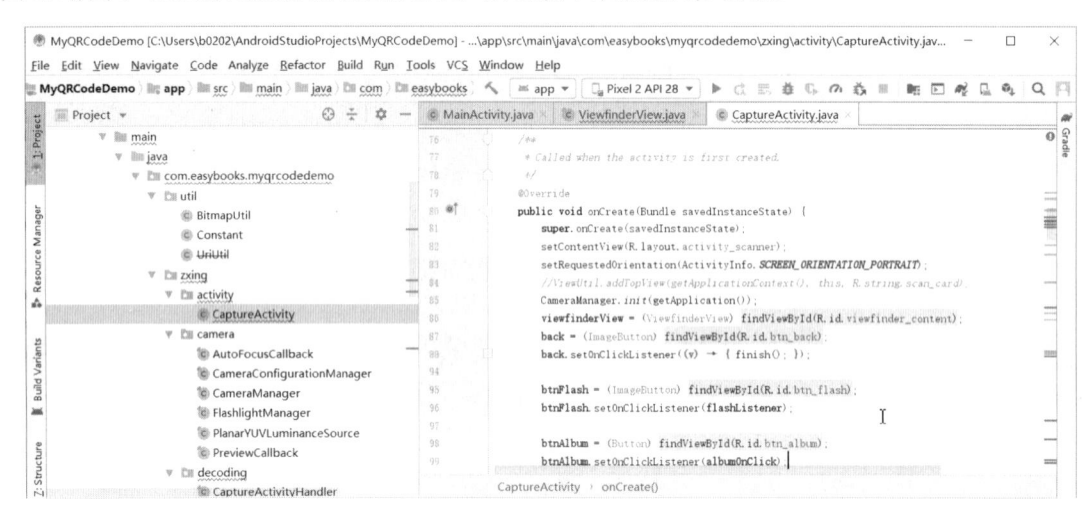

图 9.4　资源缺少引用

同步资源的大致顺序为：

复制资源目录→修改工具栏框架包→修改 ViewfinderView 包路径

① 复制资源目录。将 Demo 工程 QrScan 中 app\src\main\res 下的资源同步复制到本工程的目录中，需要同步的目录有：

- drawable：复制 btn_back.png、flash_off.png、flash_on.png。
- layout：复制 activity_capture.xml、activity_scanner.xml、toolbar_scanner.xml。
- raw：全部复制。
- values：复制/替换其中的 attrs.xml、ids.xml、colors.xml。

② 修改工具栏框架包。toolbar_scanner.xml 中的框架包要改成"androidx.appcompat.widget.Toolbar"，如图 9.5 所示。

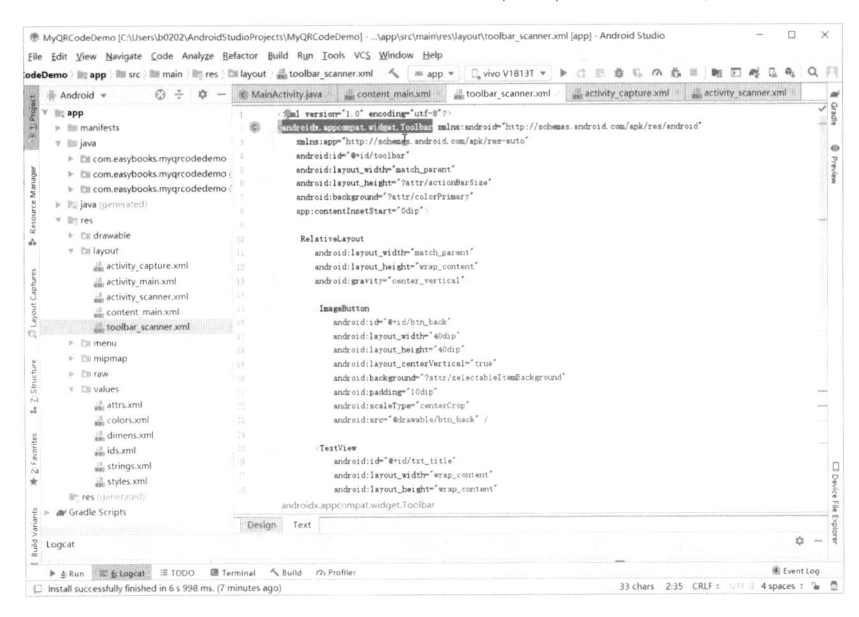

图 9.5　修改工具栏框架包

> 👀**注意：**
> 这一步千万不能忘了，否则以后开发的项目在扫描二维码时将出现程序闪退、无法启动的现象。

③ 修改 ViewfinderView 包路径。修改.xml 资源文件中 ViewfinderView 的包名为本工程的包路径名，如图 9.6 所示，工程中共有两处地方（分别位于 activity_capture.xml 和 activity_scanner.xml）要改。

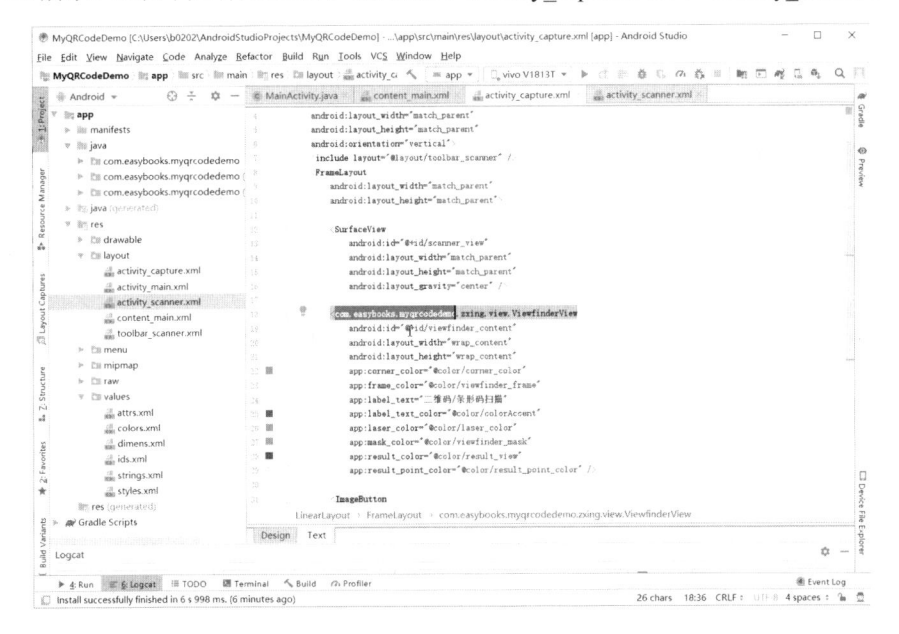

图 9.6　修改 ViewfinderView 包路径

至此，工程所有源码文件中的错误全部消失了。

（5）打开开发权限。在 AndroidManifest.xml 中打开开发权限，添加配置 activity，内容如下（加粗处为需要读者添加的内容）：

```
<manifest xmlns:android="http://schemas.android.com/apk/res/android"
    package="com.easybooks.myqrcodedemo">
    <uses-permission android:name="android.permission.VIBRATE" />      //震动权限
    <uses-permission android:name="android.permission.CAMERA" />       //摄像头权限
    <uses-permission android:name="android.permission.READ_EXTERNAL_STORAGE" />
                                                                       //读取存储空间权限
    <application
        ...
        <activity
android:name="com.easybooks.myqrcodedemo.zxing.activity.CaptureActivity"></activity>
                                                                       //添加扫描界面的activity
    </application>
</manifest>
```

最后，我们运行一下工程，如果不出现错误就表示整合成功。鉴于整合过程比较麻烦，笔者建议初学者把这个整合好 ZXing 库的工程另外备份存盘，以后凡是遇到需要开发二维码相关功能的 App 时都直接在这个已整合好框架的工程基础上进行开发，可节省不少时间和精力。

9.1.3 界面设计

在工程 content_main.xml 的设计模式下，向界面上拖曳放置一个文本编辑框（EditText）、两个按钮、一个图像视图（ImageView）和一个文本视图（TextView），如图 9.7 所示。

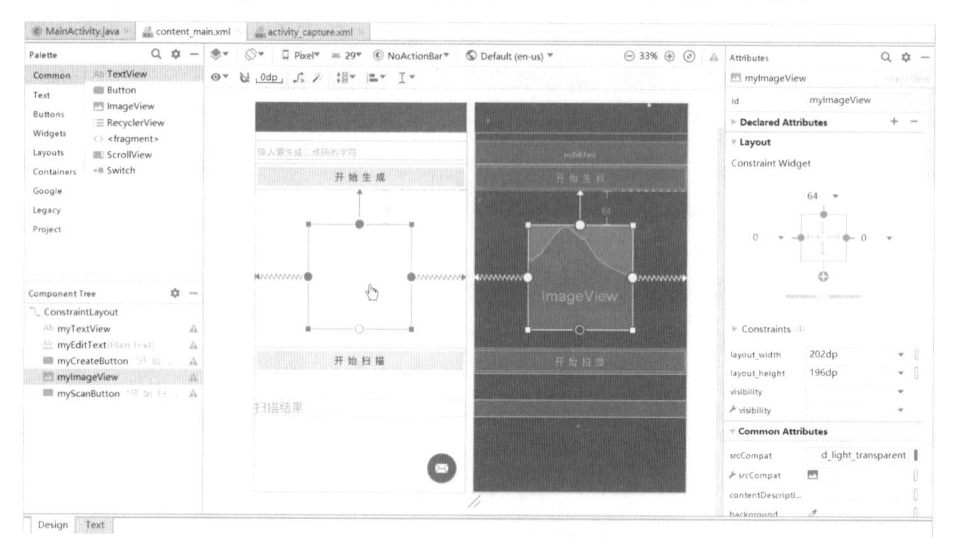

图 9.7 生成和扫描二维码程序界面设计

其中，文本编辑框提供给用户输入需要生成二维码的字符串，字符串可以是任意长度的字母、数字、空格和符号的组合；生成的二维码显示在下方图像视图中；扫描二维码得到的结果（字符串）则显示在文本视图中。

9.1.4 二维码生成

【例 9.1】 用 ZXing 库实现将用户输入文本编辑框的字符文本转换成二维码。

实现思路：

（1）我们将生成二维码的功能封装在一个自定义的 CreateUtil 类中，通过 createQRCode() 方法实现。

（2）createQRCode()方法接收 3 个参数：codestring（要生成二维码的字符串）、width（二维码图片的宽度）、height（二维码图片的高度），接收到参数后首先判断参数的合法性。

（3）createQRCode()方法将要生成二维码的字符串编码成一个 BitMatrix 对象，再创建一个像素数组 pixel，根据 BitMatrix 对象中每一个位的内容为数组 pixel 的对应元素赋值，最后由数组 pixel 生成一个 Bitmap 对象（即结果二维码图片）返回。

（4）在 MainActivity 主程序类中，从 myCreateButton（"开始生成"按钮）的点击事件监听器的 onClick()方法中调用 CreateUtil 类的 createQRCode()方法生成二维码图片，显示在界面上的图像视图中。

创建 CreateUtil 类，编写代码如下：

```java
public class CreateUtil {
    @Nullable
    public static Bitmap createQRCode(String codestring, int width, int height) {
        try {
            /**首先判断参数的合法性*/
            if (TextUtils.isEmpty(codestring) || width <= 0 || height <= 0) {
                return null;
            }                                                              //（a）
            /**然后设置二维码相关参数，生成 BitMatrix（位矩阵）对象*/
            Hashtable<EncodeHintType, String> hashtable = new Hashtable<>();
            hashtable.put(EncodeHintType.CHARACTER_SET, "utf-8");    //设置字符转码格式
            hashtable.put(EncodeHintType.ERROR_CORRECTION, "H");    //设置容错级别
            hashtable.put(EncodeHintType.MARGIN, "2");               //设置空白边距
            BitMatrix bitMatrix = new QRCodeWriter().encode(codestring, BarcodeFormat.QR_CODE, width, height, hashtable);                                                 //encode 需要抛出和处理异常
            /**再创建像素数组，并根据位矩阵为数组元素赋颜色值*/
            int[] pixel = new int[width * height];
            for (int h = 0; h < height; h++) {
                for (int w = 0; w < width; w++) {
                    if (bitMatrix.get(w, h)) {
                        pixel[h * width + w] = Color.BLACK;          //设置黑色色块
                    } else {
                        pixel[h * width + w] = Color.WHITE;          //设置白色色块
                    }
                }
            }
            /**最后创建 Bitmap 对象*/
            Bitmap qrcodemap = Bitmap.createBitmap(width, height, Bitmap.Config.ARGB_8888);
            qrcodemap.setPixels(pixel, 0, width, 0, 0, width, height);   //（b）
            return qrcodemap;
        } catch (WriterException e) {
            return null;
        }
    }
}
```

其中：

（a）**if (TextUtils.isEmpty(codestring) || width <= 0 || height <= 0) { return null; }**：参数合法性检查，包括要生成二维码的字符串内容不能为空以及二维码图片的宽和高都必须大于 0。

(b) **qrcodemap.setPixels(pixel, 0, width, 0, 0, width, height);**：根据像素数组设置 Bitmap 每个像素点的颜色值，之后返回 Bitmap 对象。

然后在 MainActivity.java 中编写代码，生成二维码：

```java
public class MainActivity extends AppCompatActivity {
    private EditText myEditText;                    //接收用户输入要生成二维码的字符串
    private Button myCreateButton;                  //点击此按钮开始生成
    private ImageView myImageView;                  //用于显示生成的二维码
    @Override
    protected void onCreate(Bundle savedInstanceState) {
        super.onCreate(savedInstanceState);
        setContentView(R.layout.activity_main);
        //生成二维码
        myEditText = findViewById(R.id.myEditText);
        myCreateButton = findViewById(R.id.myCreateButton);
        myImageView = findViewById(R.id.myImageView);
        myCreateButton.setOnClickListener(new View.OnClickListener() {
            @Override
            public void onClick(View view) {
                String codeStr = myEditText.getText().toString();    //获取用户输入的产码字符串
                myImageView.setImageBitmap(CreateUtil.createQRCode(codeStr, myImageView.getWidth(), myImageView.getHeight()));
                //调用 CreateUtil 类生成二维码后显示在图像视图中
            }
        });
        ...
    }
}
```

完成后运行程序，输入一串字符文本，点击"开始生成"按钮，生成二维码，如图 9.8 所示。

图 9.8 生成二维码

9.1.5 二维码扫描

【例 9.2】 用 ZXing 库对上面已经生成的二维码进行扫描识别,还原为字符文本显示出来。

实现思路:

(1) onCreate()方法初始化时,在 myScanButton("开始扫描"按钮)的点击事件监听器的 onClick() 方法中调用 scanQRCode()方法扫描二维码。

(2) 在 scanQRCode()方法中,首先申请相机和文件读写权限,然后通过 Intent 机制启动 zxing 框架的 CaptureActivity,请求返回结果。

(3) 在 onActivityResult()回调方法中,获取 zxing 框架扫描出的结果字符串,在界面上的文本视图中显示出来。

在 MainActivity.java 中添加代码,实现扫描二维码功能,内容如下:

```java
public class MainActivity extends AppCompatActivity {
    ...
    private Button myScanButton;                              //点击此按钮开始扫描
    private TextView myTextView;                              //扫描结果显示在这个文本视图中
    @Override
    protected void onCreate(Bundle savedInstanceState) {
        super.onCreate(savedInstanceState);
        setContentView(R.layout.activity_main);
        ...
        //扫描二维码
        myScanButton = findViewById(R.id.myScanButton);
        myTextView = findViewById(R.id.myTextView);
        myScanButton.setOnClickListener(new View.OnClickListener() {
            @Override
            public void onClick(View view) {
                scanQRCode();                                 //实现扫描二维码的方法
            }
        });
        ...
    }
    ......
    private void scanQRCode() {
        //申请相机权限
        if (ActivityCompat.checkSelfPermission(this, Manifest.permission.CAMERA) != PackageManager.PERMISSION_GRANTED) {
            ActivityCompat.requestPermissions(this, new String[]{Manifest.permission.CAMERA}, Constant.REQ_PERM_CAMERA);
            return;
        }
        //申请文件(相册)读写权限
        if (ActivityCompat.checkSelfPermission(this, Manifest.permission.READ_EXTERNAL_STORAGE) != PackageManager.PERMISSION_GRANTED) {
            ActivityCompat.requestPermissions(this, new String[]{Manifest.permission.READ_EXTERNAL_STORAGE}, Constant.REQ_PERM_EXTERNAL_STORAGE);
            return;
        }
```

```java
            //二维码扫码
            Intent intent = new Intent(this, CaptureActivity.class);
            startActivityForResult(intent, Constant.REQ_QR_CODE);
        }

    @Override
    public void onActivityResult(int requestCode, int resultCode, Intent data) {
        super.onActivityResult(requestCode, resultCode, data);
        //扫描结果回调
        if (requestCode == Constant.REQ_QR_CODE && resultCode == RESULT_OK) {
            Bundle bundle = data.getExtras();
            String scanResult = bundle.getString(Constant.INTENT_EXTRA_KEY_QR_SCAN);
            myTextView.setText(scanResult);                    //将扫描出的信息显示出来
        }
    }

    @Override
    public void onRequestPermissionsResult(int requestCode, @NonNull String[] permissions, @NonNull int[] grantResults) {
        super.onRequestPermissionsResult(requestCode, permissions, grantResults);
        switch (requestCode) {
            case Constant.REQ_PERM_CAMERA:
                //申请摄像头权限
                if (grantResults.length > 0 && grantResults[0] == PackageManager.PERMISSION_GRANTED) {        //获得授权
                    scanQRCode();
                } else {                                       //被禁止授权
                    Toast.makeText(this, "请至权限中心打开本应用的相机访问权限", Toast.LENGTH_LONG).show();
                }
                break;
            case Constant.REQ_PERM_EXTERNAL_STORAGE:
                //申请文件读写权限
                if (grantResults.length > 0 && grantResults[0] == PackageManager.PERMISSION_GRANTED) {        //获得授权
                    scanQRCode();
                } else {                                       //被禁止授权
                    Toast.makeText(this, "请至权限中心打开本应用的文件读写权限", Toast.LENGTH_LONG).show();
                }
                break;
        }
    }
}
```

完成后分别在计算机 Android Studio 环境仿真器和真实的手机上安装运行本程序，先用仿真器中的程序按 9.1.4 小节操作生成二维码，然后用手机上的程序对准二维码进行扫描，点击"开始扫描"按钮，扫描结果如图 9.9 所示。

图 9.9　扫描二维码

9.2　接入支付宝（alipaySdk 库）

9.2.1　支付接口

1. 沙箱环境

要在 App 中实现支付功能，就要接入支付宝官方提供的支付平台。目前出于安全考虑，此平台只允许拥有营业执照的企业接入使用，如果是一般的 Android 开发者想要测试 App 支付功能，只能在支付宝官方开放的沙箱中进行。沙箱是模拟的一个支付环境，里面有一对虚拟商家和买家的账号、密码，如图 9.10 所示，登录支付宝后进入"开发者中心"→"研发服务"→"沙箱环境"就可以看到。

图 9.10　测试支付功能的沙箱环境

2. App 标识

在"沙箱应用"中，可看到一个 APPID 值为"2016101300676685"，如图 9.11 所示，它是用来标识用户 App 的，稍后需要写在程序代码中，通过它来访问支付宝的 App 支付接口，请大家先记下这个值。

图 9.11　访问支付接口的 APPID

3. 获取密钥

要使程序能访问沙箱环境，除了 App 标识，用户还要获取一对密钥。可去"开发文档"→"签名专区"下载支付宝官方提供的密钥生成工具，如图 9.12 所示。

图 9.12　下载密钥生成工具

下载得到 AlipayDevelopmentAssistant-1.0.1.exe 安装包，将其安装在开发计算机上，启动程序，默认选项下直接单击"生成密钥"按钮即可生成一对私钥和公钥，如图 9.13 所示，将它们复制下来。

图 9.13 生成一对私钥和公钥

将复制下来的公钥设置在沙箱环境中，如图 9.14 所示演示了公钥加签的操作。

图 9.14 将公钥设置在沙箱环境中

而复制下来的私钥稍后将写在 App 程序代码中。

4. 下载支付 SDK

支付宝官网 SDK 下载地址为 https://docs.open.alipay.com/54/104509，如图 9.15 所示，我们下载 15.5.5 版 JAR 格式的 SDK 和 Demo。

图 9.15 下载支付 SDK

下载后解压，得到 alipaySdk-20180601.jar 以及一个名为 alipay_demo 的工程，将它们存盘待用。

经过以上一系列的操作，支付接口的配置及编程的准备工作就做好了，接下来就能进行支付功能的开发。

9.2.2 集成支付功能

我们同样也是依靠 Demo 进行支付功能的集成，具体操作如下：

（1）创建 Android Studio 工程，名为 MyEasyPayDemo。

（2）加载 alipaySdk-20180601.jar，方法同 7.2.1 小节。

（3）集成 Java 源码。

将 Demo 中的.java 文件全部复制到本工程的源码目录，修改 package 包名及 import 路径与本工程的一致。可能遇到的几种情形的处理方式参见 9.1.2 小节，不再赘述。

（4）同步资源。包括如下操作：

① 将 Demo 工程 drawable-hdpi 下的内容放入本工程 drawable 目录中。

② layout 目录下的 pay_external.xml 和 pay_main.xml 放入本工程 layout 目录中（其中包名要改成与本工程的路径包名一致）。

（5）打开开发权限。在 AndroidManifest.xml 中打开开发权限，添加配置 activity，内容如下（加粗处为需要读者添加的内容）：

```
<manifest xmlns:android="http://schemas.android.com/apk/res/android"
    package="com.easybooks.myeasypaydemo">
    <uses-permission android:name="android.permission.INTERNET" />
    <uses-permission android:name="android.permission.ACCESS_NETWORK_STATE" />
    <uses-permission android:name="android.permission.ACCESS_WIFI_STATE" />
    <uses-permission android:name="android.permission.READ_PHONE_STATE" />
    <uses-permission android:name="android.permission.WRITE_EXTERNAL_STORAGE" />
    <application
```

```xml
...
<activity
    android:name="com.easybooks.myeasypaydemo.H5PayDemoActivity"
    android:configChanges="orientation|keyboardHidden|navigation"
    android:exported="false"
    android:screenOrientation="behind" >
</activity>
<!-- alipay sdk begin -->
<activity
    android:name="com.alipay.sdk.app.H5PayActivity"
    android:configChanges="orientation|keyboardHidden|navigation|screenSize"
    android:exported="false"
    android:screenOrientation="behind"
    android:windowSoftInputMode="adjustResize|stateHidden" >
</activity>
<activity
    android:name="com.alipay.sdk.app.H5AuthActivity"
    android:configChanges="orientation|keyboardHidden|navigation"
    android:exported="false"
    android:screenOrientation="behind"
    android:windowSoftInputMode="adjustResize|stateHidden" >
</activity>
<!-- alipay sdk end -->
</application>
</manifest>
```

（6）将 Demo 工程 app 目录下 proguard-project.txt 文件中的内容全部复制到本工程的 proguard-rules.pro 文件（也位于 app 目录）中。

9.2.3 支付功能实现

【例 9.3】 用前面集成的 alipaySdk 库实现支付功能，并在沙箱环境进行测试。

在工程 PayDemoActivity.java 文件中，写入 APPID 及用户私钥：

```java
package com.easybooks.myeasypaydemo;
...
import androidx.fragment.app.FragmentActivity;

/**
 * 重要说明：
 *
 * 这里只是为方便直接向商户展示支付宝的整个支付流程；所以 Demo 中加签过程直接放在客户端完成；
 * 真实 App 里，privateKey 等数据严禁放在客户端，加签过程务必要放在服务端完成；
 * 防止商户私密数据泄露，造成不必要的资金损失，及面临各种安全风险；
 */
public class PayDemoActivity extends FragmentActivity {
    private TextView myTextView;

    /** 支付宝支付业务：入参 app_id */
    public static final String AppID = "2016101300676685";                    // (a)

    /** 支付宝账户登录授权业务：入参 pid 值 */
```

```java
        public static final String PID = "";
        /** 支付宝账户登录授权业务：入参 target_id 值 */
        public static final String TARGET_ID = "";

        /** 商户私钥，pkcs8 格式 */
        /** 如下私钥，RSA2_PRIVATE 或者 RSA_PRIVATE 只需要填入一个 */
        /** 如果商户两个都设置了，优先使用 RSA2_PRIVATE */
        /** RSA2_PRIVATE 可以保证商户交易在更加安全的环境下进行，建议使用 RSA2_PRIVATE */
        /** 获取 RSA2_PRIVATE，建议使用支付宝提供的公私钥生成工具生成 */
        /** 工具地址：https://doc.open.alipay.com/docs/doc.htm?treeId=291&articleId=106097&docType=1 */
        public static final String RSA2_PRIVATE = "MIIEv... WEmB0J4e9ywvg=";    // （b）
        public static final String RSA_PRIVATE = "";

        private static final int SDK_PAY_FLAG = 1;
        private static final int SDK_AUTH_FLAG = 2;

        @SuppressLint("HandlerLeak")
        private Handler mHandler = new Handler() {
            @SuppressWarnings("unused")
            public void handleMessage(Message msg) {
                switch (msg.what) {
                    case SDK_PAY_FLAG: {
                        @SuppressWarnings("unchecked")
                        PayResult payResult = new PayResult((Map<String, String>) msg.obj);
                        /**
                            对于支付结果，请商户依赖服务端的异步通知结果。
                            同步通知结果仅作为支付结束的通知。
                         */
                        String resultInfo = payResult.getResult();           //同步返回需要验证的信息
                        String resultStatus = payResult.getResultStatus();
                        if (TextUtils.equals(resultStatus, "9000")) {        //resultStatus 为"9000"代表支付成功
                            //该笔订单是否真实支付成功需要依赖服务端的异步通知
                            Toast.makeText(PayDemoActivity.this, "支付成功", Toast.LENGTH_SHORT).show();
                        } else {
                            //该笔订单真实的支付结果需要依赖服务端的异步通知
                            Toast.makeText(PayDemoActivity.this, "支付失败", Toast.LENGTH_SHORT).show();
                        }
                        break;
                    }
                    case SDK_AUTH_FLAG: {
                        @SuppressWarnings("unchecked")
                        AuthResult authResult = new AuthResult((Map<String, String>) msg.obj, true);
                        String resultStatus = authResult.getResultStatus();
                        if (TextUtils.equals(resultStatus, "9000") && TextUtils.equals(authResult.getResultCode(), "200")) {
                            //resultStatus 为"9000"且 result_code 为"200"代表授权成功
                            Toast.makeText(PayDemoActivity.this, "授权成功\n" + String.format("authCode:%s", authResult.getAuthCode()), Toast.LENGTH_SHORT).show();
                        } else {           //其他状态值为授权失败
                            Toast.makeText(PayDemoActivity.this,"授权失败" + String.format("authCode:%s",
```

```java
authResult.getAuthCode()), Toast.LENGTH_SHORT).show();
                    }
                    break;
                }
                default:
                    break;
            }
        };
    };

    @Override
    protected void onCreate(Bundle savedInstanceState) {
        EnvUtils.setEnv(EnvUtils.EnvEnum.SANDBOX);
        super.onCreate(savedInstanceState);
        setContentView(R.layout.pay_main);
        myTextView = findViewById(R.id.myTextView);
    }

    /**
     * 支付宝支付业务
     *
     * @param v
     */
    public void payV2(View v) {
        if (TextUtils.isEmpty(AppID) || (TextUtils.isEmpty(RSA2_PRIVATE) && TextUtils.isEmpty(RSA_PRIVATE))) {
            new AlertDialog.Builder(this).setTitle("警告").setMessage("需要配置 AppID | RSA_PRIVATE").setPositiveButton("确定", new DialogInterface.OnClickListener() {
                public void onClick(DialogInterface dialoginterface, int i) {
                    finish();
                }
            }).show();
            return;
        }
        /**
         * 这里只是为方便向商户展示支付宝的支付流程，所以 Demo 中加签过程直接放在客户端完成；
         * 真实 App 里，privateKey 等数据严禁放在客户端，加签过程必要放在服务端完成；
         * 防止商户私密数据泄露，造成不必要的资金损失，及面临各种安全风险；
         *
         * orderInfo 的获取必须来自服务端；
         */
        boolean rsa2 = (RSA2_PRIVATE.length() > 0);
        Map<String, String> params = OrderInfoUtil2_0.buildOrderParamMap(AppID, rsa2);
        String orderParam = OrderInfoUtil2_0.buildOrderParam(params);
        String privateKey = rsa2 ? RSA2_PRIVATE : RSA_PRIVATE;
        String sign = OrderInfoUtil2_0.getSign(params, privateKey, rsa2);
        final String orderInfo = orderParam + "&" + sign;
        Runnable payRunnable = new Runnable() {
            @Override
            public void run() {
```

```java
                    PayTask alipay = new PayTask(PayDemoActivity.this);
                    Map<String, String> result = alipay.payV2(orderInfo, true);
                    Message msg = new Message();
                    msg.what = SDK_PAY_FLAG;
                    msg.obj = result;
                    mHandler.sendMessage(msg);
                }
            };
            Thread payThread = new Thread(payRunnable);
            payThread.start();
        }

        /**
         * 支付宝账户授权业务
         *
         * @param v
         */
        public void authV2(View v) {
            if (TextUtils.isEmpty(PID) || TextUtils.isEmpty(AppID)
                    || (TextUtils.isEmpty(RSA2_PRIVATE) && TextUtils.isEmpty(RSA_PRIVATE))
                    || TextUtils.isEmpty(TARGET_ID)) {
                new AlertDialog.Builder(this).setTitle("警告").setMessage("需要配置 PARTNER |App_ID| RSA_PRIVATE| TARGET_ID").setPositiveButton("确定", new DialogInterface.OnClickListener() {
                    public void onClick(DialogInterface dialoginterface, int i) {}
                }).show();
                return;
            }

            /**
             * 这里只是为方便向商户展示支付宝的支付流程,所以 Demo 中加签过程直接放在客户端完成;
             * 真实 App 里, privateKey 等数据严禁放在客户端, 加签过程务必要放在服务端完成;
             * 防止商户私密数据泄露, 造成不必要的资金损失, 及面临各种安全风险;
             *
             * authInfo 的获取必须来自服务端;
             */
            boolean rsa2 = (RSA2_PRIVATE.length() > 0);
            Map<String, String> authInfoMap = OrderInfoUtil2_0.buildAuthInfoMap(PID, AppID, TARGET_ID, rsa2);

            String info = OrderInfoUtil2_0.buildOrderParam(authInfoMap);
            String privateKey = rsa2 ? RSA2_PRIVATE : RSA_PRIVATE;
            String sign = OrderInfoUtil2_0.getSign(authInfoMap, privateKey, rsa2);
            final String authInfo = info + "&" + sign;
            Runnable authRunnable = new Runnable() {
                @Override
                public void run() {
                    // 构造 AuthTask 对象
                    AuthTask authTask = new AuthTask(PayDemoActivity.this);
                    // 调用授权接口, 获取授权结果
                    Map<String, String> result = authTask.authV2(authInfo, true);
                    Message msg = new Message();
                    msg.what = SDK_AUTH_FLAG;
```

```
                msg.obj = result;
                mHandler.sendMessage(msg);
            }
        };
        //必须异步调用
        Thread authThread = new Thread(authRunnable);
        authThread.start();
    }

    /**
     * 获取 SDK 版本号
     */
    public void getSDKVersion() {
        PayTask payTask = new PayTask(this);
        String version = payTask.getVersion();
        Toast.makeText(this, version, Toast.LENGTH_SHORT).show();
    }

    /**
     * 原生的 H5（手机网页版支付转 Native 支付）【对应页面网页支付按钮】
     *
     * @param v
     */
    public void h5Pay(View v) {
        Intent intent = new Intent(this, H5PayDemoActivity.class);
        Bundle extras = new Bundle();
        /**
         * url 是要测试的网站，在 Demo App 中会使用 H5PayDemoActivity 内的 WebView 打开。
         *
         * 可以填写任一支持支付宝支付的网站（如淘宝或一号店），在网站中下订单并唤起支付宝；
         * 或者直接填写由支付宝文档提供的"网站 Demo"生成的订单地址                    （如
         * https://mclient.alipay.com/h5Continue.htm?h5_route_token=303ff0894cd4dccf591b089761dexxxx）进行测试。
         *
         * H5PayDemoActivity 中的 MyWebViewClient.shouldOverrideUrlLoading()实现了拦截 URL 唤起
         * 支付宝，可以参考它实现自定义的 URL 拦截逻辑。
         */
        String url = "http://m.taobao.com";
        extras.putString("url", url);
        intent.putExtras(extras);
        startActivity(intent);
    }
}
```

其中：

（a）**public static final String AppID = "2016101300676685";**：此处填写 9.2.1 小节获取的 APPID。

（b）**public static final String RSA2_PRIVATE = "MIIEv... WEmB0J4e9ywvg=";**：此处填写 9.2.1 小节生成的用户私钥。私钥很长，为节省篇幅，此处略去中间长串字符，只写出首尾，读者请用自己计算机上实际生成的私钥。

运行程序，点击"支付宝支付 DEMO"按钮，接下来的页面点击"支付宝账户登录"按钮，然后输入沙箱中虚拟的买家账号和密码，点击"下一步"按钮进入付款页面，确认即可，如图9.16所示。但这个过程是虚拟的，并不发生银行卡金额的实际转账。如果要使自己的App能执行实际的转账支付功能，必须向工商管理部门申请营业执照，登记企业注册上线后才能进行实际的支付操作。

图 9.16 支付功能测试

9.3 地图应用开发（高德地图开放平台）

在当今移动互联网时代，很多App都要用到地图相关的功能，例如，根据当前位置给你提供打车、最佳路线导航、搜索最近的加油站等。目前，主流的地图服务提供商有高德、百度和谷歌，本节以高德地图为例，介绍Android地图环境配置及应用开发的入门知识。

9.3.1 配置地图环境

高德地图的全部功能都是基于其官方构建的"高德开放平台"实现的，要使用该平台，首先必须获取它的一套地图SDK，将其加载到Android Studio工程中，才能在自己的App中开发地图应用功能。

1. 获取高德地图SDK

（1）下载SDK。访问高德开放平台的SDK下载页 http://lbs.amap.com/api/android-sdk/download/，如图9.17所示，单击"Android 地图SDK 一键下载"区的"下载"按钮。

图 9.17 高德地图SDK下载页

下载得到名为"AMap_Android_SDK_All"的压缩包,解压看到其中又有 6 个压缩包,如图 9.18 所示。其中,AMap2DMap、AMap3DMap 和 AMapSearch 这 3 个包中的.jar 包是开发地图必不可少的类库,将它们解压提取出来存盘待用;而另外 3 个包是 SDK 自带的文档和范例,并非构建地图开发环境所必需,读者也可将其解压后用作开发时的学习参考资料。

图 9.18 SDK 中的压缩包

(2) 加载 SDK。创建 Android Studio 工程,名称为"MyGd",切换到 Project 模式,如图 9.19 所示,将上一步解压各压缩包提取到的.jar 包(一共 3 个)复制到工程 app\libs 目录下。再在 app\src\main 目录下新建一个名为"jniLibs"的文件夹,将 AMap3DMap 解压目录下的两个子文件夹(内含.so 库)复制进去。最后选中 app\libs 目录下所有.jar 包,在右键菜单中选择"add as library"命令,在弹出的对话框中单击"OK"按钮即可。

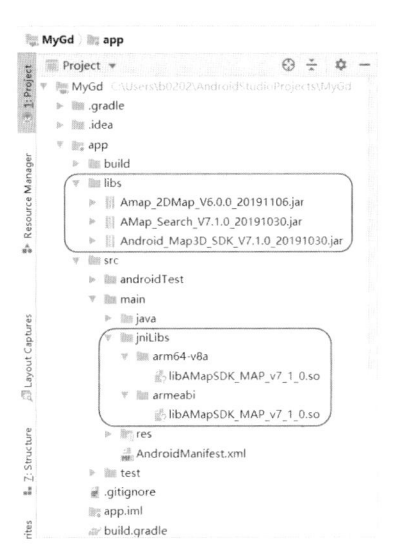

图 9.19 将 SDK 加载进工程

2. 创建和注册应用

在进行地图开发时,用户的 App 必须与地图服务提供商官网上注册的应用相对应,在获得 API 的授权访问 Key 后才能开发相应的功能。为此,读者必须首先在高德官网注册一个账号,登录后在其中创建自己的应用,而应用要获得授权还必须提供用户计算机端的本地安全码(SHA1 加密的形式)。

(1) 获取本地安全码。打开 Windows 命令行,输入"cd .android"命令并回车,进入本地计算机的 Android SDK 安装目录。再输入"keytool -v -list -keystore debug.keystore"命令并回车,出现"输入

密钥库口令"的提示,直接回车,在接下来屏幕显示的信息中就可以找到本地安全码(在"证书指纹"下)。笔者机器的安全码为"7D:3B:A9:44:89:29:06:6A:28:F3:FD:69:5E:1E:36:CC:53:AD:80:08",如图 9.20 所示。

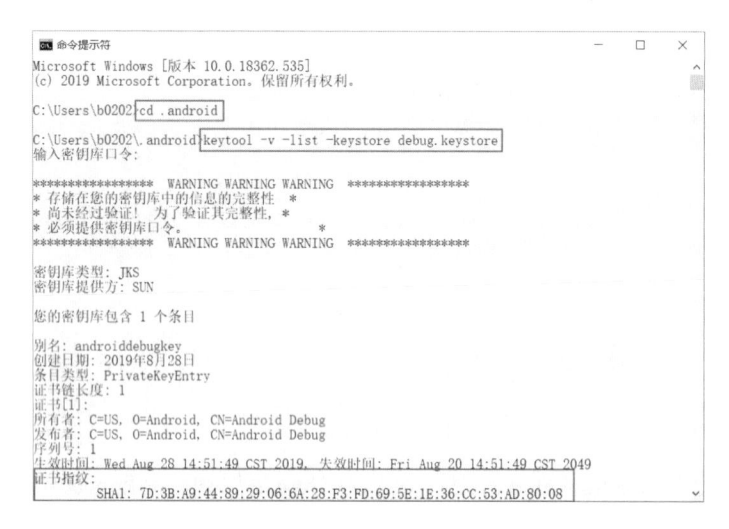

图 9.20　获取本地安全码

读者用同样的方法也可获得自己计算机的安全码,请务必将其复制记录下来以备后用。

> **注意:**
> 计算机的安全码是唯一的,如果读者在之前开发中已获取过自己机器的安全码并记录在案,之后从事地图开发就都使用这同一个码,无须再重复获取。

(2)创建应用与授权。访问高德开放平台官网 http://lbs.amap.com/,用自己的高德账号登录,单击"创建新应用"按钮,弹出"创建应用"对话框,如图 9.21 所示。填写"应用名称",就是之前创建的工程名 MyGd,根据所开发 App 的实际用途选择应用类型,这里选"导航",单击"创建"按钮。

图 9.21　新建一个高德应用

在"我的应用"页面可看到新创建的应用,如图 9.22 所示。单击"添加新 Key",弹出如图 9.23 所示的窗口。

图 9.22　新创建的应用

图 9.23　获取地图 API 的授权 Key

其中：
- **Key 名称**：填写 MyGdKey。
- **服务平台**：选 "Android 平台"。
- **发布版安全码 SHA1** 和**调试版安全码 SHA1**：均填写刚刚获取的本地安全码。
- **PackageName**：就是工程 MyGd 自身的包名，可在当前工程 AndroidManifest.xml 文件中查看，如图 9.24 所示，这里是 "com.easybooks.mygd"。

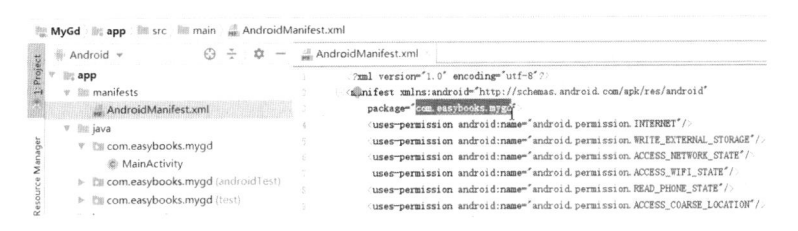

图 9.24　查看工程的包名

填写完以上各项信息后，勾选 "阅读并同意" 复选框，单击 "提交" 按钮，系统自动生成该应用的授权访问 Key，如图 9.25 所示。

图 9.25　生成应用的授权访问 Key

可以看到，本应用的授权访问 Key 值为 "b100754c1fd2cef881371ee370b842ec"（读者请以自己实际操作生成的 Key 值为准），可用记事本记录下来以备后用。

至此，高德地图开发所需的平台准备工作全部完成。

3. 显示高德地图

下面通过一个 Demo 程序来演示运行显示地图的效果。

【例 9.4】 开发一个最简单的地图 App，显示南京师范大学仙林校区的地图。

（1）配置工程环境。

① 打开工程 MyGd，在 Android Studio 主菜单栏选择"File"→"Project Structure"命令，弹出如图 9.26 所示的窗口，在其中配置工程的 keystore。

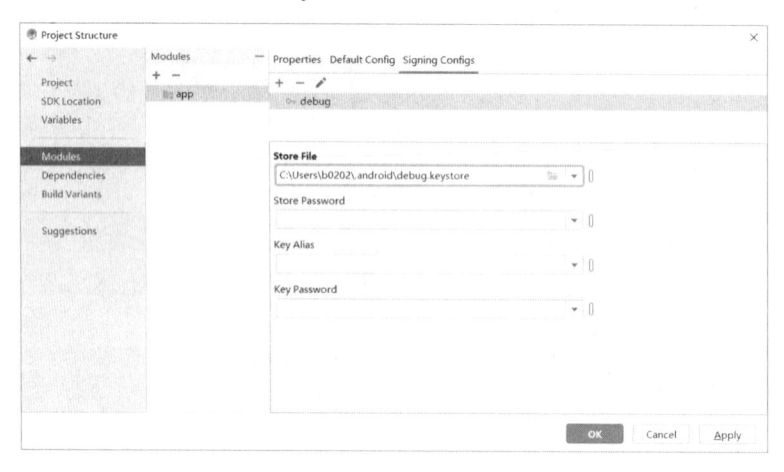

图 9.26 配置工程的 keystore

② 在 AndroidManifest.xml 中添加开发密钥、所需权限等信息（加粗处为需要添加配置的内容），代码如下：

```
<manifest xmlns:android="http://schemas.android.com/apk/res/android"
    package="com.easybooks.mygd">
    <uses-permission android:name="android.permission.INTERNET"/>
    <uses-permission android:name="android.permission.WRITE_EXTERNAL_STORAGE"/>
    <uses-permission android:name="android.permission.ACCESS_NETWORK_STATE"/>
    <uses-permission android:name="android.permission.ACCESS_WIFI_STATE"/>
    <uses-permission android:name="android.permission.READ_PHONE_STATE"/>
    <uses-permission android:name="android.permission.ACCESS_COARSE_LOCATION"/>
    <application
        android:allowBackup="true"
        android:icon="@mipmap/ic_launcher"
        android:label="@string/app_name"
        android:roundIcon="@mipmap/ic_launcher_round"
        android:supportsRtl="true"
        android:theme="@style/AppTheme">
        <meta-data android:name="com.amap.api.v2.apikey"
            android:value="b100754c1fd2cef881371ee370b842ec"/>      //应用的授权访问 Key
        <activity
            ...
        </activity>
    </application>
</manifest>
```

③ 打开工程 gradle.properties 文件，在末尾添加一行配置代码（如图 9.27 所示）：
android.injected.testOnly=false

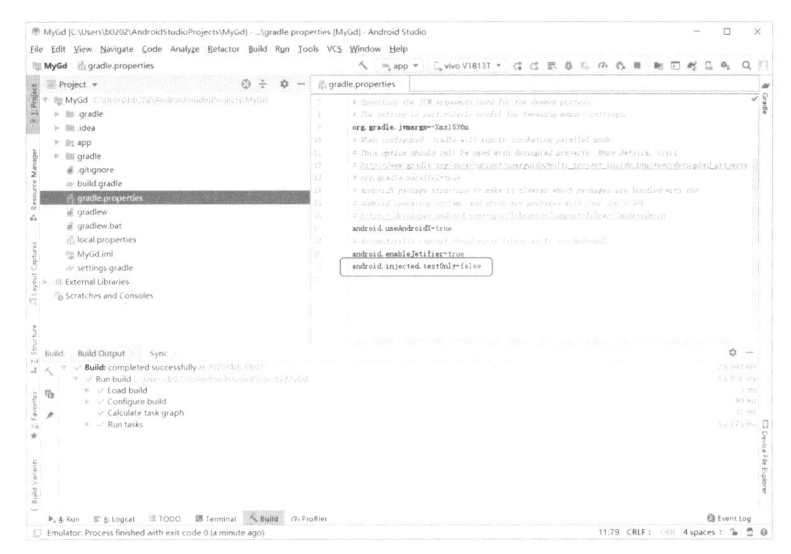

图 9.27 配置工程的 gradle.properties 文件

（2）编写代码显示地图。

经过上述一系列的操作后，终于可以在自己开发的 App 中显示高德地图了！下面实现最简单的地图显示功能。

① 引入地图控件。在工程的主显示页 content_main.xml 上放置一个 MapView（地图控件），代码如下：

```xml
<?xml version="1.0" encoding="utf-8"?>
<androidx.constraintlayout.widget.ConstraintLayout xmlns:android="http://schemas.android.com/apk/res/android"
    ...>
    <com.amap.api.maps.MapView
        android:id="@+id/myMapView"
        android:layout_width="fill_parent"
        android:layout_height="fill_parent"
        app:layout_constraintBottom_toBottomOf="parent"
        app:layout_constraintEnd_toEndOf="parent"
        app:layout_constraintStart_toStartOf="parent"
        app:layout_constraintTop_toTopOf="parent" />
</androidx.constraintlayout.widget.ConstraintLayout>
```

> **注意：**
> 由于不同地图服务提供商的地图控件互不兼容，故这里**不能**用谷歌提供的"com.google.android.gms.maps.MapView"，而**一定要**用 Android Studio 内置的"com.amap.api.maps.MapView"，否则无法显示高德地图。

② 创建地图。在 MainActivity 中创建地图并管理地图的生命周期，代码如下：

```java
......
import com.amap.api.maps.AMap;
import com.amap.api.maps.CameraUpdateFactory;
public class MainActivity extends AppCompatActivity {
    private MapView myMap = null;                    //地图控件的引用对象
    private AMap myAMap = null;                      //高德地图对象
    @Override
```

```java
protected void onCreate(Bundle savedInstanceState) {
    super.onCreate(savedInstanceState);
    setContentView(R.layout.activity_main);
    myMap = findViewById(R.id.myMapView);              //获取地图控件引用
    myMap.onCreate(savedInstanceState);                //此方法必须重写
    myAMap = myMap.getMap();                           //从地图控件获取地图对象
    myAMap.animateCamera(CameraUpdateFactory.newLatLngZoom(new  LatLng(32.10263, 118.912273),
15));                                                  //初始显示画面定位到南京师范大学
    ......
}

@Override
protected void onResume() {
    super.onResume();
    //在 Activity 执行 onResume()时执行 myMap.onResume()，实现地图生命周期管理
    myMap.onResume();
}

@Override
protected void onPause() {
    super.onPause();
    //在 Activity 执行 onPause()时执行 myMap.onPause()，实现地图生命周期管理
    myMap.onPause();
}

@Override
protected void onDestroy() {
    super.onDestroy();
    //在 Activity 执行 onDestroy()时执行 myMap.onDestroy()，实现地图生命周期管理
    myMap.onDestroy();
}
    ......
}
```

在程序初始化时，通过调用高德地图对象的 animateCamera()方法定位到南京师范大学仙林校区，其中参数对象 LatLng 封装了经纬度数据，我们这里人为地固定写入了南京师范大学仙林校区的经纬度值（32.10263，118.912273），这个值也可以通过地图的地理名检索功能获得，大家将在稍后进一步的开发中看到。

本小节我们先简单配置了高德地图应用的 App 框架，在此工程架构的基础上，再调用高德官方公开的 API 就可以实现更多丰富实用的地图功能。接下来我们将就几种常用的地图 App 功能开发展开详细介绍。

9.3.2　地图基本检索应用

高德地图基本检索应用包括按行政区检索、按地理名检索以及按经纬度检索，根据需要还可以设置地图的显示模式。

【例 9.5】　演示高德地图几种不同类型的检索方式及切换地图显示模式。

1. 设计界面

界面左上角采用 3 个单选按钮（RadioButton）组成单选按钮组，供用户选择检索方式；将一个复选框（CheckBox）定制为开关的外观，用户通过它可切换进入卫星地图模式；底部的文本视图

（TextView）则用于显示检索得到的信息详情。界面的设计视图如图 9.28 所示。

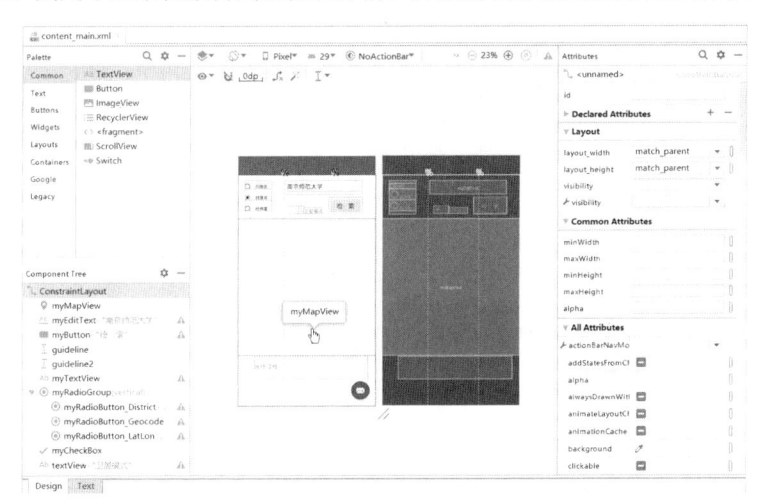

图 9.28　地图基本检索应用界面设计视图

本例界面上的单选按钮组及复选框开关的外观设计都运用了本书第 2 章所讲的状态列表图形技术，而整个界面的布局设计则用到了第 3 章所讲的约束布局来拖曳，具体操作请读者按照前面的指导自己去练习实践。

2. 功能实现

实现思路：

（1）行政区检索使用 DistrictSearch 类，用户检索的查询请求信息封装在 DistrictSearchQuery 类中，检索结果在 onDistrictSearched(DistrictResult districtResult)方法中处理和显示。

（2）地理名检索和经纬度检索本质上是根据地理编码和逆地理编码搜索特定地点的地理信息数据，为避免程序代码冗余，在本例中这两种检索方式共用同一个 GeocodeSearch 类。将查询请求信息封装在 GeocodeQuery 类中，先通过 GeocodeSearch 类的 getFromLocationNameAsyn(GeocodeQuery query)方法发起地理编码请求。

（3）在 onGeocodeSearched(GeocodeResult result, int rCode)方法中处理 GeocodeSearch 类地理编码请求的结果。若用户选择模式为地理名检索，则直接在地图上目的地点加标注并显示该点经纬度数据；若用户选择了经纬度检索模式，则还要以此地点的经纬度数据为参数，将其封装为 RegeocodeQuery 类中，通过 GeocodeSearch 类的 getFromLocationAsyn(RegeocodeQuery query)发起逆地理编码请求，然后再在 onRegeocodeSearched(RegeocodeResult result, int rCode)方法中处理和显示反向检索得到的详细中文地址信息。

（4）在"卫星模式"开关（复选框）的 OnCheckedChangeListener()监听器中，通过高德地图对象 Amap 的 setMapType()方法切换地图的显示模式。

MainActivity.java 代码如下：

```
package com.easybooks.mygd;
...
//高德地图通用库
import com.amap.api.maps.MapView;
import com.amap.api.maps.AMap;
import com.amap.api.maps.model.LatLng;
```

```java
import com.amap.api.maps.model.MarkerOptions;
import com.amap.api.maps.model.PolylineOptions;
...
//行政区检索用到的库
import com.amap.api.services.district.DistrictSearch;
import com.amap.api.services.district.DistrictSearchQuery;
import com.amap.api.services.district.DistrictSearch.OnDistrictSearchListener;
import com.amap.api.services.district.DistrictResult;
import com.amap.api.services.core.AMapException;
import com.amap.api.services.district.DistrictItem;
import com.amap.api.services.core.LatLonPoint;
import com.amap.api.maps.CameraUpdateFactory;
//地理名检索用到的库
import com.amap.api.services.geocoder.GeocodeSearch;
import com.amap.api.services.geocoder.GeocodeQuery;
import com.amap.api.services.geocoder.GeocodeSearch.OnGeocodeSearchListener;
import com.amap.api.services.geocoder.GeocodeResult;
import com.amap.api.services.geocoder.GeocodeAddress;
import com.amap.api.maps.model.Marker;
import com.amap.api.maps.model.BitmapDescriptorFactory;
//经纬度检索用到的库
import com.amap.api.services.geocoder.RegeocodeQuery;
import com.amap.api.services.geocoder.RegeocodeResult;

public class MainActivity extends AppCompatActivity implements OnDistrictSearchListener,OnGeocodeSearchListener {                                                     //（a）
    private MapView myMap = null;
    private AMap myAMap = null;
    private EditText myEditText;
    private Button myButton;
    private TextView myTextView;
    private CheckBox myCheckBox;
    private int myType;                          //检索类型（1：行政区；2：地理名；3：经纬度）
    private Marker myMarker;                     //地图上的标注（用于地理名和经纬度检索结果的标示）
    //行政区检索
    private DistrictSearch myDistrictSearch;     //区域检索对象
    //地理名检索
    private GeocodeSearch myGeocodeSearch;       //地理编码检索对象
    private String myAddrName;                   //地址名称
    //经纬度检索
    private LatLonPoint myPoint;                 //经纬度点对象

    @Override
    protected void onCreate(Bundle savedInstanceState) {
        super.onCreate(savedInstanceState);
        setContentView(R.layout.activity_main);
        myMap = findViewById(R.id.myMapView);            //获取地图控件引用
        myMap.onCreate(savedInstanceState);              //此方法必须重写
        myAMap = myMap.getMap();                         //从地图控件获取高德地图的引用对象
        myEditText = findViewById(R.id.myEditText);
        myButton = findViewById(R.id.myButton);
```

```java
        myTextView = findViewById(R.id.myTextView);
        myCheckBox = findViewById(R.id.myCheckBox);
        myCheckBox.setOnCheckedChangeListener(new CompoundButton.OnCheckedChangeListener() {
            @Override
            public void onCheckedChanged(CompoundButton compoundButton, boolean checked) {
                if (checked) myAMap.setMapType(AMap.MAP_TYPE_SATELLITE);      // (b)
                else myAMap.setMapType(AMap.MAP_TYPE_NORMAL);
            }
        });
        myType = 2;                                            //默认按地理名检索
        RadioGroup myRadioGroup = findViewById(R.id.myRadioGroup);
        myRadioGroup.setOnCheckedChangeListener(new RadioGroup.OnCheckedChangeListener() {
            @Override
            public void onCheckedChanged(RadioGroup radioGroup, int i) {
                if (i == R.id.myRadioButton_District) {
                    myType = 1;                                //行政区
                } else if (i == R.id.myRadioButton_Geocode) {
                    myType = 2;                                //地理名
                } else if (i == R.id.myRadioButton_LatLon) {
                    myType = 3;                                //经纬度
                }
            }
        });
        myMarker = myAMap.addMarker(new MarkerOptions().anchor(0.5f, 0.5f).icon(BitmapDescriptorFactory.defaultMarker(BitmapDescriptorFactory.HUE_ROSE)));      // (c)
        //行政区检索
        myDistrictSearch = new DistrictSearch(this);                                 // (d)
        myDistrictSearch.setOnDistrictSearchListener(this);
        //地理名和经纬度检索（两者共用同一个 GeocodeSearch 对象）
        myGeocodeSearch = new GeocodeSearch(this);                                   // (e)
        myGeocodeSearch.setOnGeocodeSearchListener(this);

        myButton.setOnClickListener(new View.OnClickListener() {
            @Override
            public void onClick(View view) {                   // "检索" 按钮的点击事件方法
                myTextView.setText("地理详情");
                if (myType == 1) {                             //行政区检索
                    myAMap.clear();
                    DistrictSearchQuery query = new DistrictSearchQuery();
                    query.setKeywords(myEditText.getText().toString());              // (f)
                    query.setShowBoundary(true);
                    myDistrictSearch.setQuery(query);
                    myDistrictSearch.searchDistrictAsyn();                           // (g)
                } else {                                       //地理名和经纬度检索
                    GeocodeQuery query = new GeocodeQuery(myEditText.getText().toString(), "南京");
                                                                                     // (h)
                    myGeocodeSearch.getFromLocationNameAsyn(query);
                                                   //发起异步地理编码请求
                }
            }
        });
```

```java
        ......
    }
    ......
            //行政区检索结果处理与显示
            @Override
            public void onDistrictSearched(DistrictResult districtResult) {                    // (i)
                if (districtResult == null || districtResult.getDistrict() == null) {
                    return;
                }
                if (districtResult.getAMapException() != null && districtResult.getAMapException().getErrorCode()
== AMapException.CODE_AMAP_SUCCESS) {                                                          // (j)
                    final DistrictItem item = districtResult.getDistrict().get(0);
                    if (item == null) {
                        return;
                    }
                    LatLonPoint centerLatLng = item.getCenter();           //获取结果区域中心点的经纬度
                    if (centerLatLng != null) {
                        myAMap.moveCamera(CameraUpdateFactory.newLatLngZoom(new LatLng(centerLatLng.get
Latitude(), centerLatLng.getLongitude()), 8));                             //地图画面移动至目标区域显示
                    }
                    new Thread() {                                                             // (k)
                        public void run() {
                            String[] polyStr = item.districtBoundary();    //得到区域边界点集（经纬度值字符串）
                            if (polyStr == null || polyStr.length == 0) {
                                return;
                            }
                            for (String str : polyStr) {
                                String[] lat = str.split(";");
                                PolylineOptions polylineOption = new PolylineOptions();
                                boolean isFirst = true;
                                LatLng firstLatLng = null;
                                for (String latstr : lat) {
                                    String[] lats = latstr.split(",");
                                    if (isFirst) {
                                        isFirst = false;
                                        firstLatLng = new LatLng(Double.parseDouble(lats[1]), Double.parseDou
ble(lats[0]));
                                    }
                                    polylineOption.add(new LatLng(Double.parseDouble(lats[1]), Double.parseDou
ble(lats[0])));
                                }
                                if (firstLatLng != null) {
                                    polylineOption.add(firstLatLng);
                                }
                                polylineOption.width(10).color(Color.MAGENTA);
                                myAMap.addPolyline(polylineOption);        //将结果区域边界线添加描绘在地图上
                            }
                        }
                    }.start();
                } else {
                    if (districtResult.getAMapException() != null)
```

```
            myTextView.setText("行政区检索发生错误，异常码为: " + districtResult.getAMapExcepti
on().getErrorCode());
        }
    }

    //地理名检索结果处理与显示、发起经纬度检索
    @Override
    public void onGeocodeSearched(GeocodeResult result, int rCode) {
        if (myType == 2) {                                              //地理名检索
            if (rCode == AMapException.CODE_AMAP_SUCCESS) {
                if (result != null && result.getGeocodeAddressList() != null && result.getGeocodeAddres
sList().size() > 0) {
                    GeocodeAddress address = result.getGeocodeAddressList().get(0);
                    myAMap.animateCamera(CameraUpdateFactory.newLatLngZoom(AMapUtil.convertToL
atLng(address.getLatLonPoint()), 15));                            //地图显示区切换至目标点所在区域
                    myMarker.setPosition(AMapUtil.convertToLatLng(address.getLatLonPoint()));
                                                                                    // (1)
                    myAddrName = " 纬度 " + address.getLatLonPoint().getLatitude() + ", 经度 " + ad
dress.getLatLonPoint().getLongitude() + "\n " + address.getFormatAddress();
                    myTextView.setText(myAddrName);
                } else {
                    myTextView.setText(R.string.no_result);
                }
            } else {
                myTextView.setText("地理名检索发生错误，异常码为: " + rCode);
            }
        } else if (myType == 3) {                                       //发起经纬度检索
            if (rCode == AMapException.CODE_AMAP_SUCCESS) {
                if (result != null && result.getGeocodeAddressList() != null && result.getGeocodeAddres
sList().size() > 0) {
                    GeocodeAddress address = result.getGeocodeAddressList().get(0);
                    //这里先获取经纬度值，继而以此为依据发起反向检索
                    myPoint = new LatLonPoint(Double.valueOf(address.getLatLonPoint().getLatitude()), D
ouble.valueOf(address.getLatLonPoint().getLongitude()));
                    RegeocodeQuery query = new RegeocodeQuery(myPoint, 200, GeocodeSearch.AMA
P);                                                                                  // (m)
                    myGeocodeSearch.getFromLocationAsyn(query);
                                                                    //发起逆地理编码请求
                } else {
                    myTextView.setText(R.string.no_result);
                }
            } else {
                myTextView.setText("获取经纬度值发生错误，异常码为: " + rCode);
            }
        }
    }

    //经纬度检索（反向检索）结果处理与显示
    @Override
    public void onRegeocodeSearched(RegeocodeResult result, int rCode) {
        if (rCode == AMapException.CODE_AMAP_SUCCESS) {
```

```
                    if (result != null && result.getRegeocodeAddress() != null && result.getRegeocodeAddress().g
etFormatAddress() != null) {
                        myAddrName = result.getRegeocodeAddress().getFormatAddress() + "附近";
                        myAMap.animateCamera(CameraUpdateFactory.newLatLngZoom(AMapUtil.convertToLatLn
g(myPoint), 15));
                        myMarker.setPosition(AMapUtil.convertToLatLng(myPoint));
                        myTextView.setText(myAddrName);
                    } else {
                        myTextView.setText(R.string.no_result);
                    }
                } else {
                    myTextView.setText("经纬度检索结果处理发生错误，异常码为: " + rCode);
                }
            }
        }
    }
```

其中：

（a）**public class MainActivity extends AppCompatActivity implements OnDistrictSearchListener, OnGeocodeSearchListener**：这里采用第 1 章所讲的范式四，用主程序类实现两个监听接口，其中 OnDistrictSearchListener 是处理区域检索请求的，用于行政区检索，通过"myDistrictSearch.setOnDistrictSearchListener(this);"绑定到 DistrictSearch 对象；OnGeocodeSearchListener 是处理地理编码请求的，通过"myGeocodeSearch.setOnGeocodeSearchListener(this);"绑定到 GeocodeSearch 对象。

（b）**if (checked) myAMap.setMapType(AMap.MAP_TYPE_SATELLITE);**：如果用户打开了"卫星模式"开关，则以卫星地图的形式显示地图。高德地图默认是 MAP_TYPE_NORMAL（普通模式），它同时还支持 MAP_TYPE_NIGHT（夜景模式）。

（c）**myMarker = myAMap.addMarker(new MarkerOptions().anchor(0.5f, 0.5f).icon(BitmapDescriptorFactory.defaultMarker(BitmapDescriptorFactory.HUE_ROSE)));**：addMarker()方法给高德地图上添加标注，其参数是一个实例化的 MarkerOptions 对象。anchor(float X, float Y)方法设置标注覆盖物的锚点比例；icon()方法用于设置标注覆盖物的图标，用户可以自定义用不同的图标来区分不同类型的地理位置；设置图标时要用工厂方法来实例化一个 BitmapDescriptor，为简单起见，这里我们使用高德地图默认类型的图标，颜色设定为 HUE_ROSE（玫瑰红）。

（d）**myDistrictSearch = new DistrictSearch(this);**：高德地图的行政区检索使用 DistrictSearch（区域检索类），需要事先声明，在使用前要先实例化。

（e）**myGeocodeSearch = new GeocodeSearch(this);**：高德地图的地理名和经纬度检索皆使用同一个 GeocodeSearch（地理编码检索类），也需要事先声明，在使用前也要先实例化。

（f）**query.setKeywords(myEditText.getText().toString());**：DistrictSearchQuery 类的 setKeywords() 方法接收用户输入的行政区名称，但目前只支持省（自治区/直辖市）、地级市、区县级别的行政区查询，**不支持**县以下乡镇的查询。

（g）**myDistrictSearch.searchDistrictAsyn();**：调用 DistrictSearch 对象的 searchDistrictAsyn()方法发起区域检索。

（h）**GeocodeQuery query = new GeocodeQuery(myEditText.getText().toString(), "南京");**：实例化 GeocodeQuery 对象来封装地理编码请求，第 1 个参数表示地理名称；第 2 个参数表示查询的城市（最多精确到地级市，不能是县或乡镇，也可指定行政区代码。

（i）**public void onDistrictSearched(DistrictResult districtResult) { ... }**：区域检索的结果在

DistrictResult 类型的引用参数中返回,可根据应用需要对其进行处理。

(j) **if (districtResult.getAMapException() != null && districtResult.getAMapException().getErrorCode() == AMapException.CODE_AMAP_SUCCESS) { ... }**:高德地图库将检索请求结果的成功与异常信息等同看待,都放在一个名为 AMapException 的类中统一处理,即把检索成功的返回内容也看作是一种特殊的异常,内部定义异常代码 CODE_AMAP_SUCCESS 表示成功无误。

(k) **new Thread() { ... }**:绘制区域边界的操作是耗时操作,要放在一个单独的线程中执行。

(l) **myMarker.setPosition(AMapUtil.convertToLatLng(address.getLatLonPoint()));**:将标注定位于地图上的检索结果点,这里用到了 AMapUtil 类,它是高德官方 Demo 中提供的一个通用工具类,通过它将结果地点的 LatLonPoint(经纬度点对象)转换为 LatLng 对象,以便传递进 setPosition() 进行设置。

(m) **RegeocodeQuery query = new RegeocodeQuery(myPoint, 200, GeocodeSearch.AMAP);**:实例化 RegeocodeQuery 对象来封装逆地理编码请求,第 1 个参数表示一个 LatLng 对象;第 2 个参数表示范围是多少米;第 3 个参数表示是火星坐标系还是 GPS 原生坐标系。

3. 运行效果

本程序可按照 4 种检索模式运行。

(1)运行程序,默认是地理名检索模式,输入"南京师范大学",点击"检索"按钮,程序定位到南京师范大学校园内的一个地点,并显示该点所对应的经纬度值,如图 9.29 所示(地图背景已做不可视处理)。

(2)打开"卫星模式"开关,可切换显示南京师范大学的卫星地图。

需要指出的是,初始时地图画面是以该点为中心显示的,读者可用手指触摸屏幕缩放画面到与书上图示一样的效果,并可查看南京师范大学校园里每处的细节。

(3)点选界面左上方"经纬度"单选按钮进入经纬度检索模式,点击"检索"按钮,地图仍然定位的是南京师范大学校园,但在底部文字区显示出该地具体的地址信息。

(4)点选界面左上方"行政区"单选按钮进入行政区检索模式,输入"栖霞区",点击"检索"按钮,地图上勾勒出南京市栖霞区的行政区域边界轮廓,如图 9.30 所示(地图背景已做不可视处理)。

图 9.29 地理名检索   图 9.30 行政区检索

9.3.3 GPS 定位和周边搜索

Poi 检索是地图应用最有用、最普及的功能。当人们来到一个陌生的城市，一般都急需一个旅行攻略，以便知道周边都有哪些美食店、商场、娱乐城、景点公园等。根据需求类型搜索地点，就是 Poi 的最大作用。Poi 还可以与其他检索类型（如公交路线检索）相结合，为用户提供更多的便利。本小节结合 Android 系统的 GPS 定位与高德地图的 Poi 检索功能来完成一个根据需要对用户当前位置周边搜索特定类型地点的应用。

【例 9.6】 运用 Poi 来找出用户当前位置周边方圆 3km 范围内的所有加油站。

1. 设计界面

界面很简单，如图 9.31 所示。

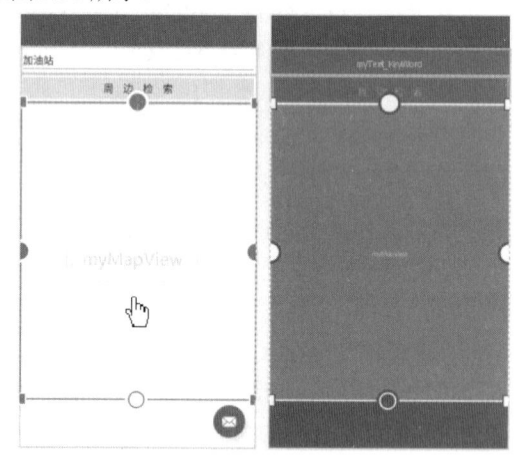

图 9.31 界面设计视图

在图 9.31 中，输入检索关键词的文本框使用了 AutoCompleteTextView 控件，第 2 章我们已经学过了，它是一个扩展功能的 TextView 控件，当用户在其中输入关键词时，能够自动弹出建议列表供用户选择匹配项。

2. 功能实现

实现思路：

（1）本程序需要开启和使用手机的 GPS 定位功能，onCreate()初始化时从系统服务中获取 LocationManager（定位管理器对象），创建 Criteria（定位条件器对象）并设置定位需求。然后，执行 locationRefresh()方法刷新用户所在位置信息。locationRefresh()方法是我们自定义的，在其中先通过 LocationManager 获取到最佳定位服务提供者，然后用 requestLocationUpdates()设置其定位监听器。定位监听器实例对象需要在外部创建，重写它的 onLocationChanged()方法获取位置信息。

（2）在 GPS 给出用户当前位置后，用户就可以输入关键词点击按钮启动 Poi 检索。以用户所在位置为圆心进行周边搜索。Poi 检索使用 PoiSearch 类与 PoiSearch.Query（Poi 查询条件类）配合完成，用 PoiSearch.Query 设置搜索条件，通过 PoiSearch 类的 searchPOIAsyn()执行搜索，在 onPoiSearched(PoiResult result, int rcode)方法中处理结果，将所有符合要求的地点在地图上标注出来。

MainActivity.java 代码如下：

```
package com.easybooks.mygd;
...
```

```java
//Poi 检索用到的库
import com.amap.api.maps.AMap;
import com.amap.api.services.poisearch.PoiSearch;
import com.amap.api.services.poisearch.PoiSearch.Query;              //Poi 查询条件类
import com.amap.api.services.poisearch.PoiResult;
import com.amap.api.services.core.PoiItem;
import com.amap.api.services.poisearch.PoiSearch.OnPoiSearchListener;
import com.amap.api.services.poisearch.PoiSearch.SearchBound;
import com.amap.api.services.core.LatLonPoint;
import com.amap.api.services.core.AMapException;
import com.amap.api.maps.model.Marker;
import com.amap.api.maps.model.BitmapDescriptorFactory;
import android.graphics.BitmapFactory;
import com.amap.api.maps.model.MarkerOptions;
import com.amap.api.maps.model.LatLng;
import com.amap.api.maps.model.CircleOptions;
//整合 GPS 用到的库
import android.location.LocationManager;
import android.location.Criteria;
import android.location.LocationListener;
import android.location.Location;
import android.os.Handler;
...
public class MainActivity extends AppCompatActivity implements OnPoiSearchListener {   //(a)
    private MapView myMap = null;
    private AMap myAMap = null;
    private AutoCompleteTextView myText_KeyWord;
    private Button myButton_Nearby;
    private PoiSearch myPoiSearch;                          //Poi 检索对象
    private Query query;                                    //Poi 查询条件类
    private LatLonPoint lpoint;                                             //(b)
    private int myRadius = 3000;                            //搜索半径
    private int currentPage = 0;                            //当前页面，从 0 开始计数
    private Marker myLastMarker;                            //之前被点击的标注
    private PoiOverlay myPoiOverlay;                        //Poi 图层
    //整合 GPS
    private LocationManager myLocationManager;              //声明定位管理器对象
    private Criteria myCriteria;                            //声明定位条件器对象
    private boolean enableLoc = false;                      //标识定位服务的可用性
    private double lat = 32.047601;                         //初始纬度
    private double lon = 118.794041;                        //初始经度

    @Override
    protected void onCreate(Bundle savedInstanceState) {
        super.onCreate(savedInstanceState);
        setContentView(R.layout.activity_main);
        myMap = findViewById(R.id.myMapView);               //获取地图控件引用
        myMap.onCreate(savedInstanceState);                 //此方法必须重写
        myAMap = myMap.getMap();
        myAMap.setOnMapClickListener(new AMap.OnMapClickListener() {
            @Override
            public void onMapClick(LatLng latLng) {
```

```java
            if (myPoiOverlay != null) {
                myPoiOverlay.removeFromMap();                    // (c)
            }
            resetLastMarker();
            doPoiSearch();                                       //重新搜索刷新图层
        }
    });
    myText_KeyWord = findViewById(R.id.myText_KeyWord);
    myButton_Nearby = findViewById(R.id.myButton_Nearby);
    myButton_Nearby.setOnClickListener(new View.OnClickListener() {
        @Override
        public void onClick(View view) {
            doPoiSearch();                                       //执行 Poi 检索
        }
    });
    //整合 GPS
    myLocationManager = (LocationManager) getSystemService(Context.LOCATION_SERVICE);
                                                                 //获取定位管理器对象
    myCriteria = new Criteria();                                 // (d)
    myCriteria.setAccuracy(Criteria.ACCURACY_FINE);              //设置定位精确度
    myCriteria.setBearingRequired(true);                         //设置是否需要方位信息
    locationRefresh();                                           //刷新位置
    myHandler.postDelayed(myRefresh, 100);                       //延迟 100 毫秒启动刷新
    ......
}
......
@Override
protected void onDestroy() {
    super.onDestroy();
    //在 activity 执行 onDestroy 时执行 myMap.onDestroy()，实现地图生命周期管理
    myMap.onDestroy();
    //整合 GPS
    if (myLocationManager != null) {                             //移除定位监听器
        myLocationManager.removeUpdates(myLocationListener);
    }
}
......
protected void doPoiSearch() {                                   // (e)
    currentPage = 0;
    query = new Query(myText_KeyWord.getText().toString(), "", "南京");  // (f)
    query.setPageSize(20);                                       //设置每页最多返回多少条 PoiItem
    query.setPageNum(currentPage);                               //设置查第 1 页
    myPoiSearch = new PoiSearch(this, query);                    //实例化 Poi 检索类
    myPoiSearch.setOnPoiSearchListener(this);
    lpoint = new LatLonPoint(lat, lon);                          //搜索中心点
    myPoiSearch.setBound(new SearchBound(lpoint, myRadius, true)); // (g)
    myPoiSearch.searchPOIAsyn();                                 //开始异步搜索
}

@Override
public void onPoiSearched(PoiResult result, int rcode) {         // (h)
    if (rcode == AMapException.CODE_AMAP_SUCCESS) {
```

```java
            if (result != null && result.getQuery() != null) {
                if (result.getQuery().equals(query)) {                    //是否为同一条
                    List<PoiItem> poiItems = result.getPois();                              // (i)
                    if (poiItems != null && poiItems.size() > 0) {
                        resetLastMarker();                         //还原上一次点击的标注样式
                        if (myPoiOverlay != null) {
                            myPoiOverlay.removeFromMap();          //清理之前搜索结果的标注
                        }
                        myAMap.clear();
                        myPoiOverlay = new PoiOverlay(myAMap, poiItems);
                        myPoiOverlay.addToMap();                   //图层添加覆盖到地图上
                        myPoiOverlay.zoomToSpan();                 //缩放到搜索区域显示
                        myAMap.addMarker(new MarkerOptions().anchor(0.5f, 0.5f).icon(BitmapDescriptorFactory.fromBitmap(BitmapFactory.decodeResource(getResources(), R.drawable.point_center))).position(new LatLng(lpoint.getLatitude(), lpoint.getLongitude())));                    //添加标注
                        myAMap.addCircle(new CircleOptions().center(new LatLng(lpoint.getLatitude(), lpoint.getLongitude())).radius(myRadius).strokeColor(Color.BLUE).fillColor(Color.argb(50, 1, 1, 1)).strokeWidth(2));
                                                                    //添加圆形覆盖区图层
                    } else {
                        ToastUtil.show(this.getApplicationContext(), R.string.no_result);    // (j)
                    }
                }
            } else {
                ToastUtil.show(this.getApplicationContext(), R.string.no_result);
            }
        } else {
            ToastUtil.showerror(this.getApplicationContext(), rcode);
        }
    }

    @Override
    public void onPoiItemSearched(PoiItem poiitem, int rcode) {
    }

    private void resetLastMarker() {                         //将之前被点击过的标注置为原来状态
        if (myLastMarker != null) {
            myLastMarker.setIcon(BitmapDescriptorFactory.fromBitmap(BitmapFactory.decodeResource(getResources(), R.drawable.marker_highlight)));
            myLastMarker = null;
        }
    }

    /** 整合 GPS */
    private void locationRefresh() {                              //自定义位置刷新方法
        String provider = myLocationManager.getBestProvider(myCriteria, true);      // (k)
        if (provider == null) {
            provider = LocationManager.NETWORK_PROVIDER;                             // (k)
        }
        if (myLocationManager.isProviderEnabled(provider)) {    //定位提供者可用
            if (ActivityCompat.checkSelfPermission(this, Manifest.permission.ACCESS_FINE_LOCATION) != PackageManager.PERMISSION_GRANTED) {                                             // (l)
                Toast.makeText(this, "定位功能未打开", Toast.LENGTH_SHORT).show();
```

```
                    return;
                }
                myLocationManager.requestLocationUpdates(provider, 200, 0, myLocationListener);
                                                                                        // (m)
                Location location = myLocationManager.getLastKnownLocation(provider);   // (n)
                if (location != null) {
                    lon = location.getLongitude();
                    lat = location.getLatitude();
                }
                enableLoc = true;
            } else {
                enableLoc = false;
            }
        }

        //创建定位监听器对象
        private LocationListener myLocationListener = new LocationListener() {         // (o)
            @Override
            public void onLocationChanged(Location loc) {                              // (o)
                lon = loc.getLongitude();                       //更新经度
                lat = loc.getLatitude();                        //更新纬度
            }

            @Override
            public void onProviderDisabled(String arg0) {                              // (o)
            }

            @Override
            public void onProviderEnabled(String arg0) {                               // (o)
            }

            @Override
            public void onStatusChanged(String arg0, int arg1, Bundle arg2) {          // (o)
            }
        };

        private Handler myHandler = new Handler();
        private Runnable myRefresh = new Runnable() {
            @Override
            public void run() {
                if (enableLoc == false) {    //若初次定位失败（服务不可用），每隔800毫秒重试一次
                    locationRefresh();
                    myHandler.postDelayed(this, 800);
                }
            }
        };
    }
```

其中：

（a）**public class MainActivity extends AppCompatActivity implements OnPoiSearchListener**：这里依然采用第1章的范式四，由主程序类实现监听接口 OnPoiSearchListener 来处理 Poi 检索请求，该接口必须同时提供实现 onPoiSearched(PoiResult result, int rcode)和 onPoiItemSearched(PoiItem poiitem, int rcode)两个方法，监听接口通过"myPoiSearch.setOnPoiSearchListener(this);"绑定到 PoiSearch 对象。

（b）**private LatLonPoint lpoint;**：这个点就是搜索中心点，最终要通过 GPS 定位得到。

（c）**myPoiOverlay.removeFromMap();**：本例程序对周边搜索区域的标注是建立在 Poi 图层的基础上的，由高德官方 Demo 提供的 PoiOverlay（Poi 图层类）负责管理和维护图层的显示状态，这里调用其 removeFromMap()方法清除上一次检索在图层上留下的旧标注。用户也可以在 PoiOverlay 类的基础上修改扩充，设计自己想要的图层外观。

（d）**myCriteria = new Criteria();**：这里创建的 Criteria（定位条件器）是用来设置（提出）定位需求的，包括定位所需精度以及是否需要提供海拔、方位信息等。GPS 系统将根据 Criteria 设定的需求为用户提供服务。

（e）**protected void doPoiSearch() { ... }**：本例自定义了这个 doPoiSearch()用于执行 Poi 检索操作。Poi 检索操作的过程代码必须封装在这个方法里，否则在实例化 PoiSearch 对象时会有 new PoiSearch(this, query)类型不兼容错误。

（f）**query=new Query(myText_KeyWord.getText().toString(), "", "南京");**：实例化 PoiSearch.Query（Poi 查询条件类）对象来封装 Poi 检索请求，第 1 个参数表示搜索关键词；第 2 个参数表示 Poi 搜索类型；第 3 个参数表示 Poi 搜索区域（空字符串代表全国）。

（g）**myPoiSearch.setBound(new SearchBound(lpoint, myRadius, true));**：设置搜索区域为以 lpoint 点为圆心，其周围半径 myRadius 米范围。

（h）**public void onPoiSearched(PoiResult result, int rcode) { ... }**：Poi 检索的结果集由 PoiResult 参数返回，判断检索是否成功的标准与前面介绍的几种检索完全一样。

（i）**List<PoiItem> poiItems = result.getPois();**：取得第 1 页的 PoiItem 数据，页数从 0 开始。

（j）**ToastUtil.show(this.getApplicationContext(), R.string.no_result);**：当搜索不到结果时，显示提示消息通知。这里用了高德官方 Demo 中的 ToastUtil 类，它可用于显示各种类型提示消息，是一个通用工具类。

（k）**String provider = myLocationManager.getBestProvider(myCriteria, true);** 和 **provider = LocationManager.NETWORK_PROVIDER;**：开始定位的时候，由 LocationManager（定位管理器）负责搜寻、获取附近最佳的定位服务提供者。定位服务提供者可以是全球卫星定位系统中的卫星，也可以是网络（附近通信基站或 WiFi 等）。本例优先使用卫星定位，若手机不巧正处在不容易收到卫星信号的地方（如密闭建筑物内），此时返回的 provider 为空，将转而使用网络定位服务（NETWORK_PROVIDER）。

（l）**if (ActivityCompat.checkSelfPermission(this, Manifest.permission.ACCESS_FINE_ LOCATION)!= PackageManager.PERMISSION_GRANTED)**：检查当前手机是否已开启定位功能（打开定位权限）。需要做两件事：

① 在工程 AndroidManifest.xml 中开放权限，代码如下：
<uses-permission android:name="android.permission.ACCESS_FINE_LOCATION" />

② 在手机上打开应用的定位权限，如图 9.32 所示。

> **注意：**
> 手机上该应用的定位权限**必须**同时打开，只在 AndroidManifest.xml 中开放权限是没用的。

（m）**myLocationManager.requestLocationUpdates(provider, 200, 0, myLocationListener);**：requestLocationUpdates()方法设置定位监听器。它有 4 个参数：第 1 个参数是之前由 LocationManager 获取的定位服务提供者；第 2 个参数为位置刷新时间间隔（此处取 200 毫秒）；第 3 个参数是位置刷新

的间距；第 4 个参数是所使用的定位监听器对象实例。

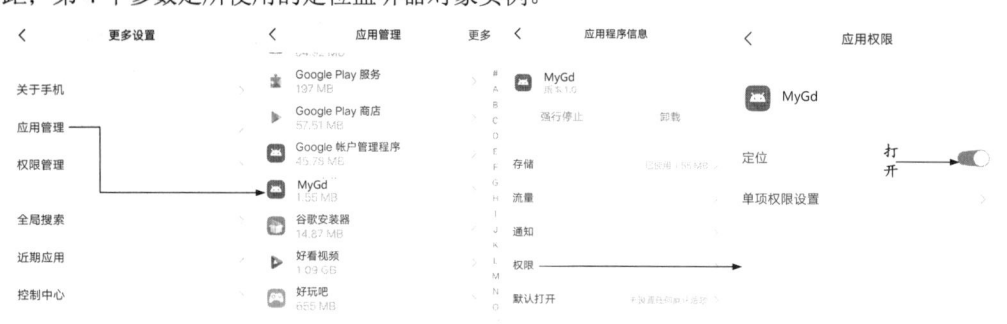

图 9.32　打开应用的定位权限

（n）**Location location = myLocationManager.getLastKnownLocation(provider);**：getLastKnownLocation()方法获取上一次成功定位的位置信息。

（o）**private LocationListener myLocationListener = new LocationListener()**：GPS 定位功能是依靠定位监听器（LocationListener）来更新用户位置信息的，为此，先要在程序中创建一个定位监听器对象实例，将其与定位管理器关联起来。程序运行时，定位监听器随时监听位置信息的变化事件从而实现位置刷新。使用定位监听器需要重写它的以下方法：

- **public void onLocationChanged(Location loc)**：用户位置发生变化时调用，在其中获取用户最新的位置信息进行处理。
- **public void onProviderDisabled(String arg0)**：当定位服务提供者被用户关闭（如关闭手机定位功能）时调用，执行一些善后处理工作。
- **public void onProviderEnabled(String arg0)**：当定位服务提供者被用户开启（如打开手机定位功能）时调用，执行一些初始化工作。
- **public void onStatusChanged(String arg0, int arg1, Bundle arg2)**：当定位服务提供者的状态发生变化（如用户走出某颗卫星的信号覆盖范围）时调用，执行变更处理，比如重新搜寻新的定位服务提供者。

本例为简单起见，只用了第一个 onLocationChanged()方法来简单地获取位置信息，其他 3 个方法虽然不用，但依 Java 规范也必须写出它们的完整方法体声明。在实际的 App 产品中，为了保证定位服务的可靠性，以上 4 个方法都要实现相应的逻辑处理。

3．运行效果

运行程序，启动初始地图就显示在用户当前所处的位置（通过 GPS 定位得到）。输入检索关键词"加油站"，点击"周边检索"按钮，程序在地图上绘出以用户所在点为中心的圆形区域，其中标识出所有的加油站所在位置。点击地图上的某个加油站标注，可查看其详细的地址信息，如图 9.33 所示（地图背景已做不可视处理）。

9.3.4　驾驶路径规划

在城市内驾车，驾驶员经常需要路径导航功能，高德地图就提供了这样的能力，能够自动检索并为驾驶员规划出一条到目的地的最佳路径。

【例 9.7】　根据输入的目的地地址规划最佳驾驶路径。

1．设计界面

界面很简单，如图 9.34 所示。

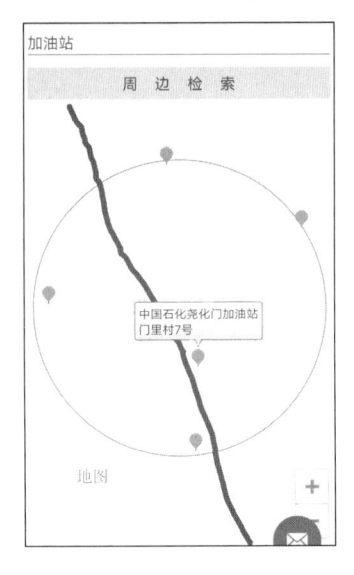

图 9.33　点击显示加油站详细地址

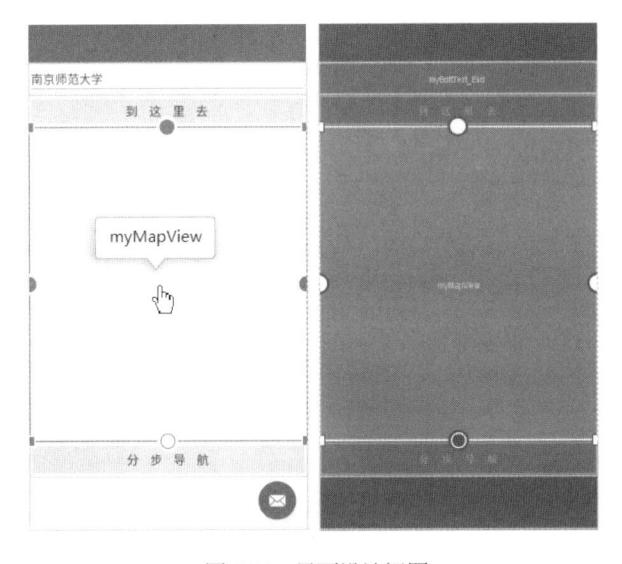

图 9.34　界面设计视图

在图 9.34 中，底部的"分步导航"按钮是供用户点击查看每一步分段路径的走法的。

2. 功能实现

实现思路：

（1）本例同样采用 GPS 定位驾驶员当前所在的位置作为路径起点。

（2）当用户输入目的地后点击"到这里去"按钮时，由程序发起地理名检索来确定目的地的具体方位。

（3）根据前两步确定的起终点，使用 RouteSearch 类实现路径规划，通过 DriveRouteQuery 设置搜索条件，使用 RouteSearch 的 calculateDriveRouteAsyn(DriveRouteQuery query)方法进行驾车规划路径计算，返回的结果在 onDriveRouteSearched(final DriveRouteResult result, int errorCode)方法中做进一步的处理。

MainActivity.java 代码如下：

```java
package com.easybooks.mygd;
...
//高德地图通用库                                                          // (a)
import com.amap.api.maps2d.model.Marker;
import com.amap.api.maps2d.MapView;
import com.amap.api.maps2d.AMap;
import com.amap.api.maps2d.model.LatLng;
...
//路径规划用到的库
import com.amap.api.services.route.RouteSearch.OnRouteSearchListener;
import com.amap.api.services.route.BusRouteResult;
import com.amap.api.services.route.DriveRouteResult;
import com.amap.api.services.route.WalkRouteResult;
import com.amap.api.services.route.RideRouteResult;
import com.amap.api.services.route.RouteSearch;
import com.amap.api.services.core.LatLonPoint;
import com.amap.api.services.route.RouteSearch.FromAndTo;
import com.amap.api.services.route.RouteSearch.DriveRouteQuery;
import com.amap.api.services.core.AMapException;
import com.amap.api.services.route.DrivePath;
import com.amap.api.services.route.DriveStep;
import com.amap.api.maps2d.CameraUpdateFactory;
import com.amap.api.maps2d.model.MarkerOptions;
import com.amap.api.maps2d.model.BitmapDescriptorFactory;
//整合 GPS 用到的库
import android.location.LocationManager;
import android.location.Criteria;
import android.location.LocationListener;
import android.location.Location;
import android.os.Handler;
import android.widget.Toast;
//整合地理名检索用到的库
import com.amap.api.services.geocoder.GeocodeSearch;
import com.amap.api.services.geocoder.GeocodeSearch.OnGeocodeSearchListener;
import com.amap.api.services.geocoder.GeocodeResult;
import com.amap.api.services.geocoder.RegeocodeResult;
import com.amap.api.services.geocoder.GeocodeAddress;
import com.amap.api.services.geocoder.GeocodeQuery;

public class MainActivity extends AppCompatActivity implements OnRouteSearchListener,OnGeocodeSearchListener {                                                                  // (b)
    private MapView myMap = null;
    private AMap myAMap = null;
    private EditText myEditText_End;
    private Button myButton_RoutePlan;
    private RouteSearch myRouteSearch;           //路径规划类
    private Context myContext;
    //整合 GPS
    private LocationManager myLocationManager;   //声明定位管理器对象
    private Criteria myCriteria;                 //声明定位条件器对象
    private boolean enableLoc = false;           //标识定位服务的可用性
```

```java
    private double lat = 39.942295;                              //默认初始纬度
    private double lon = 116.335891;                             //默认初始经度
    private LatLonPoint myStartPoint;                                         // (c)
    //整合地理名检索
    private GeocodeSearch myGeocodeSearch;                       //地理编码检索对象
    private LatLonPoint myEndPoint;                                           // (d)
    //点击翻看路径节点详细信息
    private Button myButton_PreNext;
    private DrivePath myDrivePath = null;                                     // (e)
    private int myNodeIndex = 0;                                 //当前节点索引
    private int myNodeCount;                                     //路径上的节点总数
    private MarkerOptions myMarkerOption;                        //节点标注信息显示的对象
    private Marker myMarker;

    @Override
    protected void onCreate(Bundle savedInstanceState) {
        super.onCreate(savedInstanceState);
        setContentView(R.layout.activity_main);
        myContext = this.getApplicationContext();
        myMap = findViewById(R.id.myMapView);                    //获取地图控件引用
        myMap.onCreate(savedInstanceState);                      //此方法必须重写
        myAMap = myMap.getMap();
        myEditText_End = findViewById(R.id.myEditText_End);
        myButton_RoutePlan = findViewById(R.id.myButton_RoutePlan);
        myRouteSearch = new RouteSearch(this);                                // (f)
        myRouteSearch.setRouteSearchListener(this);
        myButton_RoutePlan.setOnClickListener(new View.OnClickListener() {
            @Override
            public void onClick(View view) {
                //先根据目的地发起地理名检索
                GeocodeQuery query = new GeocodeQuery(myEditText_End.getText().toString(), "南京");
                myGeocodeSearch.getFromLocationNameAsyn(query);
                                                                 //发起地理编码请求
            }
        });
        //整合 GPS
        myLocationManager = (LocationManager) getSystemService(Context.LOCATION_SERVICE);
                                                                 //获取定位管理器对象
        myCriteria = new Criteria();                             //创建定位条件器
        myCriteria.setAccuracy(Criteria.ACCURACY_FINE);          //设置定位精确度
        myCriteria.setBearingRequired(true);                     //设置是否需要方位信息
        locationRefresh();                                       //刷新位置
        myHandler.postDelayed(myRefresh, 100);                   //延迟 100 毫秒启动刷新
        //整合地理名检索的 GeocodeSearch 对象
        myGeocodeSearch = new GeocodeSearch(this);
        myGeocodeSearch.setOnGeocodeSearchListener(this);
        //点击翻看路径节点详细信息
        myButton_PreNext = findViewById(R.id.myButton_PreNext);
        myButton_PreNext.setOnClickListener(new View.OnClickListener() {
            @Override
            public void onClick(View view) {
```

```java
                DriveStep step = myDrivePath.getSteps().get(myNodeIndex);
                LatLng nodeLocation = new LatLng(step.getPolyline().get(0).getLatitude(), step.getPolyline().get(0).getLongitude());
                                                                //获取节点的经纬度数据
                myAMap.animateCamera(CameraUpdateFactory.newLatLngZoom(nodeLocation, 15));
                                                                //地图显示区切换至节点附近区域
                String nodeTitle = step.getInstruction();       //获取在该节点上驾驶的指导信息
                myMarkerOption = new MarkerOptions().icon(BitmapDescriptorFactory.defaultMarker(BitmapDescriptorFactory.HUE_AZURE)).position(nodeLocation).title("第" + String.valueOf(myNodeIndex + 1) + "步").snippet(nodeTitle).draggable(true);
                myMarker = myAMap.addMarker(myMarkerOption);
                                                                //将指导信息作为标注覆盖物添加到地图上
                myMarker.showInfoWindow();
                if (myNodeIndex < myNodeCount - 1) myNodeIndex++;
                else myNodeIndex = 0;
            }
        });
        ......
    }
     ......
    @Override
    protected void onDestroy() {
        super.onDestroy();
        //在 activity 执行 onDestroy 时执行 myMap.onDestroy()，实现地图生命周期管理
        myMap.onDestroy();
        //整合 GPS
        if (myLocationManager != null) {                        //移除定位监听器
            myLocationManager.removeUpdates(myLocationListener);
        }
    }
    ......
    /**
     * 因为 MainActivity 声明了 implements OnRouteSearchListener，所以下面这 4 个都是必须实现的方法
     */
    @Override
    public void onBusRouteSearched(BusRouteResult result, int errorCode) {
    }

    @Override
    public void onWalkRouteSearched(WalkRouteResult result, int errorCode) {
    }

    @Override
    public void onRideRouteSearched(RideRouteResult result, int errorCode) {
    }

    @Override
    public void onDriveRouteSearched(final DriveRouteResult result, int errorCode) {
        myAMap.clear();                                         //清理地图上的所有覆盖物
        if (errorCode == AMapException.CODE_AMAP_SUCCESS) {
            if (result != null && result.getPaths() != null) {
                if (result.getPaths().size() > 0) {
```

```java
                myDrivePath = result.getPaths().get(0);                          // (g)
                myNodeCount = myDrivePath.getSteps().size();
                DrivingRouteOverLay drivingRouteOverlay = new DrivingRouteOverLay(this, myAMap,
myDrivePath, result.getStartPos(), result.getTargetPos(), null);
                drivingRouteOverlay.setNodeIconVisibility(false);
                                                                //设置节点标注是否显示
                drivingRouteOverlay.setIsColorfulline(true);
                                                //是否用颜色显示交通拥堵情况，默认为true
                drivingRouteOverlay.removeFromMap();
                drivingRouteOverlay.addToMap();         //添加到地图上
                drivingRouteOverlay.zoomToSpan();
            } else if (result != null && result.getPaths() == null) {
                ToastUtil.show(this, R.string.no_result);
            }
        } else {
            ToastUtil.show(this, R.string.no_result);
        }
    } else {
        ToastUtil.showerror(this.getApplicationContext(), errorCode);
    }
}

/** 整合GPS */
private void locationRefresh() {                            //自定义位置刷新方法
    String provider = myLocationManager.getBestProvider(myCriteria, true);
    if (provider == null) {
        provider = LocationManager.NETWORK_PROVIDER;
    }
    if (myLocationManager.isProviderEnabled(provider)) {    //定位提供者可用
        if (ActivityCompat.checkSelfPermission(this, Manifest.permission.ACCESS_FINE_LOCATION) !=
PackageManager.PERMISSION_GRANTED) {
            Toast.makeText(this, "定位功能未打开", Toast.LENGTH_SHORT).show();
            return;
        }
        myLocationManager.requestLocationUpdates(provider, 200, 0, myLocationListener);
        Location location = myLocationManager.getLastKnownLocation(provider);
        if (location != null) {
            lon = location.getLongitude();
            lat = location.getLatitude();
        }
        enableLoc = true;
    } else {
        enableLoc = false;
    }
}

//创建定位监听器对象
private LocationListener myLocationListener = new LocationListener() {
    @Override
    public void onLocationChanged(Location loc) {
        lon = loc.getLongitude();                               //更新经度
```

```java
                    lat = loc.getLatitude();                              //更新纬度
                }

                @Override
                public void onProviderDisabled(String arg0) {
                }

                @Override
                public void onProviderEnabled(String arg0) {
                }

                @Override
                public void onStatusChanged(String arg0, int arg1, Bundle arg2) {
                }
    };

    private Handler myHandler = new Handler();
    private Runnable myRefresh = new Runnable() {
        @Override
        public void run() {
            if (enableLoc == false) {          //若初次定位失败（服务不可用），每隔800毫秒重试一次
                locationRefresh();
                myHandler.postDelayed(this, 800);
            }
        }
    };

    /**
     * 整合地理名检索结果处理
     */
    @Override
    public void onGeocodeSearched(GeocodeResult result, int rCode) {
        //处理结果得到目的地的经纬度值，并由此生成路径规划终点的LatLonPoint对象
        if (rCode == AMapException.CODE_AMAP_SUCCESS) {
            if (result != null && result.getGeocodeAddressList() != null && result.getGeocodeAddressList().size() > 0) {
                GeocodeAddress address = result.getGeocodeAddressList().get(0);
                myEndPoint = new LatLonPoint(Double.valueOf(address.getLatLonPoint().getLatitude()), Double.valueOf(address.getLatLonPoint().getLongitude()));
                myStartPoint = new LatLonPoint(lat, lon);
                final FromAndTo fromAndTo = new FromAndTo(myStartPoint, myEndPoint);
                DriveRouteQuery query = new DriveRouteQuery(fromAndTo, RouteSearch.DrivingDefault, null, null, "");          // (h)
                myRouteSearch.calculateDriveRouteAsyn(query);     //计算规划驾车路径
            } else {
                ToastUtil.show(this, R.string.no_result);
            }
        } else {
            ToastUtil.show(this, "获取目的地经纬度值发生错误，异常码为：" + rCode);
        }
    }
```

```
@Override
public void onRegeocodeSearched(RegeocodeResult result, int rCode) {
}
}
```

其中：

（a）**import com.amap.api.maps2d...**：这个是高德官方发布的 2D 地图库，与前面的"com.amap.api.maps"库在功能上基本一样。但要注意，如果主程序使用了 2D 库，界面地图控件的包名也要改成对应一致的"com.amap.api.maps**2d**.MapView"。

（b）**public class MainActivity extends AppCompatActivity implements OnRouteSearchListener, OnGeocodeSearchListener**：主程序类实现监听接口 OnRouteSearchListener 来进行驾驶路径规划计算，该接口必须同时提供实现 onBusRouteSearched(BusRouteResult result, int errorCode)（公交路径规划）、onWalkRouteSearched(WalkRouteResult result, int errorCode)（步行路径规划）、onRideRouteSearched(RideRouteResult result, int errorCode)（骑行路径规划）和 onDriveRouteSearched(final DriveRouteResult result, int errorCode)（驾驶路径规划）4 个方法，本例实际使用的只有最后一个方法，但其他 3 个方法也要写出方法体，高德地图 API 设计这样的接口规范是为了使系统具有很强的功能扩展性，便于扩充二次开发。监听接口通过"myRouteSearch.setRouteSearchListener(this);"绑定到 RouteSearch 对象。OnGeocodeSearchListener 接口则是用来进行地理名检索处理的，用法见 9.3.2 小节。

（c）**private LatLonPoint myStartPoint;**：这个点就是起点，通过 GPS 得到用户当前位置作为路径规划的起点。

（d）**private LatLonPoint myEndPoint;**：这个点是终点，通过用户输入目的地名在程序内部按地理名检索得到。

（e）**private DrivePath myDrivePath = null;**：DrivePath 对象用于存储所规划路径的信息。

（f）**myRouteSearch = new RouteSearch(this);**：驾车路径规划检索使用 RouteSearch 对象，要声明和实例化。

（g）**myDrivePath = result.getPaths().get(0);**：检索到符合条件的驾车路径，通过 DriveRouteResult 类的 getPaths()方法获取。

（h）**DriveRouteQuery query = new DriveRouteQuery(fromAndTo, RouteSearch.DrivingDefault, null, null, "");**：DriveRouteQuery 设置搜索条件，第 1 个参数表示路径规划的起点和终点；第 2 个参数表示驾车模式；第 3 个参数表示途经点；第 4 个参数表示避让区域；第 5 个参数表示避让道路。

3．运行效果

运行程序，输入"南京师范大学"，点击"到这里去"按钮，程序自动规划出最佳驾驶路径并在地图上以绿色粗线标注出来，如图 9.35 所示（地图背景已做不可视处理）。点击地图底部的"分步导航"按钮，可逐步定位并查看路径上每一个节点处的驾驶指导信息。

9.3.5 百度地图应用开发

目前，比较流行的地图服务除了高德地图外，百度地图服务也占有相当的市场比重。本书第 1 版对百度地图服务进行了详细介绍，为了方便对百度地图服务感兴趣的读者，这里以二维码链接形式列出百度地图服务文档，仅供参考。

百度地图文档

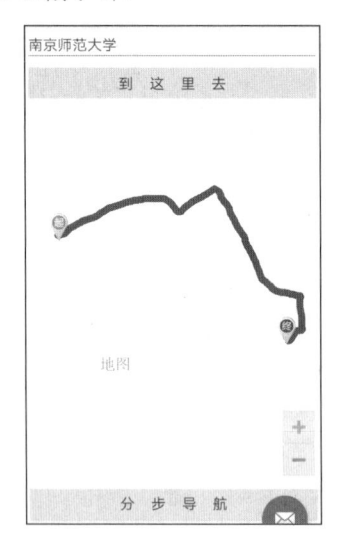

图 9.35 在地图上标注路径

9.4 Android 设备操作

Android 操作系统可以对当前智能手机上配备的几乎所有硬件设备如计步器、加速度传感器、GPS、蓝牙、手电筒等进行控制和操作，下面分别举例说明。

9.4.1 计步器

现代社会人们都崇尚健身，很多 App 都提供对运动指标的测量功能，如步多多、走步宝、猫扑运动等都是通过统计用户每日运动行走的步数向人们发放红包，以此来鼓励全民健身的健康生活方式。计步器已经成为所有手机的标配，它的原理是：通过传感器检测用户走步时带动手机的摆动次数来计量步数值。

【例 9.8】 用手机内置的计步器实现一个简单的步数统计功能。

创建 Android Studio 工程 MyStep，界面上仅放置一个文本视图用于显示步数。

实现思路：

（1）onCreate()初始化时，程序从系统服务中获取 SensorManager（传感器管理器对象），SensorManager 通过注册监听器与 TYPE_STEP_COUNTER 类型的传感器（即计步器）发生关联。

（2）主程序类 MainActivity 实现 SensorEventListener 接口，通过其中的 onSensorChanged()方法得到步数信息。

在 MainActivity.java 中调用计步器实现步数统计，代码如下：

```
public class MainActivity extends AppCompatActivity implements SensorEventListener {         // (a)
    private TextView myTextView;                                    //文本视图（显示步数）
    private SensorManager mySensorManager;                          //声明传感器管理器对象
    private int count = 0;                                          //记录步数
    @Override
    protected void onCreate(Bundle savedInstanceState) {
        super.onCreate(savedInstanceState);
        setContentView(R.layout.activity_main);
        myTextView = findViewById(R.id.myTextView);
        mySensorManager = (SensorManager) getSystemService(Context.SENSOR_SERVICE);           // (b)
```

```
            ...
        }
        ......
        @Override
        protected void onResume() {
            super.onResume();
            mySensorManager.registerListener(this,
mySensorManager.getDefaultSensor(Sensor.TYPE_STEP_COUNTER), SensorManager.SENSOR_DELAY_NORMAL);
                                                                            // (c)
        }

        @Override
        protected void onPause() {
            super.onPause();
            mySensorManager.unregisterListener(this);              //注销传感器监听器
        }

        @Override
        public void onSensorChanged(SensorEvent event) {           // (a)
            if (event.sensor.getType() == Sensor.TYPE_STEP_COUNTER) count = (int) event.values[0];
                                                                            // (d)
            myTextView.setText(String.format("您今天一共走了 %d 步", count));
        }

        @Override
        public void onAccuracyChanged(Sensor sensor, int accuracy) {    // (a)
        }
    }
```

其中：

（a）**public class MainActivity extends AppCompatActivity implements SensorEventListener { ...**、**public void onSensorChanged(SensorEvent event) { ... }** 和 **public void onAccuracyChanged(Sensor sensor, int accuracy) { ... }**：要使 App 能够读取计步器传来的步数信息，必须继承传感器事件监听器 SensorEventListener 并实现其中的 onSensorChanged() 和 onAccuracyChanged() 方法。前一个方法在检测到计步器步数变化时触发，得到实时的步数信息；而后一个方法在传感器的精度改变时触发，一般不用做任何操作，但必须写出完整的方法体声明。

（b）**mySensorManager = (SensorManager) getSystemService(Context.SENSOR_SERVICE);**：程序是通过 getSystemService() 方法从系统服务中获取传感器管理器对象的。

（c）**mySensorManager.registerListener(this, mySensorManager.getDefaultSensor(Sensor.TYPE_STEP_COUNTER), SensorManager.SENSOR_DELAY_NORMAL);**：传感器管理器对象通过 registerListener() 方法注册监听器来与特定类型的传感器发生关联，这里的传感器类型 TYPE_STEP_COUNTER 也就是手机内置的计步器。在手机内还有另一个 TYPE_STEP_DETECTOR（步行检测器）传感器是用来检测用户当前走动步数的，而 TYPE_STEP_COUNTER（计步器）给出的是用户当日行走的总步数。

（d）**if (event.sensor.getType() == Sensor.TYPE_STEP_COUNTER) count = (int) event.values[0];**：程序通过传感器事件监听器从传感器事件 SensorEvent 中获取传感器的类型以及传来的数据信息，本例也就是得到步数值。

运行程序，显示出步数，如图 9.36 所示。读者可以把手机放在口袋里走一走，再拿出来看步数是否增加了。

9.4.2 摇一摇

手机摇一摇也是今天很多 App 流行的功能，常用于社交类软件发现附近好友、抽奖领红包等类型的应用。摇一摇功能的实现原理与计步器一样，也是通过传感器来检测手机的摇晃。

图 9.36 显示步数

【例 9.9】 设计一个模拟抽奖领红包功能，当用户摇晃手机或点击按钮时进入抽奖，手机发出震动并显示抽奖信息，界面效果如图 9.37 所示。

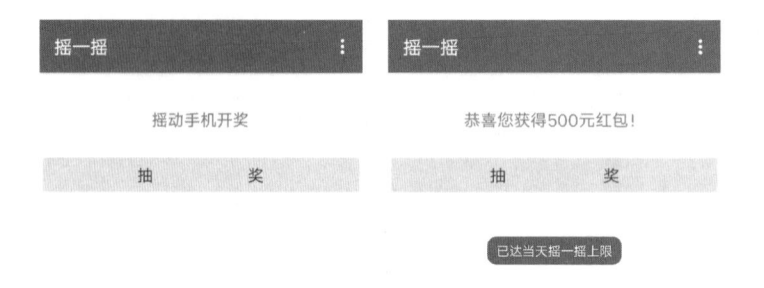

图 9.37 摇一摇运行效果

创建 Android Studio 工程 MyAcceleration，界面很简单，只有一个文本视图和一个按钮。

实现思路：

（1）从系统服务中获取 SensorManager（传感器管理器对象），SensorManager 通过注册监听器与 TYPE_ACCELEROMETER 类型的传感器（即加速度传感器）发生关联。

（2）从系统服务中获取 Vibrator（震动器）。

（3）主程序类 MainActivity 实现 SensorEventListener 接口，在 onSensorChanged()方法中通过一个 values 数组得到手机在三维空间各个方向上的摆动幅度数据。

（4）在 myButton（"抽奖"按钮）的点击事件监听器 onClick()方法中启动 Vibrator 震动手机，提示用户摇一摇上限。

在 MainActivity.java 中调用加速度传感器（TYPE_ACCELEROMETER）实现检测手机摇动，代码如下：

```java
public class MainActivity extends AppCompatActivity implements SensorEventListener {
    private TextView myTextView;                          //文本视图（显示中奖信息）
    private SensorManager mySensorManager;                //声明传感器管理器对象
    private Vibrator myVibrator;                          //（a）声明震动器对象
    private Button myButton;                              //"抽奖"按钮
    @Override
    protected void onCreate(Bundle savedInstanceState) {
        super.onCreate(savedInstanceState);
        setContentView(R.layout.activity_main);
        myTextView = findViewById(R.id.myTextView);
        myTextView.setText("摇动手机开奖");
        mySensorManager = (SensorManager) getSystemService(Context.SENSOR_SERVICE);
                                                          //获取传感器管理器对象
        myVibrator = (Vibrator) getSystemService(Context.VIBRATOR_SERVICE);
                                                          //（a）
```

```java
            myButton = findViewById(R.id.myButton);
            myButton.setOnClickListener(new View.OnClickListener() {
                @Override
                public void onClick(View view) {
                    myVibrator.vibrate(600);                                //震动手机
                    Display display = getWindowManager().getDefaultDisplay();
                    int height = display.getHeight();                       //获取屏幕高度
                    Toast toast = Toast.makeText(MainActivity.this, "已达当天摇一摇上限", Toast.LENGTH_LONG);
                    toast.setGravity(Gravity.TOP, 0, height / 3);           //（b）
                    toast.show();
                }
            });
            ...
        }
        ......
        @Override
        protected void onResume() {
            super.onResume();
            mySensorManager.registerListener(this, mySensorManager.getDefaultSensor(
                    Sensor.TYPE_ACCELEROMETER), SensorManager.SENSOR_DELAY_NORMAL);
                                                                            //注册加速度传感器监听器
        }

        @Override
        protected void onPause() {
            super.onPause();
            mySensorManager.unregisterListener(this);                       //注销传感器监听器
        }

        @Override
        public void onSensorChanged(SensorEvent event) {
            if (event.sensor.getType() == Sensor.TYPE_ACCELEROMETER) {
                float[] val = event.values;                                 //（c）
                if ((Math.abs(val[0]) > 10 || Math.abs(val[1]) > 10 || Math.abs(val[2]) > 10)) {
                    myTextView.setText("恭喜您获得 500 元红包！");
                    //系统检测到摇一摇事件后，震动手机
                    myVibrator.vibrate(600);
                }
            }
        }

        @Override
        public void onAccuracyChanged(Sensor sensor, int accuracy) {
        }
    }
```

其中：

（a）**private Vibrator myVibrator;** 和 **myVibrator = (Vibrator) getSystemService(Context.VIBRATOR_SERVICE);**：为了实现控制手机震动，这里使用了 Vibrator（震动器），它也是手机上普遍内置的标配设备，与计步器和加速度传感器一样，也是从系统服务中获取其对象，不同的是，它是对外做出可感知的物理动作而非向程序传递外部的环境信息。要使用震动器还必须在工程的 AndroidManifest.xml 中打开权限：

```
<uses-permission android:name="android.permission.VIBRATE" />
```

（b）**toast.setGravity(Gravity.TOP, 0, height / 3);**：由于本例界面简单，UI 元素很少，故需要将用户提示通知消息显示在屏幕上半部，这样看起来才紧凑。用 setGravity() 方法自定义设置系统通知消息的显示位置，方法参数中要传入显示位置的 X、Y 坐标，这里给了一个 1/3 屏幕高度的 Y 轴偏移量。

（c）**float[] val = event.values;**：从加速度传感器事件中获取到的数据信息存放在一个 values 数组中，表示手机摇动时在空间 X、Y、Z 三个坐标轴方向的幅度，其中 values[0] 对应 X 轴，values[1] 对应 Y 轴，values[2] 对应 Z 轴，程序判断只要在某一个方向上的摇动幅度超过了一定限度即为检测到一次"摇一摇"事件，并据此做出相应的处理和响应。

9.4.3 蓝牙设备发现

现在，很多智能手机上还配备了蓝牙功能，通过它可实现短距离内手机与手机或其他蓝牙设备之间的通信和数据传输。

【例 9.10】 编写程序让手机自动发现附近其他带有蓝牙功能的手机或设备。

创建 Android Studio 工程 MyBlueTooth，界面上放置一个文本视图，用于显示其他蓝牙手机或设备的名称信息。

手机蓝牙启用的过程通常为：

初始化蓝牙适配器→"蓝牙权限请求"对话框提示用户授权→打开蓝牙功能→搜索附近的设备

实现思路：

（1）onCreate() 初始化时从系统服务中获取 BluetoothManager（蓝牙管理器对象），然后根据 Android 版本的差别获取相应的蓝牙适配器。

（2）以 Intent 机制通过蓝牙适配器启动"蓝牙权限请求"对话框的 Activity，在其回调响应 onActivityResult() 方法中获得用户的选择结果。

（3）若用户授予权限，则调用适配器的 startDiscovery() 方法搜索附近的蓝牙设备。

（4）创建并注册一个 BroadcastReceiver（广播接收器），在接收器的 onReceive() 方法中解析并显示所发现蓝牙设备的名称。

在 MainActivity.java 中实现蓝牙发现功能，代码如下：

```java
public class MainActivity extends AppCompatActivity {
    private TextView myTextView;                                //文本视图（显示设备名称）
    private BluetoothAdapter myBlueAdapter;                     //声明蓝牙适配器对象
    private int myOpenCode = 1;                                 //"蓝牙权限请求"对话框返回码
    @Override
    protected void onCreate(Bundle savedInstanceState) {
        super.onCreate(savedInstanceState);
        setContentView(R.layout.activity_main);
        myTextView = findViewById(R.id.myTextView);
        if (Build.VERSION.SDK_INT >= Build.VERSION_CODES.JELLY_BEAN_MR2) {
                                                                //（a）
            BluetoothManager blueManager = (BluetoothManager) getSystemService(Context.BLUETOOTH_SERVICE);
                                                                //获取蓝牙管理器对象
            myBlueAdapter = blueManager.getAdapter();
        } else {
            myBlueAdapter = BluetoothAdapter.getDefaultAdapter();
```

```java
        }
        if (myBlueAdapter == null) {
            Toast.makeText(this, "本机不支持蓝牙功能！", Toast.LENGTH_SHORT).show();
            finish();
        }
        Intent intent = new Intent(BluetoothAdapter.ACTION_REQUEST_DISCOVERABLE);
        startActivityForResult(intent, myOpenCode);                    // (b)
    }

    protected void onActivityResult(int requestCode, int resultCode, Intent intent) {
        super.onActivityResult(requestCode, resultCode, intent);
        if (requestCode == myOpenCode) {                               // (b)
            myHandler.postDelayed(myRefresh, 100);                     //延迟100毫秒启动蓝牙设备刷新
            if (resultCode == RESULT_OK) {
                myTextView.setText("允许本机被附近的其他蓝牙设备发现。");
            } else if (resultCode == RESULT_CANCELED) {
                myTextView.setText("不允许本机被附近的其他蓝牙设备发现！");
            } else {
                myTextView.setText("未知设备！");
                if (!myBlueAdapter.isDiscovering()) {
                    myBlueAdapter.startDiscovery();                    // (c)
                }
            }
        }
    }

    protected void onStart() {
        super.onStart();
        IntentFilter intentFilter = new IntentFilter();
        intentFilter.addAction(BluetoothDevice.ACTION_FOUND);          //添加过滤新动作
        registerReceiver(broadcastReceiver, intentFilter);             // (d) 注册广播接收器
    }

    protected void onStop() {
        super.onStop();
        unregisterReceiver(broadcastReceiver);                         //注销广播接收器
    }

    private BroadcastReceiver broadcastReceiver = new BroadcastReceiver() {
        @Override
        public void onReceive(Context context, Intent intent) {
            String action = intent.getAction();
            if (action.equals(BluetoothDevice.ACTION_FOUND)) {         // (d)
                BluetoothDevice blueDevice = intent.getParcelableExtra(BluetoothDevice.EXTRA_DEVICE);
                myTextView.setText("发现蓝牙设备：" + blueDevice.getName());
            }
        }
    };

    private Handler myHandler = new Handler();
    private Runnable myRefresh = new Runnable() {
```

```
            @Override
            public void run() {
            }
        };
    }
```

其中：

(a) **if (Build.VERSION.SDK_INT >= Build.VERSION_CODES.JELLY_BEAN_MR2)**：Android 自 4.3 版开始支持蓝牙 BLE（Bluetooth Low Energy，蓝牙低能耗）技术，并且引入了专门的 BluetoothManager（蓝牙管理器），这样就使得蓝牙操作在形式上与 Android 其他设备的操作基本类同了（都是先从系统服务中获取管理器，再执行接下来的操作）。而在此之前，蓝牙管理器的角色一直是由 BluetoothAdapter（蓝牙适配器）兼任的。所以，这里要先判断 Android 的版本，根据版本差异采用不同的操作方式：若是 4.3 之后的新版本，就由管理器调用 getAdapter()方法获取适配器；若是 4.3 之前的老版本 Android，则直接由适配器的 getDefaultAdapter()方法获取系统默认的适配器。

(b) **startActivityForResult(intent, myOpenCode);** 和 **if (requestCode == myOpenCode)**：弹出"蓝牙权限请求"对话框，提示用户是否让其他蓝牙设备能够检测到本手机，如图 9.38 所示。默认情况下，蓝牙适配器是不允许外部发现本手机设备的，本例程序运行时点击"是"按钮授予权限。

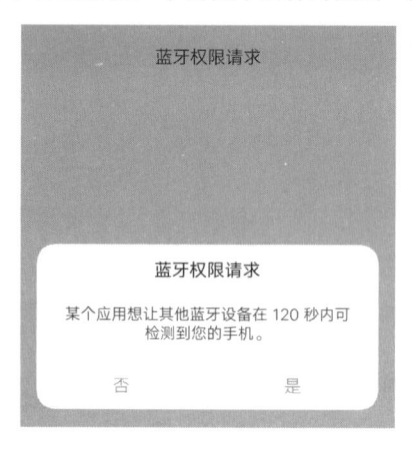

图 9.38 "蓝牙权限请求"对话框

程序中通过重写 onActivityResult()方法判断"蓝牙权限请求"对话框的用户选择结果，根据对话框返回码（myOpenCode 的值）进行判断。

(c) **myBlueAdapter.startDiscovery();**：在蓝牙功能打开（即用户在"蓝牙权限请求"对话框授予权限）之后，就可以调用适配器的 startDiscovery()方法搜索附近的蓝牙设备。

(d) **registerReceiver(broadcastReceiver, intentFilter);** 和 **if (action.equals(BluetoothDevice.ACTION_FOUND))**：由于搜索和发现外部设备需要一定时间，这是个异步的过程，故适配器并不是直接返回搜索结果，而是通过 Android 的广播机制，在 BluetoothDevice.ACTION_FOUND 中返回发现的蓝牙设备信息。为此，需要创建并注册一个 BroadcastReceiver（广播接收器），在接收器中对发现的蓝牙设备信息进行解析。

最后，不要忘了在工程 AndroidManifest.xml 文件中开放蓝牙相关的权限，代码如下：

```
<uses-permission android:name="android.permission.BLUETOOTH_ADMIN" />
<uses-permission android:name="android.permission.BLUETOOTH" />
<uses-feature android:name="android.hardware.bluetooth_le" android:required="true"/>
```

运行程序，附近刚好有一部开启了蓝牙功能的华为 Mate 手机被我们搜索到了，如图 9.39 所示。

```
设备发现
发现蓝牙设备：HUAWEI Mate 20 Pro (UD)
```

图 9.39　搜索到蓝牙功能的华为 Mate 手机

9.4.4　手电筒

手机上的手电筒实际与摄像头共用一套照明系统，所以打开手电筒实质上就是开启照相机的闪光灯。

【例 9.11】　编写程序控制手机上手电筒的打开和关闭。

创建 Android Studio 工程 MyTorch。由于 Android 相机类一般要与表面视图（SurfaceView）配合使用，所以本例界面上放置一个表面视图和一个按钮。

实现思路：

（1）使用 Android 5.0 以上的版本，通过控制照相机闪光灯来实现手电筒功能。

（2）在 onCreate() 方法初始化时，通过表面视图持有者 SurfaceHolder 的回调函数打开相机闪光灯。

（3）在"打开/关闭手电筒"按钮点击事件的 onClick() 方法中，通过 setFlashMode() 方法设置相机闪光灯状态为 FLASH_MODE_TORCH，进入手电筒模式。

在 MainActivity.java 中实现开、关手电筒的功能，代码如下：

```java
public class MainActivity extends AppCompatActivity {
    private SurfaceView mySurfaceView;                              //表面视图对象
    private Button myButton;                                         //"打开/关闭手电筒"按钮
    private Camera myCamera;                                         //照相机对象
    private Camera.Parameters myCameraParas;                         //照相机参数
    private boolean flash = false;                                   //指示闪光灯当前状态
    private SurfaceHolder myHolder;                                  //（a）
    private static String[] PERMISSION_STORAGE = {                   //（b）
            Manifest.permission.CAMERA
    };
    private static int REQUEST_PERMISSION_CODE = 1;                  //（c）
    @Override
    protected void onCreate(Bundle savedInstanceState) {
        super.onCreate(savedInstanceState);
        setContentView(R.layout.activity_main);
        mySurfaceView = findViewById(R.id.mySurfaceView);
        myButton = findViewById(R.id.myButton);
        //判断版本号与棒棒糖版本号的等级
        if (Build.VERSION.SDK_INT > Build.VERSION_CODES.LOLLIPOP) { //（d）
            if (ActivityCompat.checkSelfPermission(this, Manifest.permission.CAMERA) != PackageManager.PERMISSION_GRANTED) {
                ActivityCompat.requestPermissions(this, PERMISSION_STORAGE, REQUEST_PERMISSION_CODE);
            }
        }
        myCamera = Camera.open();                                    //打开相机
        myCameraParas = myCamera.getParameters();                    //获取相机参数
```

```java
                    myHolder = mySurfaceView.getHolder();                    //获取表面持有者
                    myHolder.addCallback(new SurfaceHolder.Callback() {
                        @Override
                        public void surfaceCreated(SurfaceHolder holder) {
                            try {
                                myCamera.setPreviewDisplay(myHolder);        //（e）设置预览界面
                            } catch (Exception e) {
                                myCamera.release();                          //释放照相机对象
                            }
                            myCamera.startPreview();                         //开始预览
                        }

                        @Override
                        public void surfaceChanged(SurfaceHolder holder, int format, int width, int height) {
                        }

                        @Override
                        public void surfaceDestroyed(SurfaceHolder holder) {
                        }
                    });

                    myButton.setOnClickListener(new View.OnClickListener() {
                        @Override
                        public void onClick(View view) {
                            if (!flash) {
                                flash = true;                                //闪光灯状态切换
                                myCameraParas.setFlashMode(Camera.Parameters.FLASH_MODE_TORCH);
                                                                             //（f）
                                myCamera.setParameters(myCameraParas);
                                myButton.setText("关 闭 手 电 筒");
                            } else {
                                flash = false;
                                myCameraParas.setFlashMode(Camera.Parameters.FLASH_MODE_OFF);
                                                                             //关闭闪光灯
                                myCamera.setParameters(myCameraParas);
                                myButton.setText("打 开 手 电 筒");
                            }
                        }
                    });
                }
            }
```

其中：

（a）**private SurfaceHolder myHolder;**：表面视图持有者 SurfaceHolder 是一个接口，它的作用就相当于表面视图的监听器，为用户提供了访问和控制表面视图相关的方法。

（b）**private static String[] PERMISSION_STORAGE = { ... };**：权限 PERMISSION_STORAGE，允许访问手机存储器。

（c）**private static int REQUEST_PERMISSION_CODE = 1;**：REQUEST_PERMISSION_CODE 是请求许可代码。

（d）**if (Build.VERSION.SDK_INT > Build.VERSION_CODES.LOLLIPOP)**：由于 Android 自 5.0

以上对手电筒的操作机制发生了变化，通过 SurfaceHolder 的回调机制来控制相机的闪光灯，比之老版本大大精简了代码。但在操作之前要保证 Android 系统版本在 Android 5.0 以上，故这里要检查版本号是否大于 VERSION_CODES.LOLLIPOP（棒棒糖，即 Android 5.0 的代号）。操作之前还要用 checkSelfPermission()方法进行权限自检。

（e）**myCamera.setPreviewDisplay(myHolder);**：由于本例使用表面视图只是为了能控制照相机的闪光灯作为手电筒用，而非要用它来拍照，故在设计界面的时候将表面视图高度设为 1（即隐去不显示），代码如下：

```xml
<?xml version="1.0" encoding="utf-8"?>
<androidx.constraintlayout.widget.ConstraintLayout xmlns:android="http://schemas.android.com/apk/res/android"
    ...>
    <SurfaceView
        android:id="@+id/mySurfaceView"
        android:layout_width="0dp"
        android:layout_height="1dp"                               //将表面视图高度设为1
        ... />
    <Button
        android:id="@+id/myButton"
        ...
        android:text="打 开 手 电 筒"
        android:textSize="24sp"
        ...
        app:layout_constraintTop_toBottomOf="@+id/mySurfaceView" />
</androidx.constraintlayout.widget.ConstraintLayout>
```

这样在运行程序时界面上就只看到一个按钮，如图 9.40 所示。

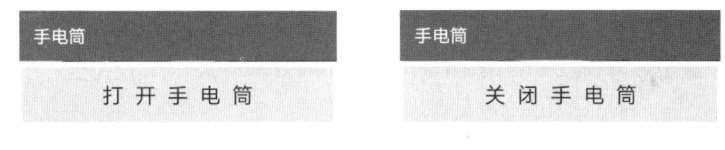

(a) 关闭状态　　　　　　(b) 开启状态

图 9.40　运行程序时界面上只看到一个按钮

（f）**myCameraParas.setFlashMode(Camera.Parameters.FLASH_MODE_TORCH);**：用 setFlashMode()方法设置闪光灯的模式为 FLASH_MODE_TORCH（火炬）即可当手电筒使用。

最后，不要忘了在 AndroidManifest.xml 中开放对照相机硬件及其闪光灯的操作权限，代码如下：

```xml
<uses-permission android:name="android.permission.CAMERA"/>
<uses-permission android:name="android.permission.FLASHLIGHT"/>
<uses-feature android:name="android.hardware.Camera"/>
```

将程序安装到真实的手机上运行，点击"打开手电筒"按钮，界面如图 9.40 所示，读者可以试着用自己设计的程序控制手电筒的照明效果。

习题和实验

第 1 章 Android 开发入门

一、填空题

1. Android 是 Google 公司基于_____平台开发的手机及平板电脑的操作系统。
2. Android 的分层架构中,应用框架层使用_____语言开发。
3. 程序员编写 Android 应用程序时,主要调用_____层提供的接口实现。
4. Android Studio 是 Google 官方推荐的 Android 开发环境,目前已经发布到 3.x 系列,与 2.x 相比,Android Studio 3.0 往后的版本不再默认集成_____,为的是缩小其安装包的体积。
5. Android Studio 工程创建后,默认生成一个布局文件 activity_main.xml,该文件位于工程的_____目录中;默认生成一个 Activity 文件 MainActivity.java,该文件中会自动导入_____类。
6. Android 应用程序界面布局文件采用_____格式,若要使自己创建的工程默认生成两个布局文件(activity_main.xml 和 content_main.xml),那么在创建工程时要选择 Activity 的类型为_____。
7. Android 应用程序的配置文件名称为_____。
8. 如果需要开发支持多国语言的 App,通常会将程序中用到的全部界面文字字符串集中定义在 res\values 目录下的_____文件中。
9. 为了让程序员更加方便地运行调试程序,Android Studio 提供了_____。

二、选择题

1. Android Studio 安装运行的软、硬件要求包括(多选题)(   )。
   A. Windows 7(32 位)   B. 8GB 以上内存
   C. i7CPU   D. 固态硬盘
2. Android Studio 开发环境运行需要(多选题)(   )。
   A. JDK   B. 配置 JDK 的环境变量
   C. AVD   D. Android SDK

三、简答题

1. Android Studio 开发环境安装的步骤有哪些?
2. Android Studio 平台应用程序创建和运行的步骤有哪些?
3. 创建第一个 Android Studio 工程,说明它由哪几部分构成。
4. Android 程序事件处理有哪几种通行的编程范式?各自的适用场合是什么?

四、说明题

1. 请说明在创建 Android Studio 工程时以下名称的意义:
(1) Name;
(2) Package name;

（3）Save location；

（4）Language；

（5）Minimum API level。

2．在创建 Android Studio 工程时为什么需要选择应用程序将要运行的平台？

3．模拟运行和真机运行有什么差别？

4．在编辑 XML 界面文件时，Design 和 Text 方式有什么不同？

五、实验题

1．安装 JDK（参考教材 1.2.1 小节）。

（1）建议从 Oracle 官网下载最新版本的 JDK，网址为 https://www.oracle.com/technetwork/java/javase/downloads/index.html。

（2）执行 JDK 安装向导（注意：新版 JDK 已经不再需要用户专门指定 JRE 了）。

（3）配置环境变量，使后面安装的 Android Studio 能够找到 JDK。

2．安装 Android Studio（参考教材 1.2.2 小节）。

（1）下载 Android Studio。

（2）安装 Android Studio，并完成第一次启动的配置工作。

3．开发第一个 Android 应用程序。

（1）创建 Android Studio 工程（参考教材 1.3.1 小节）

（2）设计应用程序界面（参考教材 1.3.2 小节），效果如图 T1.1 所示（读者可自行设计制作自己喜爱的图片）。

图 T1.1　应用程序界面

（3）熟悉 Android Studio 开发环境。

（4）创建仿真器，模拟运行程序；再尝试将程序安装到自己的手机运行（参考教材 1.3.4 小节）。

（5）初步理解 Android Studio 工程结构（参考教材 1.3.5 小节）。

（6）用 4 种不同的编程范式分别实现点击"开始课程"按钮后的事件处理功能（参考教材 1.3.7 小节）。

4．安装和管理 Android SDK（参考教材 1.4 节）。

（1）用 AVD 映像方式安装 SDK。

（2）用 SDK Manager 下载方式安装 SDK。

（3）找到自己计算机的"C:\Users\<用户名>\AppData\Local\Android\Sdk\platforms"路径，如图 T1.2 所示，查看自己所安装的 SDK 目录。

图 T1.2　查看安装的 SDK 目录

（4）试着安装多个不同版本的 SDK，测试 HelloWorld 程序在它们各自映像系统上的运行效果。

第 2 章　Android 用户界面

一、选择题

1. Android 中有许多控件，这些控件都继承自（　　）类。
A．Control　　　　　　　B．Window　　　　　　　C．TextView　　　　　　　D．View

2. 指定"将控件右部贴于另一控件左部"的属性是（　　）。
A．app:layout_constraintRight_toAlignLeftOf
B．app:layout_constraintRight_toAlignParentLeftOf
C．app:layout_constraintRight_toLeftOf
D．app:layout_constraintLeft_toRightOf

3. 如图 T2.1 所示展示的是支付宝提现界面，要实现图中带 LOGO 的银行卡选项（图中圈出），最合适的控件是（　　）。

图 T2.1　支付宝提现界面

A. ImageView　　　　B. TextView　　　　C. EditText　　　　D. ImageButton
4. 以下哪个控件**无法**同时显示图片和文字？（　　）
A. ImageButton　　　B. TextView　　　　C. EditText　　　　D. Button
5. （　　）属性用来表示引用图片的资源。
A. text　　　　　　　B. img　　　　　　　C. id　　　　　　　　D. srcCompat
6. Spinner 是（　　）控件。
A. 浮动菜单　　　　　B. 组合框　　　　　　C. 下拉框　　　　　　D. 下拉列表
7. 如果需要捕捉某个控件的事件，需要为该控件创建（　　）。
A. 属性　　　　　　　B. 方法　　　　　　　C. 监听器　　　　　　D. 事件

二、简答题

1. 在 Android 的布局文件 activity_main.xml 中，"@+id/username" 与 "@android:id/username" 两者有何区别？
2. 控件的只读属性和是否可用属性有什么不同？
3. 文本视图（TextView）和文本编辑框（EditText）控件功能有哪些差别？
4. 按钮（Button）和图像按钮（ImageButton）控件功能有哪些差别？这种差别又源于什么？
5. 下拉框（Spinner）控件在手机 App 上通常显示为"浮动菜单"的外观，这是通过其哪个属性控制的？用户选择下拉框的值又是如何得到的？
6. 单选按钮（RadioButton）为什么需要放在它的容器中？

三、实验题

1. 熟悉 Android 中基本的界面控件（参考教材 2.2 节）。
（1）按照本节教材实例试设计程序，并在仿真器中模拟运行测试。
（2）将【例 2.5】的质感单选按钮界面修改成如图 T2.2 所示的简约风格界面，自己动手试试看。

江苏GDP排名第一的城市是：苏州
○ 南京
◉ 苏州
○ 无锡

图 T2.2　简约风格界面

2. 应用设计。
（1）运用本章所学的状态列表图形技术，将 CheckBox 控件打造成如图 T2.3 所示的形似开关按钮的外观，设计一个程序获取开关的状态信息并用文字说明的形式通过文本视图显示出来。

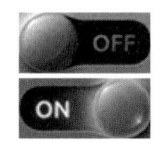

图 T2.3　CheckBox 打造的开关按钮

（2）本章【例 2.7】的下拉框中带有图标的选项（图 T2.4）是用图像视图与文本视图组合在线性布局中实现的，代码位于 myspinner_item.xml 中，内容如下：

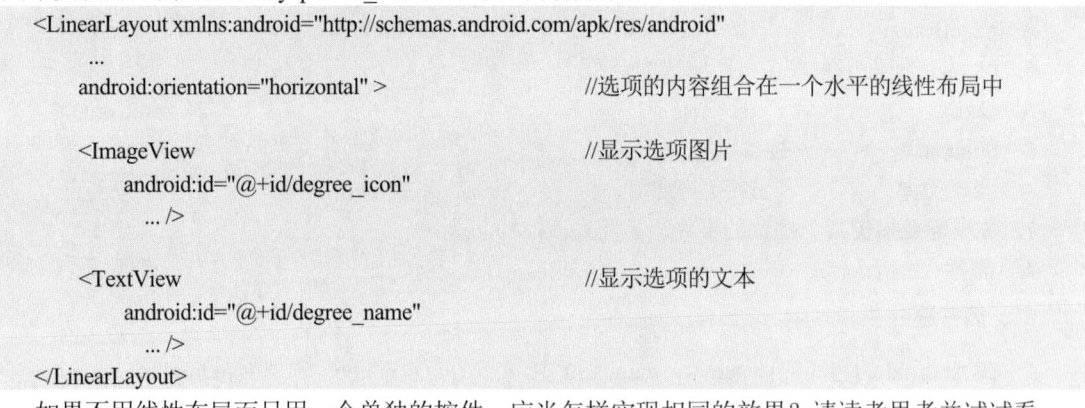

```
<LinearLayout xmlns:android="http://schemas.android.com/apk/res/android"
    ...
    android:orientation="horizontal" >           //选项的内容组合在一个水平的线性布局中

    <ImageView                                    //显示选项图片
        android:id="@+id/degree_icon"
        ... />

    <TextView                                     //显示选项的文本
        android:id="@+id/degree_name"
        ... />
</LinearLayout>
```

如果不用线性布局而只用一个单独的控件，应当怎样实现相同的效果？请读者思考并试试看。

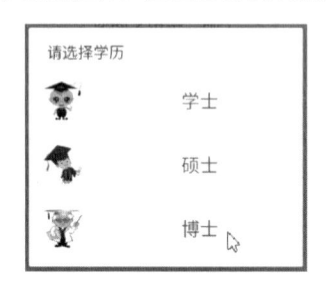

图 T2.4　带有图标的选项

提示：Android 中的控件 TextView、Button、EditText 都具备同时显示图像和文字的能力，试着用一用。

（3）将【例 2.8】的搜索框（图 T2.5）改为使用下拉框实现，设定其模式属性，并编程实现关键词匹配功能，使之呈现出与原自动完成文本视图 AutoCompleteTextView 一模一样的效果。

图 T2.5　搜索框外观

（4）现在的手机普遍都有手势解锁功能，进入 App 之前需要在屏幕上用手指划出预定的径迹才被允许登录。Android 中专门提供有 GestureDetector 类，它里面有一个 OnGestureListener 接口就是用来检测用户手指在屏幕上划出的方向的。请读者参考教材 2.3 节关于界面事件的原理和编程机制，通过自己查阅 Android 官方文档或网络资料，设计一个模拟开机手势验证的程序，界面可参考图 T2.6。

提示：OnGestureListener 接口无法被 MainActivity 主类继承和实现，只能由 GestureDetector 类来实现和使用。

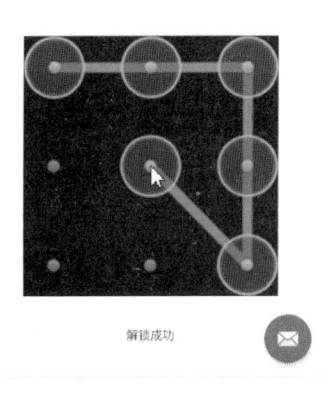

图 T2.6　手势验证程序参考界面

第 3 章　界面布局与活动页

一、填空题

1. Android Studio 3.x 开始创建 Android 程序时，默认使用的布局是_____。
2. 在相对布局中，_____属性确定横向相对布局；_____属性表示"是否跟父布局左对齐"；_____属性表示"与指定控件右对齐"。
3. 表格布局可以包含多行，_____代表是一行。
4. 板块布局中可以添加多个控件，这些控件会重叠在屏幕_____显示。
5. 如果控件的 id 设置为 myButton，那么在调用方法 findViewById()获取控件的对象时，引用该控件的参数应为_____。
6. 一般采用_____和_____两种形式来使用 Intent。
7. Intent 的构成包括 Action、_____、_____、_____、_____和_____。
8. Activity 控制的页面从产生到结束，会经历_____个阶段。
9. Toast 的作用是_____。

二、选择题

1. (　　) 属性可以指定"在指定控件左边"。
   A．android:layout_alignleft　　　　　B．android:layout_alignParentLeft
   C．android:layout_left　　　　　　　D．android:layout_toLeftOf
2. 在相对布局中，(　　) 属性确定"是否跟父布局底部对齐"。
   A．android:layout_alignBottom　　　B．android:layout_alignParentBottom
   C．android:layout_alignBaseline　　　D．android:layout_below
3. 在相对布局中，(　　) 属性指定一个控件位于布局水平方向中央的位置。
   A．android:layout_alignBaseline　　　B．android:layout_centerVertical
   C．android:layout_centerInParent　　　D．android:layout_centerHorizontal

4. 在表格布局中，android:layout_column 属性指定（　　）。
   A. 行数   B. 列数   C. 总行数   D. 总列数
5. 一个 Android 应用程序，系统默认包含（　　）Activity。
   A. 0 个   B. 1 个   C. 2 个   D. 多个
6. （　　）方法，Activity 从启动到关闭不会执行。
   A. onCreate()   B. onStart()   C. onResume()   D. onRestart()
7. 不能使用 Intent 启动的是（　　）。
   A. Activity   B. 服务   C. 广播   D. 内容提供者
8. startActivityForResult 方法接收两个参数，第一个是 Intent，第二个是（　　）。
   A. resultCode   B. requestCode   C. 请求码   D. data
9. 在 Activity 的生命周期中，当 Activity 处于栈顶时，处于（　　）状态。
   A. 活动   B. 暂停   C. 停止   D. 销毁

10. 如图 T3.1 所示的是某 App 网站春节期间发起的"集鼠卡抽奖领红包"活动，需要通过做任务来增加抽奖机会，对于抽奖机会已用完的用户，当点转盘中央"开始抽奖"按钮后，页面下半部会弹出"做任务领抽奖机会"的对话框，与此同时页面的上半部分会变暗隐去。运用本章所学有关 Activity 生命周期的知识，推想一下此时的抽奖转盘页正处于（　　）状态。

图 T3.1　"集鼠卡抽奖领红包"活动 App 界面

A. 活动   B. 暂停   C. 停止   D. 销毁

三、简答题

1. 为什么需要使用布局嵌套？
2. 线性布局能否实现网格布局功能？
3. 表格布局和网格布局的相同点与不同点有哪些？
4. 绝对布局的优点和缺点有哪些？

5. Android 的三个基本组件（Activity、Service 和 Broadcast Receiver）传递 Intent 的方式分别是什么？

6. 通过 Bundle 在 Activity 之间传递数据的过程是怎样的？

四、实验题

1. 约束布局（参考教材 3.1.1 小节）。

根据如图 T3.2（b）所示的操作指导，完成如图 T3.2（a）所示的登录界面设计。

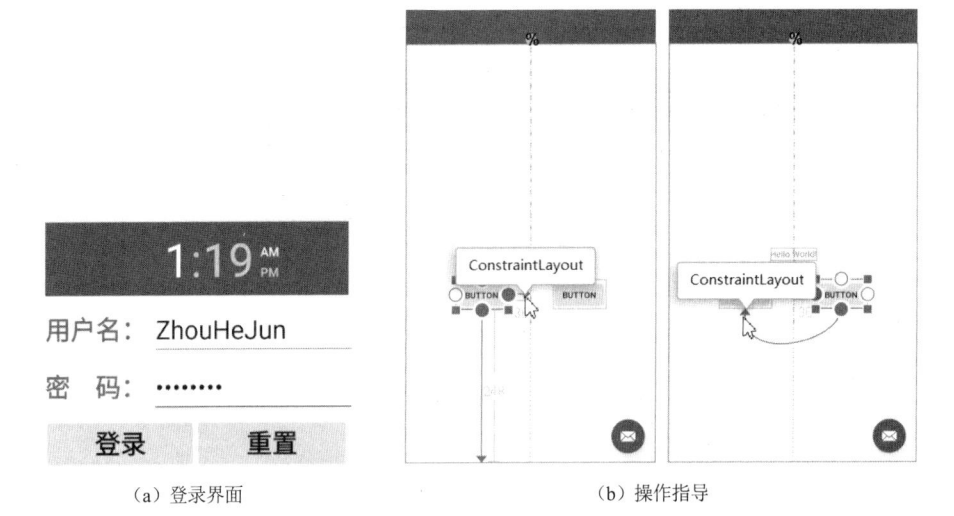

（a）登录界面　　　　　　　　　　　　　（b）操作指导

图 T3.2　约束布局界面设计及操作指导

2. 线性布局和相对布局（参考教材 3.1.2 小节）。

（1）根据教材线性布局和相对布局代码创建工程实例，在仿真器中查看运行效果。

（2）用嵌套的线性布局设计如图 T3.3 所示的图书展示页面。

图 T3.3　图书展示页面

3. 表格布局（参考教材 3.1.2 小节）。

（1）根据教材表格布局代码创建工程实例，在仿真器中查看运行效果。

（2）用表格布局设计一个输入学生基本信息的表单界面。

4．网格布局（参考教材 3.1.2 小节）。

用网格布局显示如图 T3.4 所示的界面。

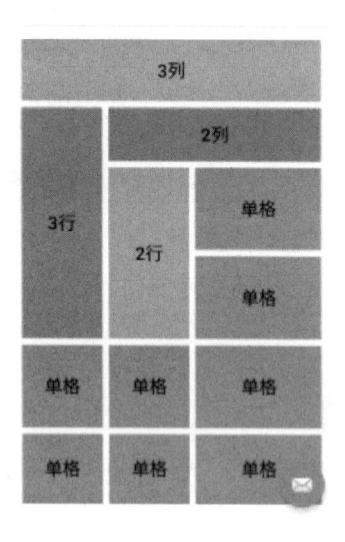

图 T3.4　网格布局界面

5．页面间数据交互（Intent 机制）。

（1）参考教材 3.2.2 小节内容，完成【例 3.2】的登录响应功能。

（2）设计一个程序，计算方程 $ax^2+bx+c=0$ 的根，第一个页面输入系数，通过 Intent 传递给第二个页面，由第二个页面执行求根运算并将结果返回给第一个页面显示。

6．页面生命周期。

（1）参考教材 3.2.3 小节内容，观察【例 3.3】程序的生命周期。

（2）在上题的 $ax^2+bx+c=0$ 求根应用中加入显示生命周期提示的信息。

第 4 章　移动 App 高级界面开发技术

一、填空题

1．移动 App 通用的界面元素包括（列举 3 个）：_____、_____、_____。
2．互联网电商的 App 内容呈现形式有（列举 3 个）_____、_____、_____。
3．要实现标签栏按钮的图标和背景随用户的点选而改变颜色和背景，通常采用_____技术。
4．Android 控件的自定义风格写在工程 values 目录下的_____文件中。
5．TabActivity 标签栏前端布局代码中，根元素必须是_____。
6．设置标签按钮的文字及图标资源用_____方法。
7．轮播条的图片翻页功能通常使用 Android 中的_____控件实现。
8．RecyclerView 有多种布局管理器，要实现列表展示商品信息选用_____布局管理器。

二、简答题

1．App 界面开发中为什么常常需要定义控件风格？
2．轮播条翻页时将对应图片的指示器圆点点亮为实心，这是通过什么原理实现的？

3. RecyclerView（循环视图）是个功能极其强大的 Android 控件，它可以实现的界面元素或内容呈现形式有哪些？

4. 视图类组件开发中为什么要定义数据模型？

5. BaseAdapter 有什么作用？

6. 开发互联网电商的 App 商品展示页为什么经常需要自定义文本视图？

7. TabHost 与基于 TabActivity 的标签栏是同一种控件吗？为什么？

8. Fragment 是什么？与 Activity 又是什么关系？为什么有时候开发页面要使用 Fragment 而不是 Activity？

三、实验题

1. 参考教材【例 4.2】（轮播条）和【例 4.6】（类别标签列表），制作如图 T4.1 所示的某电商 App 首页头部效果。

图 T4.1　某电商 App 首页头部效果

2. 参考教材【例 4.4】（列表视图），制作如图 T4.2 所示的图书展示列表。

图 T4.2　图书展示列表

3. 改用 RecyclerView（循环视图）实现图 T4.2 一样的界面效果，比较一下与 ListView（列表视图）相比哪个在实现上更方便。

4. 改用多维数组的数据结构重新创建【例 4.6】的数据模型，修改 Fruit.java 的代码，看能不能在完全**不改变**前端 XML 代码的情况下实现与图 T4.3 一模一样的类别标签列表的界面效果。从实践中体会 MVC 设计模式的优越性所在。

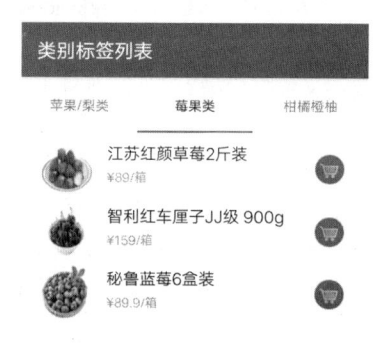

图 T4.3　类别标签列表的界面效果

5．为本章"易果鲜超市"App"我的"标签页开发内容，可参考如图 T4.4 所示界面的样式。想一想：其中用到哪些学过的界面元素？

图 T4.4　提供参考的界面样式

6．完善本章"易果鲜超市"App 的通知消息计数功能，当消息数大于 99 时，消息通知圆点内的数字变为省略号。

7．综合运用本章学到的知识，实现一个完整的酒店订餐系统 App 界面。

第 5 章　Android 服务与广播程序设计

一、选择题

1．每一次启动服务都会调用（　　）方法。
A．onCreate()　　　　　B．onStart()　　　　　C．onResume()　　　　　D．onStartCommand()

2．（　　）方法不属于 Service 生命周期。

A．onResume()　　　B．onStart()　　　C．onStop()　　　D．onDestroy()

3．继承 BroadcastReceiver 会重写（　　）方法。

A．onReceive()　　　B．onUpdate()　　　C．onCreate()　　　D．onStart()

4．下列方法中，用于发送一条有序广播的方法是（　　）。

A．startBroadcastReceiver()　　　　　　B．sendOrderedBroadcast()
C．sendBroadcast()　　　　　　　　　　D．sendReceiver()

5．在清单文件中，注册广播时使用的节点是（　　）。

A．<activity>　　　B．<broadcast>　　　C．<receiver>　　　D．<broadcastreceiver>

6．属于绑定服务特点的是（　　）。

A．以 bindService()方法开启　　　　　　B．调用者关闭后服务关闭
C．必须实现 ServiceConnection()　　　　D．使用 stopService()方法关闭服务

7．Service 与 Activity 的共同点是（　　）。

A．都是四大组件之一　　　　　　　　　B．都有 onResume()方法
C．都可以被远程调用　　　　　　　　　D．都可以定义界面

8．关于 Service 生命周期的 onCreate()和 onStart()方法，正确的是（　　）。

A．如果 Service 已经启动，将先后调用 onCreate()和 onStart()方法
B．当第一次启动的时候先后调用 onCreate()和 onStart()方法
C．当第一次启动的时候只会调用 onCreate()方法
D．如果 Service 已经启动，只会执行 onStart()方法，不再执行 onCreate()方法

二、简答题

1．什么情况下需要使用线程？
2．Service 的使用方式及其特点有哪些？
3．多个 Service 交互的特点有哪些？
4．什么情况下需要使用广播？普通广播和有序广播的应用场合有哪些？

三、实验题

1．参考教材 5.1.2 小节的内容，完成启动方式使用 Service 的实例。
2．参考教材 5.1.3 小节的内容，完成绑定方式使用 Service 的实例。
3．参考教材 5.1.4 小节的内容，完成多个 Service 交互及生命周期实例。
4．参考教材 5.2.2 小节的内容，完成普通广播应用实例。
5．参考教材 5.2.3 小节的内容，完成有序广播应用实例。

第 6 章　Android 数据存储与共享

一、填空题

1．SharedPreferences 本质上是一个 XML 文件，存储的数据是以_____的格式保存的。
2．SharedPreferences 用于存储应用程序的_____。
3．指定文件内容可以追加的文件操作权限是_____。
4．可以通过使用_____方法创建数据库。另外还可以通过写一个继承_____类的方式创建数据库。

5. 查询 SQLite 数据库中的信息使用_____接口，使用完毕后调用_____关闭。
6. ContentProvider 对数据进行操作的方法包括_____。
7. ContentResolver 可以通过 ContentProvider 提供的_____进行数据操作。
8. ContentProvider 创建时首先会调用_____方法。

二、简答题

1. 采用什么方式可以减少不同的存储数据的方法对其他程序的影响？
2. Android 系统文件访问为什么需要采用两个方法打开文件？
3. 如何查看采用 SQLite 数据库方式存储的数据文件内容？
4. 说明 ContentProvider 的调用关系。

三、实验题

1. 参考教材 6.1 节的内容，完成【例 6.1】：采用 SharedPreferences 方式，存取注册信息实例。
2. 参考教材 6.2 节的内容，完成【例 6.2】：采用文件存储方式，存取注册信息实例。
3. 参考教材 6.3 节的内容，完成【例 6.3】：采用 SQLite，存取注册信息实例。
4. 参考教材 6.4 节的内容，另创建一个工程，通过 ContentResolver 调用 ContentProvider 的添加（insert()）方法向 SQLite 数据库中注册新用户，然后调用更新（update()）方法修改用户密码，最后调用删除（delete()）方法删除用户注册信息。

第 7 章　Android 数据库和网络编程

一、填空题

1. 目前能支持 Android 程序直连 MySQL 的最高版本驱动是_____。
2. 为保证移动用户的使用体验，数据库操作必须放在一个_____中来执行。
3. 当前大多数互联网应用系统都采用_____、_____、_____的 3 层架构方式。
4. Android 为 HTTP 编程提供了_____类，可用它连接 Java/Java EE、.NET、PHP 等几乎所有主流平台的 Web 服务器。
5. Android 程序与 WebService 通信使用的是_____协议。
6. 目前移动互联网通行的信息传输格式是_____。

二、选择题

1. 以下哪一款 DBMS 目前尚不支持 Android 程序的 JDBC 直连？（　　）
   A. SQL Server 2019　　B. MySQL 8.0　　C. Oracle 19c　　D. Oracle 11g
2. 以下哪种方式可以用主机名访问数据库？（　　）
   A. HTTP 通过 Servlet 访问 MySQL　　　B. JDBC 直连本地 MySQL
   C. JDBC 直连远程 MySQL　　　　　　D. JDBC 直连 SQL Server
3. 以下说法不正确的是（　　）。
   A. 基于.NET 4.0 创建的 WebService 只能通过.NET 4.0 兼容的驱动访问数据库
   B. MySQL 8 最新驱动 mysql-connector-java-8.0.17.jar 不支持 Android 程序访问数据库
   C. 一个 Android 程序可访问多种 DBMS，它们的驱动包都添加依赖在同一个工程中，互不干扰
   D. JDBC 直连本地 MySQL 程序连接字符串中的 IP 地址可写成 127.0.0.1

三、简答题

1. 简述访问数据库过程中的 UI 界面刷新机制。
2. 实际开发中为什么不推荐以 JDBC 直连的方式访问数据库？
3. 移动 App 访问电商网站，为什么多数情况下都是在与服务器上的 WebService 交互？
4. 简述一个 Android 程序访问 WebService 的典型流程。

四、实验题

1. 准备本章所用的数据库（参考教材 7.1 节）。

（1）将 3 个数据库（MySQL、SQL Server 和 Oracle）分别安装在 3 台不同的计算机上，为每台计算机命名主机名、配置 IP 地址。

（2）熟悉 Navicat Premium 工具的使用。

（3）创建 emarket（电子商城）数据库、表，准备样本数据，如图 T7.1 所示。

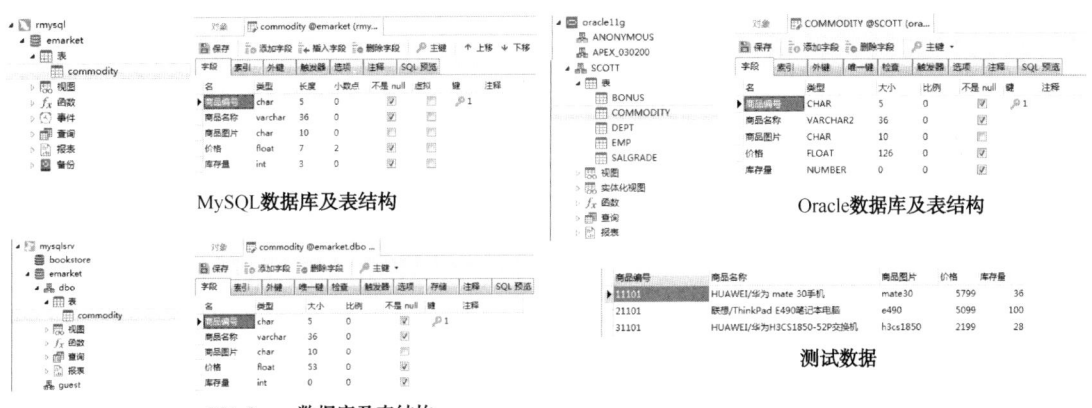

图 T7.1　创建 emarket（电子商城）数据库、表，准备样本数据

2. JDBC 直连数据库（参考教材 7.2 节）。

（1）按照书中指导，试做【例 7.1】。

（2）将程序改为直连 SQL Server、Oracle 数据库，并分别试验远程和本地连接两种情况。

3. HTTP 访问数据库。

（1）搭建"移动端—Web 服务器—后台数据库"的 3 层架构开发环境（参考教材 7.3.2 小节），并测试可用性。

（2）按照书中指导，试做【例 7.2】，掌握 Servlet 程序的编写、Web 项目的打包发布操作（参考教材 7.3.3 小节）。

（3）将服务器 Servlet 程序的功能改为用当前互联网应用开发流行的 SSM 框架（Spring+SpringMVC+MyBatis）实现，通过网络和相关专业书籍自学，完成项目试验。

4. WebService 访问数据库。

（1）学会在 Windows 上配置 IIS 服务器（参考教材 7.4.2 小节）。

（2）按照书中指导，试做【例 7.4】，开发一个 WebService（参考教材 7.4.3 小节），并发布测试（参考教材 7.4.4 小节）

（3）按照书中指导，试做【例 7.5】，开发与 WebService 交互的移动端 Android 程序（参考 7.4.5 小节），重点掌握 SOAP 协议编程以及谷歌 ksoap2 库的使用。

5．JSON 数据操作。

（1）按照书中指导，试做【例 7.7】程序，开发"网上商城商品管理"App 界面，如图 T7.2 所示（参考教材 7.5 节）。

图 T7.2　"网上商城商品管理"App 界面

（2）为"网上商城商品管理"App 添加开发一个商品入库页，其上表单录入新进商品信息，提交后包装成 JSON 格式发给服务器，再存入后台数据库，并在如图 T7.2 所示页面上查询到新添加的商品信息。

第 8 章　Android 多媒体和图形图像编程

一、填空题

1．Android 程序播放 App 内置视频时，视频文件要预先存放在工程的_____目录下。

2．打开视频时，在_____方法中获得 Intent 传递过来的视频 URI，通过_____方法从 Intent 数据中解析出 URI，用_____方法将 URI 转化为路径名字符串。

3．要使用 SeekBar 进度条需要实现_____监听接口；要实现当用户拖曳进度条停止后在其释放的位置处继续播放视频，需要重写该接口中的_____方法。

4．Android 程序是根据页面_____方法的_____参数来判断当前 App 是否处于画中画模式的。

5．播放音频时，要设置 Intent 的类型为_____。

6．Android 中对图形图像的处理实质上是对图片在内存中_____的处理，为此 Android 专门提供了_____类来辅助用户进行图像处理。

7．实现手机相册需要结合使用 Android 的_____和_____控件。

二、简答题

1. 使用 SeekBar 要想在视频播放过程中看到进度条随播放进程的动态变化，在编程中应当怎么处理？

2. MediaController（媒体控制条）与 VideoView（视频视图）绑定需要分哪两个步骤？

3. 录像编程中用到的 SurfaceHolder 是什么？有什么作用？

4. 在图形图像处理过程中，当需要刷新视图时，究竟用 invalidate()还是 postInvalidate()方法好？说说两者的区别与适应场合。

5. 简要说说 OpenGL 绘图的一般流程。

三、实验题

1. 视频、音频与录像（参考教材 8.1 节）。

（1）按照书中指导，试做【例 8.1】，用两种方式打开并播放视频。

（2）按照书中指导，试做【例 8.2】，分别用 SeekBar 和 MediaController 两种控件实现播放器的控制条。

（3）按照书中指导，试做【例 8.3】，实现画中画特效。

（4）按照书中指导，试做【例 8.4】，用两种方式打开并播放音频。

（5）参考【例 8.5】，在录像功能的基础上增加弹出对话框，让用户选择录像视频的存放路径，以及打开已存盘录像视频文件播放的功能。

2. 图形图像处理（参考教材 8.2 节）。

结合【例 8.6】和【例 8.7】的功能，开发一个完善的图片处理器，可打开指定路径下的图片文件，对其内容进行缩放、旋转任意角度、扭曲、裁剪等各种常用的变换处理。

3. 手机相册（参考教材 8.3 节）。

（1）按照书中指导，试做【例 8.8】的相册 App，完成效果如图 T8.1 所示。

图 T8.1 相册 App 界面

（2）修改程序，调整相框底部小图的尺寸，增加同时预览的照片张数。

（3）给画廊中的小照片加上漂亮的边框。

4．OpenGL 初步（参考教材 8.4 节及相关书籍和网络资料）。

学会构建 OpenGL 环境和 OpenGL ES 2.0 库的基本使用，尝试用 OpenGL 绘制正方形、圆、多边形等不同形状的基本几何图形，并做出简单的动画效果来。

第 9 章　Android 第三方开发与设备操作

一、填空题

1．在整合二维码 zxing 框架时，对 Demo 的源文件需要修改_____、_____、_____。

2．在进行二维码程序开发时，需要在 AndroidManifest.xml 中打开的权限有_____、_____、_____。

3．为了访问支付宝的支付接口，需要在程序代码中提供_____来标识用户的 App。

4．为使 App 能够读取到计步器的步数信息，必须继承_____监听器，通过其中的_____方法得到实时的步数信息。

5．手机"摇一摇"功能是通过_____传感器实现的。

6．手机上手电筒与_____共用一套照明系统。

二、简答题

1．什么是"沙箱"？它的作用是什么？

2．在工程中如何引入高德地图功能？

3．什么是"定位服务提供者"？举例说明。

4．简述智能手机上蓝牙启用的过程。

三、实验题

1．生成和扫描二维码（参考教材 9.1 节）。

（1）按照书中指导，在工程中整合 zxing 框架。

（2）试做【例 9.1】和【例 9.2】，完成二维码生成和扫描功能。

（3）给第 4 章"易果鲜超市"App 页面顶部开发一个工具栏，其中集成二维码功能，界面可参考图 T9.1。

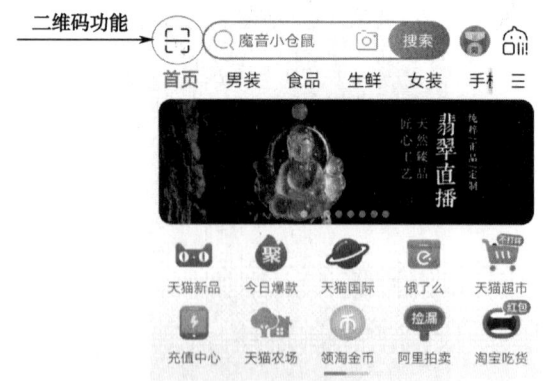

图 T9.1　带二维码功能的工具栏

2．接入支付宝（参考教材 9.2 节）。

（1）按照书中指导，熟悉支付宝沙箱的使用。

（2）集成支付功能，完成【例 9.3】基本的支付应用。

3．地图应用开发（参考教材 9.3 节）。

（1）下载高德地图 SDK，安装高德地图应用环境。

（2）完成【例 9.4】显示一张地图。

（3）将【例 9.5】地图改用夜景模式显示，查看效果。

（4）结合第 2 章所学 AutoCompleteTextView 控件的相关知识，为【例 9.6】程序增加检索关键词提示列表功能，并检索周边 5km 范围内的所有游泳馆。

（5）扩充【例 9.7】程序 OnRouteSearchListener 接口中的 onRideRouteSearched(RideRouteResult result, int errorCode)方法，增加骑行路径规划功能。

4．设备操作（参考教材 9.4 节）。

（1）按照书中的指导，完成【例 9.8】～【例 9.11】程序，学会用 Android 程序操作手机计步器、加速度传感器、GPS 定位、蓝牙、手电筒等各种设备。

（2）设计开发一个"走路赚钱"App，界面可参考图 T9.2。根据用户走路步数兑换现金红包（10000 步=1 元），并通过支付宝（沙箱）提现到用户账户。

图 T9.2 "走路赚钱"App 界面（参考）

（3）结合高德地图和 GPS 开发一个定位 App，界面如图 T9.3 所示。打开 App 后程序自动通过 GPS 获取用户当前所在位置的经、纬度值，显示在界面上方，并在地图上标示出用户所在地点的名称信息。

图 T9.3 定位 App 界面

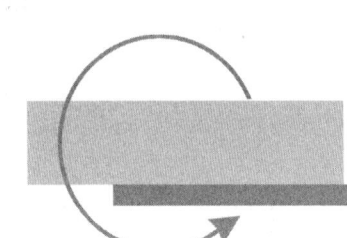

习题参考答案

第 1 章　Android 开发入门

一、填空题

1．Linux
2．Java
3．应用框架层
4．Android SDK
5．layout，AppCompatActivity
6．XML，Basic Activity
7．AndroidManifest.xml
8．strings.xml（字符串资源文件）
9．AVD（Android Virtual Device）仿真器

二、选择题

1．B C D　　2．A B

三、简答题

1．Android Studio 安装步骤：
（1）准备高性能的计算机；
（2）安装 JDK 并配置环境变量；
（3）下载 Android Studio，执行安装向导；
（4）初次启动，忽略 SDK 检查，选择界面主题风格，下载必需的 Android SDK 组件。

2．创建和运行应用程序的步骤：
（1）启动 Android Studio，单击"Start a new Android Studio project"项来创建新的 Android Studio 工程；
（2）选择 Activity（活动页面）类型；
（3）填写应用程序相关的信息；
（4）在仿真器管理器中创建 AVD 及其系统映像；
（5）安装所选映像对应版本的 SDK；
（6）启动仿真器运行程序。

3．Android Studio 工程由以下 3 个部分构成：
（1）**manifests** 目录：AndroidManifest.xml 是应用程序配置文件。
（2）**java** 目录：存放 Java 源程序文件。
（3）**res** 目录：存放资源文件，其中的 layout 子目录专门存放用户界面的布局文件。

4. Android 程序事件处理有 4 种通行的编程范式：

（1）**UI 组件直接创建监听器**：适用于事件处理的代码量不大的简单情形。

（2）**UI 组件实例化监听类**：事件处理代码量大，且程序中用到的 UI 组件和事件类型极其丰富多样。

（3）**layout 界面设置事件方法**：操作简单，无须程序员了解 Java 事件机制，但只适用于按钮的点击事件。

（4）**主程序类实现监听接口**：应用最为广泛，但对于某些 UI 组件所特有的事件类型不适用。

四、说明题

1.
（1）应用程序名（也就是工程的名称）；
（2）工程中 Java 程序所在的包名；
（3）工程存放的路径；
（4）Android 开发所使用的编程语言；
（5）要安装此应用程序的目标设备，其操作系统最低要求的版本。

2. 因为现在智能手机更新换代很快，Android 操作系统版本也随之持续升级，且使用 Android 的硬件设备也多种多样，选择应用程序要运行的平台就是为了保证所开发的程序在目标平台上的兼容性。

3. 所谓"模拟运行"就是使用 Android Studio 的仿真器运行应用程序；而"真机运行"则是在真实的物理手机上运行 App，需要安装驱动并打开手机的开发者权限进入调试模式。对于一般的 Android 应用程序，这两种方式运行是等效的，但如果需要测试自己的 App 在某一款特定手机上的兼容性和性能，或者所开发的 App 需要操作摄像头、手电筒、计步器等真实手机才有的硬件设备的话，则必须使用真机运行程序。

4. Design 方式是界面设计模式，可在设计区直接看到界面效果和进行鼠标拖曳放置控件的操作；Text 方式是界面编码模式，只能看到 XML 源码，程序员通过在其中编写控件元素标签代码来设计界面。通常情况下这两种方式是等效的，用户可视方便程度和个人习惯来切换使用。但是，对于某些特殊（或引自第三方库）的控件，Android Studio 工具箱面板里并未提供，这个时候就只能采用 Text 方式编写设计界面的代码。

第 2 章　Android 用户界面

一、选择题

1. D　　2. C　　3. B　　4. A　　5. D　　6. C　　7. C

二、简答题

1. @+id/username 表示标识 id 为 username 的 UI 元素；而@android:id/username 是属于 Android 框架的资源，必须添加其 Android 包的命名空间。

2. 只读属性表示控件（如文本编辑框）内的文字不可修改，但内容仍然正常显示（不会变灰）；是否可用属性则设置了控件的可用性，被置为不可用的控件一般会呈现灰色无法操作的状态。

3. 文本视图（TextView）是显示类视图，通常用于显示 App 界面的文字信息，用户不能对信息内容进行编辑；而文本编辑框（EditText）则可提供用户光标输入和编辑文字内容，但也可以设为只读或不可用，只显示信息而不允许编辑。

4. 按钮（Button）既可以显示文字，也可以带图片，且图片在按钮上的方位还可以根据用户需要进行设定；图像按钮（ImageButton）却只能显示图片而不能在其上配文字。出现这种差别的根源是：按钮（Button）在 Android 的类体系中与文本视图（TextView）直接继承自同一父类，都具备同时显示图片和文字的功能；而图像按钮（ImageButton）则与图像视图（ImageView）在类体系中的亲缘关系更近，其直接父类原本只是专门用来显示图像的，只是在后来才派生出按钮的功能而已。

5. 通过 spinnerMode 属性设为 dialog 就会以浮动菜单的样式显示下拉框中的选项，而用户选择下拉框的值是通过其 OnItemSelectedListener 监听器接口中的 onItemSelected(AdapterView<?> arg0, View arg1, int arg2, long arg3)方法的第 3 个参数 arg2 索引得到的。

6. 单选按钮（RadioButton）放在它的容器按钮组 RadioGroup 中，就可以在编程时通过绑定于 RadioGroup 上的 OnCheckedChangeListener 监听器统一检查和响应容器中按钮的选中状态，保证在同一个组中选择的唯一性。如果不放在容器中，那么就要在每一个单选按钮上都逐一绑定 OnCheckedChangeListener 或 OnClickListener 监听器，并且要由程序员自己编写程序来检查和确保用户选择的唯一性，会使程序代码极其冗长，故不建议这么做。

第 3 章　界面布局与活动页

一、填空题

1. 约束布局（ConstraintLayout）
2. android:orientation = "horizontal"，android:layout_alignParentLeft，android:layout_alignRight
3. <TableRow>……</TableRow>
4. 左上角
5. R.id.myButton
6. 显示，隐式
7. Data，Category，Type，Component，Extras
8. 7
9. 显示提示通知消息

二、选择题

1. D　2. B　3. D　4. B　5. B　6. A　7. D　8. C　9. A　10. B

三、简答题

1. 用户可根据需要选用特定类型的组件拖曳至界面上已有的布局中作为其子布局，在子布局中再放入控件，如此可设计出更为丰富复杂的界面来。某些布局类型，如线性布局，必须使用嵌套才能实现多行布局效果。

2. 可以。采用嵌套多行布局的方式可实现类似网格布局的效果。

3. 相同点：都是将屏幕划分成行和列，向其中添加控件实现布局。不同点：网格布局中一个控件可占用多个网格（跨多行多列），比表格布局要灵活。

4. 优点：指定控件的绝对坐标位置，针对特定目标手机的设计非常完美。缺点：无法根据屏幕尺寸的改变对控件位置进行调整，无法适应不同类型的手机。

5. 三大组件传递 Intent 方式：

（1）**Activity**：通过调用 startActivity()或 startActivityForResult()方法。

（2）**Service**：通过调用 startService()方法或 bindService()方法与服务绑定。

（3）**BroadcastReceiver**：通过 sendBroadcast()、sendOrderBroadcast()或 sendStickBroadcast()方法发送广播 Intent，广播发出后所有注册的拥有与之匹配的 Intent 过滤器的 BroadcastReceiver 都会被激活，并运行各自的处理代码。

6. 通过 Bundle 在 Activity 之间传递数据的过程如下：
（1）发送方 Activity 获得页面交互信息（数据）；
（2）定义 Intent 对象；
（3）定义数据的载体 Bundle 对象；
（4）将数据绑定在 Bundle 上；
（5）将 Bundle 放入 Intent 的传输机制，然后启动页面间的数据传输；
（6）接收方 Activity 解析 Intent 消息，得到 Bundle 中的数据信息，处理后的结果通过 resultCode 返回给发送方 Activity；
（7）发送方 Activity 再对返回码 resultCode 进行处理，决定接下来的动作。

第 4 章　移动 App 高级界面开发技术

一、填空题

1. 轮播条，频道栏，标签栏（当然也有其他一些元素）
2. 列表，网格，类别标签（读者也可举一些不常见的例子）
3. 状态列表图形
4. styles.xml
5. TabHost
6. setIndicator()
7. ViewPager（翻页视图）
8. LinearLayoutManager

二、简答题

1. 对于很多具有相同风格（如字体、字号和颜色）的界面控件元素，如果逐一设置它们的相同属性，会很烦琐且造成代码冗余。自定义风格后一次性地统一应用到所有相同风格的控件上，可以极大地简化开发工作量，精简代码，易于维护和修改。

2. 指示器圆点实为单选按钮，ViewPager 的 SimpleOnPageChangeListener（简单页面变更监听器）执行图片翻页切换时的处理动作。通过重写该监听器的 onPageSelected()方法，在页面变更时将对应页的单选按钮设为选中状态，就实现了将对应图片的指示器圆点点亮为实心的效果。

3. 有频道栏、类别标签列表、网格等效果，都可以用 RecyclerView 实现。

4. 当所要显示的数据项比较多而复杂时，定义一个数据模型类来专门集中存储和管理数据，使得软件结构清晰、易于维护，也符合 MVC 的设计理念。

5. BaseAdapter 是 Android 为具有复杂内容项的视图提供的一种适应性很强的基本适配器，用户开发的适配器继承自它，重写 getView()方法，通过其 convertView 参数获取布局填充器生成的内容项视图，然后进一步操作其中的各控件元素。

6. 当前互联网应用中某些高级特效，例如带有醒目边框和图片的商品返利信息、带边框的淡化文字消息等，在 Android 中都没有现成的控件，而 Android 的文本视图控件不仅可以显示文字信息，还能够在上下左右各个方位显示图片，甚至滚动字幕，对其进行扩展定制也十分方便，所有对于很多复杂的显示特效往往通过自定义文本视图来实现。

7. 不是。TabHost 支持用户自定义名称；而 TabActivity 只能使用内部名称 tabhost，且编程时必须继承 TabActivity 父类。

8. Fragment 是板块，实际的 App 往往一个 Activity 对应有很多个 Fragment。它们可以挂载到 Activity 上，也可以从 Activity 剥离销毁，有着类似于 Activity 的生命周期，但比之 Activity，Fragment 占用的系统资源更少、更灵活。所以，对于一个页面中的各个子功能模块，通常优先考虑用 Fragment 实现而非开启一个新的 Activity，为的是节省系统资源和降低运行开销，提高性能。

第 5 章　Android 服务与广播程序设计

一、选择题

1. B　　2. A　　3. A　　4. B　　5. C　　6. AB　　7. A　　8. BD

二、简答题

1. 在以启动方式使用服务时，主程序无法直接与服务进行交互，这个时候就必须使用线程。

2. Service 有两种使用方式：

（1）**启动方式**：Service 组件**不能够**获取自身的对象实例，因此无法调用 Service 的方法获取到 Service 中的任何状态和数据信息。Service 具备自管理能力，当操作完成时能够停止它本身。

（2）**绑定方式**：绑定 Service 的组件能够获取 Service 的对象实例，因此可以调用 Service 的方法直接获取 Service 中的状态和数据信息。这种方式下的 Service 只有在与调用者组件绑定后才能运行，调用者组件一旦退出，Service 也会随之退出。

3. 多个 Service 交互的特点：

（1）**被启动的服务**。如果一个 Service 被某个 Activity 调用 startService()方法所启动，那么不管是否有 Activity 使用 bindService()绑定或 unbindService()解绑该服务，它都会始终在后台运行。只要调用一次 stopService()方法便可以停止服务，而不考虑调用了多少次启动服务的方法。

（2）**被绑定的服务**。如果一个 Service 被某个 Activity 调用 bindService()方法绑定启动，不管调用 bindService()几次，onCreate()方法都只会调用一次，同时 onStart()方法始终不会被调用。

（3）**被启动又被绑定的服务**。如果一个 Service 既被启动又被绑定，则它将会一直在后台运行。并且不管如何调用，onCreate()始终只会调用一次，而对应 startService()调用了多少次，onStart()便会调用多少次。调用 unbindService()将不会停止服务，而必须调用 stopService()或 Service 自身的 stopSelf()来停止服务。

（4）**当服务被停止时清除服务**。当一个 Service 被停止时，onDestroy()方法将会被调用，在其中完成清理工作。

4. 当产生了 Android 系统事件，该事件发生的时间事先是不确定的（如开机完成、电池电量即将耗尽、网络连接状态改变等），且系统中很多组件都需要对此做出响应，但每个组件的处理方式又都不一样时，会使用广播。

（1）**普通广播**：对于多个接收者来说完全异步，接收者之间互不影响。一个接收者也无法阻止其他接收者的接收动作。应用场合是各类源于 Android 系统自身的消息和事件，通过普通广播告知系统中所有组件，有需要即可响应处理。

（2）**有序广播**：同步执行，优先级高的接收器可以把执行结果传播到优先级低的接收器，且有能力终止这个广播。主要应用于黑名单、骚扰短信和诈骗电话的拦截。

第 6 章　Android 数据存储与共享

一、填空题

1. 名称/值对（NVP）
2. 共享
3. MODE_AppEND
4. getWritableDatabase()，SQLiteOpenHelper
5. Cursor，close()
6. delete()、insert()、query()和 update()
7. URI
8. onCreate()

二、简答题

1. 通过 URI 确定要访问的 ContentProvider 数据集。在发起一个请求的过程中，系统根据 URI 确定处理这个查询的 ContentProvider，然后初始化 ContentProvider 所需的资源。ContentProvider 提供一组标准的数据操作接口，使用者只要调用接口中的方法即可完成所有的数据操作。

2. 为了简化读写流式文件的过程，其中，openFileOutput()方法为写入数据做准备而打开文件；openFileInput()方法为读取数据做准备而打开文件。

3. 通过 Android 自带的 sqlite3 进行，在 Android SDK 的 tools 目录中有 sqlite3 工具，可通过 Windows 命令行启动。输入 ".tables" 命令并回车，查看数据库中的表，用 select 语句直接查询即可看到表的内容。

4. 不能直接调用 ContentProvider 的接口方法，需要使用 ContentResolver 对象，通过 URI 间接调用。

第 7 章　Android 数据库和网络编程

一、填空题

1. mysql-connector-java-5.1.48.jar
2. 子线程
3. 移动端，Web 服务器，后台数据库
4. HttpURLConnection
5. SOAP
6. JSON

二、选择题

1．C　　2．A　　3．D

三、简答题

1. 用 Android 系统的 Message 对象来存储要向主线程传递的消息，创建一个 Handler 句柄，其 handleMessage()方法与主线程是公用的，故可以在其中直接操作界面上的 UI 对象,完成对 UI 界面的刷新。

2．基于两点：

（1）不安全。数据库连接字符串（含用户名和密码）很容易被截获；

（2）加重数据库系统负荷。

3．App 访问电商网站很多情况下只是为了获取所需要的数据和处理数据的服务，并非为了显示网页界面，故与它交互的服务器程序不一定非要有可视的 UI 界面，只须完成传递数据或者某种处理任务即可，这个时候使用 WebService 是最经济的做法。另一个重要原因是，WebService 高度的兼容性和扩展性，可以用 Java、C#、PHP 等各种不同编程语言来实现，后台数据库也可以是 MySQL、SQL Server、Oracle 等多种多样异构类型的 DBMS。

4．Android 程序访问 WebService 的典型流程如下：

（1）开启一个子线程；

（2）创建一个 SOAP 请求对象；

（3）创建一个信封，将请求对象装入信封；

（4）指定 WebService 的 WSDL 文档的 URL；

（5）调用 WebService 中的方法；

（6）获得 WebService 返回的结果；

（7）根据返回结果刷新主线程 UI。

第 8 章　Android 多媒体和图形图像编程

一、填空题

1．raw

2．onActivityResult()，getData()，getPath()

3．OnSeekBarChangeListener，onStopTrackingTouch()

4．onPictureInPictureModeChanged()，isInPicInPicMode

5．audio/*

6．像素矩阵，Matrix

7．ImageSwitcher（图像切换器），Gallery（画廊）

二、简答题

1．用 Handler 机制创建进度条刷新任务，在其中以线程实时刷新进度条的当前值。然后在 VideoView 启动后调用 post()方法启动刷新任务。

2．绑定步骤：

（1）调用 VideoView 的 setMediaController()方法设置与之相关联的 MediaController；

（2）调用 MediaController 的 setAnchorView()或 setMediaPlayer()方法设置与之绑定的 VideoView。

3．SurfaceHolder 是一个接口，通过它可以编辑和控制 SurfaceView 录像视图，每个 SurfaceView 都包含一个 SurfaceHolder，通过它可调用操作录像视图参数的 API。

4．invalidate()仅用于本 UI 线程内部的刷新；而 postInvalidate()支持跨线程刷新，postInvalidate()在底层调用的也是 invalidate()，只是在其上封装了跨线程消息机制。如果我们自定义的视图本身就是在 UI 线程，没有用到多线程，直接用 invalidate()刷新就可以了。跨线程的刷新则必须用 postInvalidate()。

5．OpenGL 绘图的一般流程：

（1）预置要绘制图形的顶点坐标数据；
（2）初始化字节缓存并清屏；
（3）初始化视图；
（4）编写着色器代码；
（5）加载并编译着色器；
（6）创建绘图程序；
（7）链接两个着色器，添加程序到 GPU 运行；
（8）连接顶点属性；
（9）绘制图形。

第 9 章　Android 第三方开发与设备操作

一、填空题

1. package 包名，import 路径，类包名
2. 震动，摄像头，读取存储空间
3. APPID 值
4. SensorEventListener，onSensorChanged()
5. 加速度传感器（TYPE_ACCELEROMETER）
6. 摄像头

二、简答题

1. 沙箱是支付宝官方为开发者提供的一个模拟的支付环境。出于安全考虑，支付宝真实的支付平台只允许拥有营业执照的企业接入使用，而一般的 Android 开发者想要测试 App 支付功能，就必须借助沙箱环境，里面有一对虚拟商家和买家的账号、密码，可模拟各种常用的支付应用。App 在沙箱环境调试稳定后，开发者可申请营业执照，再将 App 移植到支付宝真正的支付平台运行。

2. 引入高德地图功能的步骤：
（1）去高德开放平台下载 SDK，加载到自己的 Android Studio 工程中；
（2）在高德开放平台官网注册一个账号，登录后创建一个应用；
（3）用自己计算机的安全码对应用授权，得到访问 Key 值；
（4）在自己的工程中配置 keystore；
（5）在 AndroidManifest.xml 中打开权限、写入访问 Key 值；
（6）在工程界面设计文件中引入地图控件；
（7）在程序中创建地图并编写管理地图生命周期的代码。

3. "定位服务提供者"就是向用户提供位置服务的第三方系统，可以是全球卫星定位系统中的卫星，也可以是网络（附近通信基站或 WiFi 等）。

4. 手机蓝牙启动过程：
（1）初始化蓝牙适配器；
（2）"蓝牙权限请求"对话框提示用户授权；
（3）打开蓝牙功能；
（4）搜索附近的设备。

欢迎广大院校师生**免费**注册应用

华信SPOC官方公众号

www.hxspoc.cn

华信SPOC在线学习平台
专注教学

- 数百门精品课 数万种教学资源
- 教学课件 师生实时同步
- 多种在线工具 轻松翻转课堂
- 电脑端和手机端（微信）使用
- 测试、讨论、投票、弹幕…… 互动手段多样
- 一键引用，快捷开课 自主上传，个性建课
- 教学数据全记录 专业分析，便捷导出

登录 www.hxspoc.cn 检索 华信SPOC 使用教程 获取更多

华信SPOC宣传片

教学服务QQ群：1042940196
教学服务电话：010-88254578/010-88254481
教学服务邮箱：hxspoc@phei.com.cn

电子工业出版社　华信教育研究所